Imperial College London
2407013531

Sediment-Hosted Gas Hydrates: New Insights on Natural and Synthetic Systems

The Geological Society of London
Books Editorial Committee

Chief Editor
BOB PANKHURST (UK)

Society Books Editors
JOHN GREGORY (UK)
JIM GRIFFITHS (UK)
JOHN HOWE (UK)
PHIL LEAT (UK)
NICK ROBINS (UK)
JONATHAN TURNER (UK)

Society Books Advisors
MIKE BROWN (USA)
ERIC BUFFETAUT (FRANCE)
JONATHAN CRAIG (ITALY)
RETO GIERÉ (GERMANY)
TOM MCCANN (GERMANY)
DOUG STEAD (CANADA)
RANDELL STEPHENSON (UK)

Geological Society books refereeing procedures

The Society makes every effort to ensure that the scientific and production quality of its books matches that of its journals. Since 1997, all book proposals have been refereed by specialist reviewers as well as by the Society's Books Editorial Committee. If the referees identify weaknesses in the proposal, these must be addressed before the proposal is accepted.

Once the book is accepted, the Society Book Editors ensure that the volume editors follow strict guidelines on refereeing and quality control. We insist that individual papers can only be accepted after satisfactory review by two independent referees. The questions on the review forms are similar to those for *Journal of the Geological Society*. The referees' forms and comments must be available to the Society's Book Editors on request.

Although many of the books result from meetings, the editors are expected to commission papers that were not presented at the meeting to ensure that the book provides a balanced coverage of the subject. Being accepted for presentation at the meeting does not guarantee inclusion in the book.

More information about submitting a proposal and producing a book for the Society can be found on its web site: www.geolsoc.org.uk.

It is recommended that reference to all or part of this book should be made in one of the following ways:

LONG, D., LOVELL, M. A., REES, J. G. & ROCHELLE, C. A. (eds) 2009. *Sediment-Hosted Gas Hydrates: New Insights on Natural and Synthetic Systems*. Geological Society, London, Special Publications, **319**.

KLAPP, S. A., KLEIN, H. & KUHS, W. F. 2009. Gas hydrate crystallite size investigations with high-energy synchrotron radiation. *In*: LONG, D., LOVELL, M. A., REES, J. G. & ROCHELLE, C. A. (eds) *Sediment-Hosted Gas Hydrates: New Insights on Natural and Synthetic Systems*. Geological Society, London, Special Publications, **319**, 161–170.

GEOLOGICAL SOCIETY SPECIAL PUBLICATION NO. 319

Sediment-Hosted Gas Hydrates: New Insights on Natural and Synthetic Systems

EDITED BY

D. LONG
British Geological Survey, Edinburgh, UK

M. A. LOVELL
University of Leicester, UK

J. G. REES
British Geological Survey, Keyworth, UK

and

C. A. ROCHELLE
British Geological Survey, Keyworth, UK

2009
Published by
The Geological Society
London

THE GEOLOGICAL SOCIETY

The Geological Society of London (GSL) was founded in 1807. It is the oldest national geological society in the world and the largest in Europe. It was incorporated under Royal Charter in 1825 and is Registered Charity 210161.

The Society is the UK national learned and professional society for geology with a worldwide Fellowship (FGS) of over 9000. The Society has the power to confer Chartered status on suitably qualified Fellows, and about 2000 of the Fellowship carry the title (CGeol). Chartered Geologists may also obtain the equivalent European title, European Geologist (EurGeol). One fifth of the Society's fellowship resides outside the UK. To find out more about the Society, log on to www.geolsoc.org.uk.

The Geological Society Publishing House (Bath, UK) produces the Society's international journals and books, and acts as European distributor for selected publications of the American Association of Petroleum Geologists (AAPG), the Indonesian Petroleum Association (IPA), the Geological Society of America (GSA), the Society for Sedimentary Geology (SEPM) and the Geologists' Association (GA). Joint marketing agreements ensure that GSL Fellows may purchase these societies' publications at a discount. The Society's online bookshop (accessible from www.geolsoc.org.uk) offers secure book purchasing with your credit or debit card.

To find out about joining the Society and benefiting from substantial discounts on publications of GSL and other societies worldwide, consult www.geolsoc.org.uk, or contact the Fellowship Department at: The Geological Society, Burlington House, Piccadilly, London W1J 0BG: Tel. +44 (0)20 7434 9944; Fax +44 (0)20 7439 8975; E-mail: enquiries@geolsoc.org.uk.

For information about the Society's meetings, consult *Events* on www.geolsoc.org.uk. To find out more about the Society's Corporate Affiliates Scheme, write to enquiries@geolsoc.org.uk.

Published by The Geological Society from:
The Geological Society Publishing House, Unit 7, Brassmill Enterprise Centre, Brassmill Lane, Bath BA1 3JN, UK

(*Orders*: Tel. +44 (0)1225 445046, Fax +44 (0)1225 442836)
Online bookshop: www.geolsoc.org.uk/bookshop

The publishers make no representation, express or implied, with regard to the accuracy of the information contained in this book and cannot accept any legal responsibility for any errors or omissions that may be made.

© The Geological Society of London 2009. All rights reserved. No reproduction, copy or transmission of this publication may be made without written permission. No paragraph of this publication may be reproduced, copied or transmitted save with the provisions of the Copyright Licensing Agency, 90 Tottenham Court Road, London W1P 9HE. Users registered with the Copyright Clearance Center, 27 Congress Street, Salem, MA 01970, USA: the item-fee code for this publication is 0305-8719/09/$15.00.

British Library Cataloguing in Publication Data

A catalogue record for this book is available from the British Library.
ISBN 978-1-86239-279-3

Typeset by Techset Composition Ltd, Salisbury, UK
Printed by MPG Books Ltd, Bodmin, UK

Distributors

North America
For trade and institutional orders:
The Geological Society, c/o AIDC, 82 Winter Sport Lane, Williston, VT 05495, USA
Orders: Tel. +1 800-972-9892
         Fax +1 802-864-7626
         E-mail: gsl.orders@aidcvt.com

For individual and corporate orders:
AAPG Bookstore, PO Box 979, Tulsa, OK 74101-0979, USA
Orders: Tel. +1 918-584-2555
         Fax +1 918-560-2652
         E-mail: bookstore@aapg.org
         Website: http://bookstore.aapg.org

India
Affiliated East-West Press Private Ltd, Marketing Division, G-1/16 Ansari Road, Darya Ganj, New Delhi 110 002, India
Orders: Tel. +91 11 2327-9113/2326-4180
         Fax +91 11 2326-0538
         E-mail: affiliat@vsnl.com

Contents

Obituary	vi
LONG, D., LOVELL, M. A., REES, J. G. & ROCHELLE, C. A. Sediment-hosted gas hydrates: new insights on natural and synthetic systems	1
RIEDEL, M., COLLETT, T., MALONE, M. J. & IODP EXPEDITION 311 SCIENTISTS. Gas hydrate drilling transect across northern Cascadia margin – IODP Expedition 311	11
KLEINBERG, R. L. Exploration strategy for economically significant accumulations of marine gas hydrate	21
MCGEE, T., MACELLONI, L., LUTKEN, C., BOSMAN, A., BRUNNER, C., ROGERS, R., DEARMAN, J., SLEEPER, K. & WOOLSEY, J. R. Hydrocarbon gas hydrates in sediments of the Mississippi Canyon area, Northern Gulf of Mexico	29
MAZURENKO, L. L., MATVEEVA, T. V., PRASOLOV, E. M., SHOJI, H., OBZHIROV, A. I., JIN, Y. K., POORT, J., LOGVINA, E. A., MINAMI, H., SAKAGAMI, H., HACHIKUBO, A., SALOMATIN, A. S., SALYUK, A. N., PRILEPSKIY, E. B. & CHAOS 2003 SCIENTIFIC TEAM. Gas hydrate forming fluids on the NE Sakhalin slope, Sea of Okhotsk	51
SWART, R. Hydrate occurrences in the Namibe Basin, offshore Namibia	73
CAMPS, A. P., LONG, D., ROCHELLE, C. A. & LOVELL, M. A. Mapping hydrate stability zones offshore Scotland	81
MINSHULL, T. A. & CHAND, S. The pore-scale distribution of sediment-hosted hydrates: evidence from effective medium modelling of laboratory and borehole seismic data	93
TINIVELLA, U., LORETO, M. F. & ACCAINO, F. Regional versus detailed velocity analysis to quantify hydrate and free gas in marine sediments: the South Shetland Margin case study	103
EATON, M. W., JONES, K. W. & MAHAJAN, D. Mimicking natural systems: methane hydrate formation–decomposition in depleted sediments	121
KVAMME, B., GRAUE, A., BUANES, T., KUZNETSOVA, T. & ERSLAND, G. Effects of solid surfaces on hydrate kinetics and stability	131
ANDERSON, R., TOHIDI, B. & WEBBER, J. B. W. Gas hydrate growth and dissociation in narrow pore networks: capillary inhibition and hysteresis phenomena	145
KLAPP, S. A., KLEIN, H. & KUHS, W. F. Gas hydrate crystallite size investigations with high-energy synchrotron radiation	161
ROCHELLE, C. A., CAMPS, A. P., LONG, D., MILODOWSKI, A., BATEMAN, K., GUNN, D., JACKSON, P., LOVELL, M. A. & REES, J. Can CO_2 hydrate assist in the underground storage of carbon dioxide?	171
Index	185

Dr Leonid L. Mazurenko, 1976–2007

Dr Leonid Leonidovich Mazurenko was an enthusiastic participant of the International Conference 'Sediment-Hosted Gas Hydrates: New Insights on Natural and Synthetic Systems', upon which this volume is based. The work he presented at the meeting dealt with gas hydrates and hydrate-forming fluids from gas venting sites in the Sea of Okhotsk as a part of the CHAOS Project. His paper on the subject in this volume is his final publication.

Leonid Mazurenko was born on 4 December, 1976 in Priozersk, Kazakhstan (USSR). He studied in the Department of Lithology at Moscow State University. In 1998, he went to St Petersburg, where he continued his education at St Petersburg State University and then was employed as an 'engineer-geologist' by the All Russia Research Institute for Geology and Mineral Resources of the Ocean (VNIIOkeangeologia). Leonid's career progressed as a 'engineer-geologist', scientist, senior scientist and head of the Laboratory for Gas Hydrate Geology in VNIIOkeangeologia.

Leonid was interested in and actively involved with research in submarine gas hydrates and related geochemical systems. The successful defence of his dissertation entitled 'Gas hydrate formation in submarine fluid discharge areas' earned him the Russian scientific degree of Candidate of Science in 2004 (equivalent to a Western PhD). He was one of the most prominent specialists in the Russian gas hydrate and fluid venting geochemical community, and was recognized internationally for his understanding of gas hydrate dynamics and formation mechanisms. He was at the beginning of a bright scientific career, and had already made important contributions to the understanding of gas hydrates and pore water chemistry for numerous of fluid venting areas. Amongst these were parts of the Black Sea, the Gulf of Cadiz (NE Atlantic), the Sea of Okhotsk and Lake Baikal.

Leonid was consistently highly energetic and cheerful, and always fully enjoyed life. He had an outstanding sense of humour and combined his research responsibilities and laboratory leadership roles with building friendships and taking good care of his family. His sudden death at the tragically young age of 30 came as a shock to his family and colleagues. He will be greatly missed by them, and his death is a big loss to the gas hydrate scientific community.

Tatania Matveeva, All-Russia Research Institute for Geology and Mineral Resources of the World Ocean, is thanked for submitting this obituary.

Sediment-hosted gas hydrates: new insights on natural and synthetic systems

D. LONG[1]*, M. A. LOVELL[2], J. G. REES[3] & C. A. ROCHELLE[3]

[1]*British Geological Survey, West Mains Road, Edinburgh EH9 3LA, UK*

[2]*Department of Geology, University of Leicester, Leicester LE1 7RH, UK*

[3]*British Geological Survey, Kingsley Dunham Centre, Keyworth, Nottingham NG12 5GG, UK*

*Corresponding author (e-mail: dal@bgs.ac.uk)

Abstract: In the public's imagination, hydrates are seen as either a potential new source of energy to be exploited as the world uses up its reserves of oil and gas or as a major environmental hazard. Scientists, however, have expressed great uncertainty as to the global volume of hydrates and have reached little agreement on how they might be exploited. Both of these uncertainties can be reduced by a better understanding of how hydrates are held within sediments. There are conflicting ideas as to whether hydrates are disseminated within selected lithologies or trapped within fractures comparable to mineral lodes. To resolve this, hydrates have to be examined at all scales ranging from using seismics to microscopic studies. Their position within sediments also influences the stability of methane hydrate in responding to pressure and temperature and how the released gas might transfer to the ocean, atmosphere, or to a transport mechanism for recovery. These results also run parallel with the studies of carbon dioxide hydrate, which is being considered as a potential sequestion medium.

Over recent decades hydrates have been gradually making their way up the scientific agenda, receiving correspondingly greater wider societal interest with time. However, what is most notable to the casual observer is not the fact that it is broadly recognized that hydrates have the potential to be a major environmental hazard or a major new energy source, but that nobody is exactly shouting about it. There are several reasons for this. Geological investigations suggest that hydrates were at least partners in crime in many climatic disasters, such as the great methane outbursts that caused the mass extinctions at the end of the Permian (Erwin 1994; Krull & Retallack 2000) or Paleocene (Dickens *et al.* 1997; Zachos *et al.* 2005). Whilst people are interested in such events, the immediate instability of possibly 'life-threatening' natural hydrates does not seem to be their immediate concern; it is clear that they occurred a long time ago, since when environments have greatly changed, and it is also apparent that they were largely precipitated by external triggers (White 2002; Maclennan & Jones 2006). Likewise, although they see the potential to produce energy-providing methane from hydrates, initial difficulties in doing so have dampened interest. In both cases – as a hazard or resource – the lack of societal focus largely stems from the body language of the scientific community, which is itself highly uncertain of the importance of hydrates. Scientists are not certain enough of their ground to allow a clear direction to be mapped in relation to minimization of the risk of mass hydrate destabilization or widespread exploitation. To a large degree this uncertainty centres on our poor fundamental understanding of the occurrence and stability of sediment-hosted hydrates. There is no better illustration of this than the widely fluctuating predictions of global hydrate reserves we have seen in recent years, where estimates vary between 10^{15} and 10^{19} m^3 of methane gas at STP. Milkov (2003) describes how improved understanding of the distribution and concentration of gas hydrates in marine sediments has led to a readjustment of global estimates downwards over each subsequent decade, although the estimate of 10^{15} m^3 is challenged by Klauda & Sandler (2005), who suggest that a total volume of 10^{17} m^3 is likely with 10^{16} m^3 located on continental margins. Although there is starting to be a consensus in the order of 10^{16} m^3 (e.g. Kvenvolden 1998, 2000; Makogon *et al.* 2007), a range of values are still used and such apparent uncertainty in hydrate abundance hardly conveys the message to the wider community that the scientists know what they are talking about.

To move this debate along, we considered the source of our greatest uncertainties and found that these largely centre around how hydrates are physically stored in sediments at a range of scales. At present our understanding is extremely crude. We have very little knowledge about how hydrates are stored in sediments of different grain size or texture, whether they dominantly are separated by water films in inter-grain pores, or whether they coat grains, how bacteria control authigenic fixation, and whether the mineralogy of host-sediment influences the microscale sediment–hydrate association. At a larger scale, we have a very poor understanding of the distribution of hydrates within individual beds, let alone complex heterolithic sequences. Big questions prevail. Do hydrates form pods that are little influenced by lithology as in many ore bodies, or are they dominantly stratiform and follow rules of behaviour that are analogous to hydrocarbons in reservoirs? How much are sediment-hosted distributions controlled by structural settings, whether tectonic, gravitational or through diagenetic changes in the sediment (for instance the development of cavities or veins through shrinkage)? It is clear that our ability to describe such relationships is very immature, even before we start to make sense of changing pressure, temperature and salinity regimes, as well as the variation in natural supply of methane from external sources.

To review our current understanding, and in order to encourage debate about these issues, we decided to convene a meeting to which interested scientists could look at these many challenges in a fairly relaxed setting. Consequently, with the support of the Geological Society Hydrocarbons Group, a two-day meeting 'Sediment-Hosted Gas Hydrates: New Insights on Natural and Synthetic Systems' took place at Burlington House, Piccadilly, London on 25–26 January 2006. This was based on 35 presentations and posters and brought together over about 100 international hydrate scientists spanning the hazard and resource communities, as well as those with very different experience, for instance those involved principally with laboratory experimentation, mixing with those of geophysical field studies or geochemical mapping. The main theme of the meeting was the nature of the primary hydrate–sediment relationships that control hydrate stability. This largely addressed the distribution of natural sediment-hosted hydrates, but also covered research into synthetic systems. The latter are of interest as they provide analogues of natural environments, but in very well calibrated, controlled laboratory settings where textural relationships and processes can be mapped and measured. Synthetic sediment-hosts are also likely to be of interest in the future as a possible store of greenhouse gases; the possibility that hydrates could be used to store large volumes of human-generated carbon dioxide was discussed at some length in the meeting. The structure and content of this volume largely reflects the structure and interests of the meeting and addresses sediment-hosted hydrates in natural and synthetic systems separately.

Natural systems

Setting of natural hydrates

The study of hydrates began as that of a curiosity in the nineteenth century (Davy, 1811) and then as a practical solution to difficulties in transporting gas by pipeline in the early twentieth century (Hammerschmidt 1934). Once it was realized that the pressure and temperature conditions necessary for methane hydrate formation exist naturally, the hunt was on to locate examples. In the 1960s and 1970s hydrate was identified in wells drilled through the Siberian and Alaskan permafrost (Collett 1983), and samples of hydrate were recovered from the Black Sea in 1974 (Yeframova & Zhizhchenko 1975) and then many other deep water settings. A few years previously, bottom simulating reflectors (BSRs) were identified and attributed to the occurrence of methane hydrates (Lancelot & Ewing 1972); hydrates have subsequently been recognized on many continental slopes around the world. Recent evidence suggests however, that the presence of BSRs alone is not a reliable indicator of the presence or not of methane hydrates (Finley & Kranson 1986). In particular BSRs form at the interface between hydrate and free gas, but in the absence of free gas a BSR will not develop, even if hydrates are present (Haq 1998). Similarly other cross cutting reflectors can be formed by diagenetic processes (Davies & Cartwright 2002). Therefore careful analysis needs to be conducted of BSRs before attributing extensive areas of continental slopes as hydrate bearing. Because of the unreliability of much of the BSR evidence, even local estimates of hydrate volume are highly uncertain. For example, in the Gulf of Mexico, where hydrates are of both biogenic and petrogenic origin and have been found even at the seabed, Collet & Kuuskraa (1998) estimated that up to 500 Gton of carbon occurred as hydrate but Milkov & Sassen (2001) estimated only 5 Gton of carbon. Even if the relationship between BSRs and hydrates were well established, the fact that the reported occurrence is strongly influenced by the identification of BSRs in geophysical surveys conducted for hydrocarbon exploration, suggests that we are likely to have a poor understanding of the real distribution of hydrates. Although many areas can be shown to have the appropriate temperature and pressure conditions, they are lacking physical samples or geophysical evidence. Many of these problems are discussed by **Kleinberg** who discusses the problems of exploring for hydrates and suggests adaptations to geological and geophysical survey methods.

Characteristic hydrate domains

The primary source of methane in hydrate deposits is biogenic decay of organic matter, as demonstrated by the isotopic composition and the near absence of higher hydrocarbons such as ethane and propane (Kvenvolden & McDonald 1985). Although hydrates of a petrogenic (thermogenic) origin have been encountered in many hydrocarbon basins (e.g. the Gulf of Mexico or Caspian Sea), they appear to be less common (Brooks & Bryant 1985; Soloviev & Ginsberg 1994). In the former, a greater range of hydrates are present, including ethane and propane hydrates, which result in different hydrate structures and different interactions with the associated sediments. Hydrate can occur in various forms. The main type is stratigraphic-type hydrate deposits (Milkov & Sassen 2001) formed at the base of the hydrate stability zone (HSZ), where there is the greatest concentration of hydrate due to methane generation within sedimentary units. Structural-type hydrate deposits form where methane migration through faults allows hydrate to occur at all points within the HSZ, including near the seabed where it is vulnerable to changes in bottom water conditions. **Mazurenko et al.** describe one such seabed site in the Sea of Okhotsk where hydrate is forming rapidly at the present time in areas of focused fluid flow. They used isotopic and chemical approaches to ascertain that fluids from depth have interacted with porewaters within shallow sediments and led to the formation of hydrates. The establishment of seabed observatories, as described by **McGee et al.** will allow detailed monitoring of the growth and decay of these features and the interaction with the biosphere. It was originally presumed that hydrate occurred as thick layers often within fine-grained sediments. Increasingly the evidence suggests that hydrate preferentially occurs as layers within coarse-grained sediments due to the greater permeability of these horizons (Clennell et al. 1999). This is amply shown in the detailed work on Hydrates Ridge (**Riedel et al.**) where hydrates have been located in coarse grained sedimentary units at several levels within the HSZ implying a complex history of methane migration including via fractures identified by X-rays of the cores collected under in situ pressure (Schultheiss pers. comm.) using HYACE equipment.

A concern raised as a consequence of the increasing interest in climate change is the presence of hydrate occurring under permafrost. With rising temperatures the breakdown of permafrost systems is predicted; these are likely to have an increased potential to release methane, thereby providing positive feedbacks to climate change. The Arctic coastlines and shelf seas of Siberia, Alaska and Canada are the focus of this concern, although this assumes that methane hydrate is trapped beneath the permafrost. For such circumstances to prevail it is expected that a relatively thick permafrost will produce low temperatures at sufficient depth for high enough pressures to occur for hydrate to be stable below it. However, there is some evidence (Yakushev 2004) that methane hydrate can exist within permafrost at shallower depths, in regions where it is only metastable, and thus especially vulnerable to climate change. In some cases, it is likely that great volumes of methane as free gas may occur under a 'cap' of permafrost or hydrate, and there is the potential for this to be released should increasing temperatures melt the permafrost or disassociate hydrate. Climate change is also likely to impact on hydrate stability in other settings. The warming of oceans has been identified as a potential concern as increases in bottom water temperatures will lead to the displacement of the underlying geothermal profile and a reduction in the dimensions of the hydrate stability zone. Recent warming of the oceans has been attributed to anthropogenic greenhouse gas emissions (Levitus et al. 2005). However it is expected that changes in water temperature will be greatest in areas of shallower seas, thereby making submarine hydrates in polar shelf regions more vulnerable than elsewhere.

Submarine slope failures have been attributed to former changes in the stability regime. Examples of these mass movements are reported by **Swart** (2009) from the Namibian margin where hydrates are closely associated with the slump deposits. With future changes in seabed conditions now locked-in due to recent and predicted changes in global temperatures, further failures may occur.

Sediment–hydrate interaction

Malone (1985) describes four possible hydrate morphologies with the terminology disseminated, nodular, vein and massive. Disseminated hydrate occurs within the pore space of the sediment, while the other three occur where the sediment is disturbed either by regional tectonic stresses or through the stress resulting from hydrate crystal growth – for instance Cook and Goldberg (2007) found hydrate bearing fractures to be oriented with respect to regional tectonic stresses offshore India. Whilst such observational classifications may have had some merit (even though they have commonly been based on limited observations of natural hydrate samples recovered intact in sediment samples, and often these were in the process of undergoing dissociation by the time of the observation), they have done little to further our understanding of sediment–hydrate interactions.

Theoretical work on hydrate formation (Clennell et al. 1999; Henry et al. 1999) explored the influence of capillary pressure and thermodynamics on hydrate growth and provided some real physical constraints to hydrate morphology. It concluded that hydrate growth in fine-grained muds would be unlikely, and that coarser-grained sediments, exhibiting larger pores, would act as more likely hosts. Thus theoretical consideration suggested that hydrate morphology is controlled by the nature of the sediment host as much as by the supply of the necessary 'ingredients' and conditions (water, gas, nucleation sites, temperature and pressure). This theoretical work is also supported by experimental studies (Kleinberg et al. 2003; Camps 2007) and by observations (e.g. Tréhu et al. 2002; Riedel et al. 2006) where disseminated hydrate is limited to coarser-grained sediments and the other forms tend to occur in finer-grained sediments where the sediment fabric is disturbed. Similar observations were found through investigations of hydrate dissociation conditions (Anderson et al. 2003; Llamedo et al. 2004). These revealed that dissociation is more readily achieved within small pores compared with large pores, suggesting the possibility of hydrate breakdown in small pores though not in adjacent large pores. Tréhu et al. (2004) discuss how the distribution of gas hydrate is controlled both structurally and stratigraphically, with thick sections of hydrate in the upper 100 m of the Cascadia Margin (Riedel et al. 2009). The role of the subsea biosphere on hydrate formation and distribution is largely unknown, though Inagaki et al. (2006) found that, for two separate locations in the Pacific, prokaryotic communities in methane hydrate-bearing sediment cores are distinct from those in hydrate-free cores.

Significant advances in characterization of the relationship between sediments and hydrates have been recently developed. For instance, **Tinivella et al.** have quantified the concentrations of gas hydrate in pore space by travel-time inversion modelling of the acoustic properties of these sediments. Such analysis has allowed the identification of free gas distribution in pore spaces, likely patterns of fluid migration, the physical properties of sediments and the consequent origin of the BSR offshore the Antarctic Peninsula. At other sites, where other geophysical approaches have been applicable, hydrate volumes have often been estimated using electrical or acoustic measurements and relating these parameters (electrical resistivity and acoustic velocity) to the hydrate concentration or saturation in the pore space. Such transforms usually assume the hydrate is disseminated in the pore space and that the sediment remains wet, comparable to petroleum-bearing reservoirs. However, Ecker et al. (2000) and Dvorkin et al. (2000) demonstrate that knowledge of the interaction between hydrate and sediment grains is crucial in achieving well-constrained volume estimates of hydrate. It is also recognized that the use of Archie's equations assumes the hydrate does not completely block off the pore space at low saturations, treating hydrate as a hydrocarbon fluid. **Minshull & Chand** further refine the self-consistent approximation/differential effective medium approach relating seismic properties of sediment to its hydrate content. Their results suggest that the inferred proportion of load-bearing hydrate appears to decrease with increasing hydrate saturation for gas-rich laboratory environments, but increase when hydrate is formed from solution.

Koh & Sloan (2007) suggest there has been a recent paradigm shift from addressing the thermodynamics (i.e. time-independent properties) to hydrate formation and dissociation kinetics. Thus improved understanding of these processes is crucial if there is to be any control of gas recovery from hydrates in-situ or in assessing hydrate dissolution and ensuing environmental impact. Understanding the size and morphology of hydrates is central to modelling these processes. The paper by **Klapp et al.** adds to our knowledge in this area through the use of high-energy synchrotron diffraction to determine grain sizes of six natural gas hydrates in samples retrieved from the Bush Hill region in the Gulf of Mexico and from ODP Leg 204 at the Hydrate Ridge offshore Oregon.

Stability of methane hydrates

It is clear that naturally changing conditions over geologic timescales have forced the formation and disassociation of vast amounts of methane hydrate on Earth, with individual volumes of rock possibly having experienced several episodes of conditions favouring hydrate stability. A detailed knowledge of hydrate stability is thus important in terms of understanding and quantifying carbon cycling and methane release in the geologic record (e.g. events in the PETM; Zachos et al. 2005; Sluijs 2006), and especially so in the light of anthropogenic influences on the global climate.

Many studies have made detailed measurements of the stability of methane hydrates under different pressure and temperature conditions, and with different fluid composition. Such investigations have often been driven by the needs of the hydrocarbon industry, as hydrate formation within boreholes or surface infrastructure can impact production and the safety of operations. Compilations of data resulting from such studies (e.g. Sloan

1998) have led to the production of computer codes that can be used to model hydrate stability over a broad range of conditions. They can also be used in scoping calculations to 'map out' hydrate stability zones within sediments.

It is not necessarily straightforward however, to apply data from relatively open systems (effectively having 'huge' pore spaces), to fine-grained sediments. Indeed, previous studies (e.g. Clennell *et al.* 1999; Henry *et al.* 1999; Llamedo *et al.* 2004) show that the pressure–temperature stability field of methane hydrate is reduced in very narrow pores. The study by **Anderson *et al.*** takes these observations further, by considering the effect of narrow pores on hydrate growth, as well as dissociation. For narrow pores they find a distinct hysteresis over a range of pressure–temperature conditions during cycles of hydrate growth and disassociation, with hydrate formation occurring at significantly lower temperatures (or higher pressures) compared with dissociation.

There are other processes, however, that occur even when samples of hydrate are kept within their stability zone. Although these may not alter the overall stability of the total mass of hydrate present, they may act to alter its distribution. One such process described by Klapp *et al.* (2009) is where larger hydrate crystals grow at the expense of smaller ones. A consequence of this is that over time the average size of hydrate crystals within a sample increases; Klapp *et al.* (2009) suggest that advanced X-ray techniques may be one way to assess such size changes. Given further development and calibration against hydrate samples of known ages, it may even be possible to use rate-dependant information such as this to make estimates of the age of hydrate samples. Some of these approaches have been developed by **Eaton *et al.*** who explore two hydrate formation methods, based on depleted sediment samples from Blake Ridge using a flexible unit that records temperatures, pressures and changes in gas volume during absorption/evolution: (1) under continuous methane gas-flow conditions; and (2) where hydrates are formed from the dissolved gas phase by diffusion.

Studies such as those outlined above are very important when considering sediment-hosted gas hydrates as they will impact upon our predictions of hydrate stability; such work needs to continue, to advance our knowledge in this area. Indeed, accurate assessments of hydrate stability will have to consider the characteristics of the host sediment as well as pressure, temperature and fluid composition. Not only could such analysis alter estimates of natural methane hydrate abundance in (or recovery from) fine-grained sediments, but it could also enhance the estimate of other hydrates in anthropogenic systems – such as during the underground storage of carbon dioxide.

Synthetic systems

Carbon dioxide hydrates

As a consequence of the abundance and importance of methane hydrate, be it formed naturally in sediments or within engineered structures such as pipelines, most studies of sediment-hosted hydrates have focused on methane hydrate systems. However, other sediment-hosted hydrates can also be important in certain circumstances, and one example is CO_2 hydrate. Although the occurrence of natural CO_2 hydrate is rare (e.g. Sakai *et al.* 1990), its controlled formation and storage could provide a mechanism with which to trap and store waste CO_2 rather than emit it to the atmosphere. This approach is very relevant to current discussions about the degree of anthropogenic influence on climate and how it might be reduced (e.g. IPCC 1990, 1995, 2001, 2007; RCEP 2000). Indeed, reducing overall anthropogenic releases of CO_2 may be vital to limit the extent of global warming, and hence reduce the potential for climate-induced breakdown of vast amounts of natural methane hydrate from permafrost areas or below the seafloor.

Ongoing industrial-scale projects at places such as Sleipner (North Sea) (Baklid *et al.* 2006) and Weyburn (Canada) (Malk & Islam 2000; White *et al.* 2004; Wilson & Monea 2004) demonstrate the practicality of capturing large amounts of CO_2 and injecting it within deep, warm sediments to be eventually trapped either dissolved in solution or as carbonate minerals (e.g. Bachu *et al.* 1994; Gunter *et al.* 1993, 1997, 2000). However, relatively few studies have investigated the role that CO_2 hydrate could play in trapping CO_2 in cooler sediments. Some previous studies have considered that the relative stability of methane and CO_2 hydrates might facilitate the trapping of CO_2 as a hydrate whilst at the same time liberating methane (IEA GHG 2000; Goel 2006), whereas other studies have concentrated on just CO_2 injection followed by CO_2 hydrate formation (e.g. Kiode *et al.* 1997; House *et al.* 2006). Both approaches raise questions about our understanding of the processes involved and their inherent uncertainties. Where might storage of CO_2 by these methods be possible? How will the hydrate interact with sediment grains and pore fluids? What are the rates and magnitudes of the trapping processes involved? How safe and secure will the CO_2 hydrate be in the long term?

The study by **Rochelle *et al.*** takes a broad view of the possibilities for underground CO_2 storage in cool sediments. It considers some of the beneficial trapping mechanisms that might be enhanced relative to more conventional CO_2 storage at higher temperatures, such as the role of CO_2 hydrate as both an immobile trapping phase and as a potential cap above a store of free-phase CO_2. This approach is extended to consider where CO_2 hydrate might be stable in sediments, with **Camps et al.** calculating the extent and thickness of CO_2 hydrate and methane hydrate stability zones offshore western Scotland. Below the bed of the Faroe–Shetland Channel CO_2 hydrate is predicted to be stable to a maximum depth of 345 m, whereas methane hydrate has a greater maximum stability depth of 650 m. Rochelle et al. (2009) use a similar approach to present preliminary modelling of regions offshore western Europe, which shows that large areas may have the potential for CO_2 hydrate formation in deep-water sediments. These studies indicate that the storage of CO_2 as a hydrate within sediments may present a viable future CO_2 storage technology for some parts of Europe. However, they also highlight the importance of certain basic information when making predictions, such as the lack of detailed geothermal gradient data, which limits the extent to which detailed models and predictions can be made.

Much work remains to be undertaken to fully understand how CO_2 hydrate can best contribute to underground storage. **Kvamme et al.** explore the possibilities of replacing original hydrate-bound hydrocarbons, such as methane, by CO_2 as hydrates of the latter are considerably more stable thermodynamically than methane hydrates. Technologies allowing two goals to be accomplished at the same time – safe storage of carbon dioxide in hydrate reservoirs, and in situ release of hydrocarbon gas – would offer enormous economic potential. It certainly would provide an opportunity for hydrate scientists to utilize their detailed knowledge to provide another novel technique to add to the portfolio of strategies that could help reduce emissions of anthropogenic CO_2 to the atmosphere.

Overview and conclusions

What is encouraging about recent developments in hydrate science is the fact that, whilst great uncertainties about the potential environmental and economic impacts of hydrates remain, much new research is focusing on the key issues relating to hydrate sediment interaction. It now appears that we have left the period when hydrate research was dominated by studies of the distribution and setting of BSRs, and are now moving into one in which the primary physical questions about the physical setting of hydrates, and the processes that generated these, are uppermost. This is well illustrated by the papers in this volume. In the quest for a more detailed understanding of the distribution and stability of hydrates in sediments, a wide range of approaches have been developed, addressing questions from the microscopic to the global in scale. These have developed not only new methodologies, but also novel technologies, as may be seen in several of the papers here. As the science evolves, so do the questions and re-focusing of subsequent research. As may be seen in these papers (e.g. Kleinberg 2009), those with an interest in the economic potential of hydrates are already re-prioritizing their exploration foci. Likewise, those whose interest lies in hydrates and climate are recognizing the great importance that relatively shallow, circum-polar hydrates, may have in controlling climate, and thus are fast considering how to address the research challenges in these areas.

The hydrate community have long-recognized the importance of inter-national, -institutional and -disciplinary research. However, whilst this may have been largely stimulated by the high costs of field campaigns in the past, it is increasingly apparent that there will be many benefits in developing integrated programmes addressing issues such as hydrate–sediment relationships. Many parties have the potential to unlock part, but only a part, of a large puzzle. By coming together, researchers should be able to answer some of the key questions about sediment-hosted gas hydrates, and their stability, more rapidly than if they had done this alone. Today there are many opportunities, for instance with the IODP, and establishment of observatories, to participate in joint programmes. It is important to recognize that, whilst the top-level objectives of different scientists are often different, many of the physical observations and models that come out of the research are generic, and will have widespread applicability. It is highly likely that many advances being developed by those establishing the economic development of hydrates (e.g. in the Nankai Trough), will also assist scientists who want to address some climate-related questions about methane release, or the potential for climate change mitigation through CO_2 storage.

The meeting 'Sediment-Hosted Gas Hydrates: New Insights on Natural and Synthetic Systems' was useful in bringing together many leading edge scientists who are actively studying hydrate–sediment interactions and continuing the analysis of where the major uncertainties lie in relation to hydrate stability. The meeting marked a small, but definite, landmark in hydrate science, demonstrating the maturity of the science in some areas, and how measurements may be used to better quantify hazard or resource, as well as to new research trends. It is clear that hydrate research has a long, long, way to go before we can even confidently predict the global volume of hydrate. We look forward to watching what happens over the next decade – within which we expect to again focus on the primary interactions between hydrates and their host sediments.

The authors acknowledge the presenters and attendees of the Sediment-Hosted Gas Hydrates: New Insights on Natural and Synthetic Systems conference held at the Geological Society of London in February 2006 whose active participation encouraged the production of this volume. The authors thank Peter Jackson for reviewing this paper. DL, JGR and CAR publish with permission of the Executive Director of the British Geological Survey (NERC).

References

ANDERSON, R., LLAMEDO, C., TOHIDI, B. & BURGASS, R. W. 2003. Experimental measurement of methane and carbon dioxide clathrate hydrate equilibria in mesoporous silica. *Journal of Physical Chemistry B*, **107**, 3507–3514.

BACHU, S., GUNTER, W. D. & PERKINS, E. H. 1994. Aquifer disposal of CO_2: Hydrodynamic and mineral trapping. *Energy Conversion and Management*, **35**, 269–279.

BAKLID, A., KORNØ, L. R. & OWREN, G. 1996. Sleipner Vest CO_2 disposal, CO_2 injection into a shallow underground aquifer. *Society of Petroleum Engineers*, **36600**, 269–277.

BROOKS, J. M. & BRYANT, W. R. 1985. *Geological and Geochemical Implications of Gas Hydrates in the Gulf of Mexico*. Final report to Department of Energy, Morgantown Energy Technology Centre, West Virginia.

CAMPS, A. P. 2007. *Hydrate formation in near surface ocean sediments*. Unpublished PhD, Department of Geology, University of Leicester.

CLENNELL, M. B., HOVLAND, M., BOOTH, J. S., HENRY, P. & WINTERS, W. J. 1999. Formation of natural gas hydrates in marine sediments. Part 1: Conceptual model of gas hydrate growth conditioned by host sediment properties. *Journal of Geophysical Research B*, **104**, 22985–23003.

COLLETT, T. S. 1983. Detection and evaluation of natural gas hydrates from well logs, Prudhoe Bay, Alaska. *Proceedings of the 4th International Conference on Permafrost*, Fairbanks, AL, 169–174.

COLLETT, T. S. & KUUSKRAA, V. A. 1998. Hydrates contain vast store of world gas resources. *Oil and Gas Journal*, **96**, 90–95.

COOK, A. E. & GOLDBERG, D. 2007. Gas hydrate filled fracture distribution, eastern Indian continental margin. *American Geophysical Union Fall Meeting*, San Francisco, CA, AN: OS11C-04.

DAVIES, R. J. & CARTWRIGHT, J. A. 2002. A fossilised Opal A to C/T transformation on the northeast Atlantic margin. *Basin Research*, **14**, 467–486.

DAVY, H. 1811. On some of the combinations of oxy-muriatic gas and oxygen, and on the chemical relations of the principles to inflammable bodies. *Philosophical Transactions of the Royal Society, London*, **101**, 1–35.

DICKENS, G. R., CASTILLO, M. M. & WALKER, J. C. G. 1997. A blast of gas in the latest Paleocene: Simulating first-order effects of massive dissociation of oceanic methane hydrate. *Geology*, **25**, 259–262.

DVORKIN, J., HELGERUD, M., WAITE, W., KIRBY, S. & NUR, A. 2000. Introduction to physical properties and elasticity models. *In*: MAX, M. D. (ed.) *Natural Gas Hydrate in Oceanic and Permafrost Environments*, Kluwer Academic, Dordrecht, 245–260.

ECKER, C., DVORKIN, J. & NUR, A. 2000. Estimating the amount of gas hydrate and free gas from marine seismic data. *Geophysics*, **65**, 565–573.

ERWIN, D. H. 1994. The Permo-Triassic extinction. *Nature*, **367**, 231–236.

FINLEY, P. & KRASON, J. 1986. *Geological Evolution and Analysis of Confirmed or Suspected Gas Hydrate Localities: Basin Analysis, Formation and Stability of Gas Hydrates in the Middle America Trench*. US Department of Energy, DOE/MC/21181–1950, **9**.

GOEL, N. 2006. *In situ* methane hydrate dissociation with carbon dioxide sequestration: Current knowledge and issues. *Journal of Petroleum Science and Engineering*, **51**, 169–184.

GUNTER, W. D., PERKINS, E. H. & HUTCHEON, I. 2000. Aquifer disposal of acid gases: Modelling of water-rock reactions for trapping of acid wastes. *Applied Geochemistry*, **15**, 1085–1095.

GUNTER, W. D., PERKINS, E. H. & MCCANN, T. J. 1993. Aquifer disposal of CO_2-rich gases: Reaction design for added capacity. *Energy Conversion Management*, **34**, 941–948.

GUNTER, W. D., WIWCHAR, B. & PERKINS, E. H. 1997. Aquifer disposal of CO_2-rich greenhouse gases: Extension of the time scale of experiment for CO_2-sequestering reactions by geochemical modelling. *Mineralogy and Petrology*, **59**, 121–140.

HAMMERSCHMIDT, E. G. 1934. Formation of gas hydrates in natural gas transmission lines. *Industrial Engineering Chemistry*, **26**, 851–855.

HAQ, B. U. 1988. Natural gas hydrates: Searching for the long-term climatic and slope-stability records. *In*: HENRIET, J.-P. & MIENERT, J. (eds) *Gas Hydrates: Relevance to World Margin Stability and Climate Change*. Geological Society, London, Special Publications, **137**, 303–318.

HENRY, P., THOMAS, M. & CLENNELL, M. B. 1999. Formation of natural gas hydrates in marine sediments. Part 2: Thermodynamic calculations of stability conditions in porous sediments. *Journal of Geophysical Research B*, **104**, 23005–23022.

HOUSE, K. Z., SCHRAG, D. P., HARVEY, C. F. & LACKNER, K. S. 2006. Permanent carbon dioxide storage in deep-sea sediments. *Proceedings of the National Academy of Sciences*, **103**, 12291–12295.

IEA GHG. 2000. *Issues Underlying the Feasibility of Storing CO_2 as Hydrate Deposits*. IEA Greenhouse Gas R&D Programme Report **PH3/25**.

INAGAKI, F., NUNOURA, T. *ET AL.* 2006. Biogeographical distribution and diversity of microbes in methane hydrate-bearing deep marine sediments on the Pacific

Ocean Margin. *Proceedings of the National Academy of Sciences*, **103**, 2815–2820.

IPCC. 1990. First assessment report. Published by the IPCC. World Wide Web Address: http://www.ipcc.ch/ipccreports/assessments-reports.htm.

IPCC. 1995. Second assessment report: Climate change. World Wide Web Address: http://www.ipcc.ch/ipccreports/assessments-reports.htm.

IPCC. 2001. Third assessment report: Climate change. World Wide Web Address: http://www.ipcc.ch/ipccreports/assessments-reports.htm.

IPCC. 2007. Fourth assessment report: Climate change. World Wide Web Address: http://www.ipcc.ch/ipccreports/assessments-reports.htm.

KIODE, H., TAKAHASHI, M. ET AL. 1997. Hydrate formation in sediments in the sub-seabed disposal of CO_2. *Energy*, **22**, 279–283.

KLAUDA, J. B. & SANDLER, S. I. 2005. Global distribution of methane hydrate in ocean sediment. *Energy and Fuels*, **19**, 459–470.

KLEINBERG, R. L., FLAUM, C. ET AL. 2003. Deep sea NMR: Methane hydrate growth habit in porous media and its relationship to hydraulic permeability, deposit accumulation, and submarine slope stability. *Journal Geophysical Research*, **108**, 2508.

KOH, C. A. & SLOAN, D. E. 2007. Natural gas hydrates: Recent advances and challenges in energy and environmental applications. *American Institute of Chemical Engineers Journal*, **53**, 1636–1643.

KRULL, S. J. & RETALLACK, J. R. 2000. ^{13}C depth profiles from paleosols across the Permian–Triassic boundary: Evidence for methane release. *GSA Bulletin*, **112**, 1459–1472.

KVENVOLDEN, K. A. 1998. A primer on the geological occurrence of gas hydrate. In: HENRIET, J.-P. & MEINERTS, J. (eds) *Gas Hydrates: Relevance to World Margin Stability Change*. Geological Society, London, Special Publications, **137**, 9–30.

KVENVOLDEN, K. A. 2000. Gas hydrate and humans. In: HOLDER, G. D. & BISHNOI, P. R. (eds) *Gas Hydrates: Challenges for the Future*. Annals of the New York Academy of Sciences, **912**, 17–22.

KVENVOLDEN, K. A. & MCDONALD, T. J. 1985. *Gas Hydrates of the Middle America Trench DSDP Leg 84*. Initial Reports, **DSDP 84**. US Government Printing Office, Washington, DC, 367–375.

LANCELOT, Y. & EWING, J. I. 1972. Correlation of natural gas zonation and carbonate diagenesis in Tertiary sediments from the north-west Atlantic. In: HOLLISTER, C. D., EWING, J. I. ET AL. (eds) *Initial Reports*, **DSDP 11**. US Government Printing Office, Washington, DC, 791–799.

LEVITUS, S., ANTONOV, J. & BOYER, T. 2005. Warming of the world ocean, 1955–2003. *Geophysical Research Letters*, **32**, L02604.

LLAMEDO, M., ANDERSON, R. & TOHIDI, B. 2004. Thermodynamic prediction of clathrate hydrate dissociation conditions in mesoporous media. *American Mineralogist*, **89**, 1264–1270.

MACLENNAN, J. & JONES, S. M. 2006. Regional uplift, gas hydrate dissociation and the origins of the Paleocene–Eocene thermal maximum. *Earth and Planetary Science Letters*, **245**, 65–80.

MAKOGON, Y. F., HOLDITCH, S. A. & MAKOGON, T. Y. 2007. Natural gas-hydrates – a potential energy source for the 21st century. *Journal of Petroleum Science and Engineering*, **56**, 14–31.

MALK, Q. M. & ISLAM, M. R. 2000. CO_2 Injection in the Weyburn Field of Canada: Optimization of Enhanced Oil Recovery and Greenhouse Gas Storage with Horizontal Wells. Society of Petroleum Engineers, **59327**.

MALONE, R. 1985. *Gas Hydrates Topical Report*, **DOE/METC/SP-218 (DE85001986)**. Department of Energy, Morgantown Energy Technology Center, USA.

MILKOV, A. V. 2003. Global estimates of hydrate-bound gas in marine sediments: How much is really out there? *Earth-Science Reviews*, **66**, 183–197.

MILKOV, A. V. & SASSEN, R. 2001. Estimate of gas hydrate resource, northwestern Gulf of Mexico continental slope. *Marine Geology*, **179**, 71–83.

RCEP. 2000. *Energy – the Changing Climate*. Twenty-second report of the Royal Commission on Environmental Pollution, Cm. **4749**.

RIEDEL, M., COLLETT, T. S., MALONE, M. J. & THE EXPEDITION 311 SCIENTISTS. 2006. *Proceedings of IODP 311*. Integrated Ocean Drilling Program Management International, Inc., Washington, DC.

SAKAI, H., GAMO, T. ET AL. 1990. Venting of carbon dioxide-rich fluid and hydrate formation in mid-Okinawa trough backarc basin. *Science*, **248**, 1093–1096.

SLOAN, E. D. JR. 1998. *Clathrate Hydrates of Natural Gases*. Marcel Dekker, New York.

SLUIJS, A. 2006. *Global Change During the Paleocene–Eocene Thermal Maximum*. PhD thesis, Utrecht University.

SOLOVIEV, V. & GINSBURG, G. D. 1994. Formation of submarine gas hydrates. *Bulletin of the Geological Society of Denmark*, **41**, 86–94.

SWART, R. 2009. Hydrate occurrences in the Namibe Basin, offshore Namibia. In: LONG, D., LOVELL, M. A., REES, J. G. & ROCHELLE, C. A. (eds) *Sediment-Hosted Gas Hydrates: New Insights on Natural and Synthetic Systems*. Geological Society, London, Special Publications, **319**, 73–80.

TRÉHU, A. M., BOHRMANN, G., TORRES, M. E. & COLWELL, F. S. (eds). 2002. *Proceedings of the Ocean Drilling Program, Scientific Results*, **204**.

TRÉHU, A. M., LONG, P. E. ET AL. 2004. Three dimensional distribution of gas hydrate beneath southern hydrate ridge: Constraints from ODP Leg 204. *Earth and Planetary Science Letters*, **222**, 845–862.

WHITE, D. J., BURROWES, G. ET AL. 2004. Greenhouse gas sequestration in abandoned oil reservoirs: The International Energy Agency Weyburn pilot project. *GSA Today*, **14**, 4–10.

WHITE, R. V. 2002. Earth's biggest 'whodunnit': Unravelling the clues in the case of the end. *Philosophical Transactions of the Royal Society, London*, **360**, 2963–2985.

WILSON, M. & MONEA, M. (eds). 2004. IEA GHG Weyburn CO_2 Monitoring & Storage Project Summary Report 2000–2004. *Proceedings of the 7th International Conference on Greenhouse Gas*

Control Technologies, 5–9 September 2004, Vancouver, Canada, **III**. Petroleum Technology Research Centre, Regina.

YAKUSHEV, V. 2004. Intrapermafrost gas hydrates at the north of west Siberia. *AAPG Hedberg Conference, Gas Hydrates: Energy Resource Potential and Associated Geologic Hazards*, 12–16 September 2004, Vancouver.

YEFRAMOVA, A. G. & ZHIZHCHENKO, B. P. 1975. Occurrence of crystal hydrates of gases in the sediments of modern marine basins. *Doklady, Earth Sciences Section*, **214**, 219–220.

ZACHOS, J. C., RÖHL, U. ET AL. 2005. Rapid acidification of the ocean during the Paleocene–Eocene thermal maximum. *Science*, **308**, 1611–1615.

Gas hydrate drilling transect across northern Cascadia margin – IODP Expedition 311

M. RIEDEL[1]*, T. COLLETT[2], M. J. MALONE[3] & IODP EXPEDITION 311 SCIENTISTS[§]

[1]*Natural Resources Canada, Geological Survey of Canada – Pacific, 9860 West Saanich Road, Sidney, BC, V8L 4B2, Canada*

[2]*US Geological Survey, Denver Federal Center, Box 25046, MS-939 Denver, CO 80225, USA*

[3]*Integrated Ocean Drilling Program, Texas A&M University, 1000 Discovery Drive, College Station, TX 77845-9547, USA*

*Corresponding author (e-mail: mriedel@NRCan.gc.ca)

Abstract: A transect of four sites (U1325, U1326, U1327 and U1329) across the northern Cascadia margin was established during Integrated Ocean Drilling Program Expedition 311 to study the occurrence and formation of gas hydrate in accretionary complexes. In addition to the transect sites, a fifth site (U1328) was established at a cold vent with active fluid flow. The four transect sites represent different typical geological environments of gas hydrate occurrence across the northern Cascadia margin from the earliest occurrence on the westernmost first accreted ridge (Site U1326) to the eastward limit of the gas hydrate occurrence in shallower water (Site U1329). Expedition 311 complements previous gas hydrate studies along the Cascadia accretionary complex, especially ODP Leg 146 and Leg 204 by extending the aperture of the transect sampled and introducing new tools to systematically quantify the gas hydrate content of the sediments. Among the most significant findings of the expedition was the occurrence of up to 20 m thick sand-rich turbidite intervals with gas hydrate concentrations locally exceeding 50% of the pore space at Sites U1326 and U1327. Moreover, these anomalous gas hydrate intervals occur at unexpectedly shallow depths of 50–120 metres below seafloor, which is the opposite of what was expected from previous models of gas hydrate formation in accretionary complexes, where gas hydrate was predicted to be more concentrated near the base of the gas hydrate stability zone just above the bottom-simulating reflector. Gas hydrate appears to be mainly concentrated in turbidite sand layers. During Expedition 311, the visual correlation of gas hydrate with sand layers was clearly and repeatedly documented, strongly supporting the importance of grain size in controlling gas hydrate occurrence. The results from the transect sites provide evidence for a structurally complex, lithology-controlled gas hydrate environment on the northern Cascadia margin. Local shallow occurrences of high gas hydrate concentrations contradict the previous model of gas hydrate formation at an accretionary prism. However, long-lived fluid flow (part of the old model) is still required to explain the shallow high gas hydrate concentrations, although it is most likely not pervasive throughout the entire accretionary prism, but rather localized and focused by the tectonic processes. Differences in the fluid flow regime across all of the transect drill sites indicate site-specific and probably disconnected (compartmented) deeper fluid sources in the various parts of the accretionary prism. The data and future analyses will yield a better understanding of the geologic controls, evolution and ultimate fate of gas hydrate in an accretionary prism as an important contribution to the role of gas hydrate methane gas in slope stability and possibly in climate change.

The area of investigation for Expedition 311 is the accretionary prism of the Cascadia subduction zone (Fig. 1). The Juan de Fuca plate converges nearly orthogonally to the North American plate at a present rate of *c.* 45 mm/a^{-1} (e.g. Riddihough 1984). Seaward of the deformation front, the Cascadia Basin consists of pre-Pleistocene hemipelagic sediments overlain by rapidly deposited

---

[§]**IODP Expedition 311 Scientists**: Michael Riedel (Co-chief Scientist), Timothy S. Collett (Co-chief Scientist), Mitchell Malone (Expedition Project Manager/Staff Scientist), Gilles Guèrin, Fumio Akiba, Marie-Madeleine Blanc-Valleron, Michelle Ellis, Yoshitaka Hashimoto, Verena Heuer, Yosuke Higashi, Melanie Holland, Peter D. Jackson, Masanori Kaneko, Miriam Kastner, Ji-Hoon Kim, Hiroko Kitajima, Philip E. Long, Alberto Malinverno, Greg Myers, Leena D. Palekar, John Pohlman, Peter Schultheiss, Barbara Teichert, Marta E. Torres, Anne M. Tréhu, Jiasheng Wang, Ulrich G. Wortmann, Hideyoshi Yoshioka.

From: LONG, D., LOVELL, M. A., REES, J. G. & ROCHELLE, C. A. (eds) *Sediment-Hosted Gas Hydrates: New Insights on Natural and Synthetic Systems.* The Geological Society, London, Special Publications, **319**, 11–19. DOI: 10.1144/SP319.2   0305-8719/09/$15.00 © The Geological Society of London 2009.

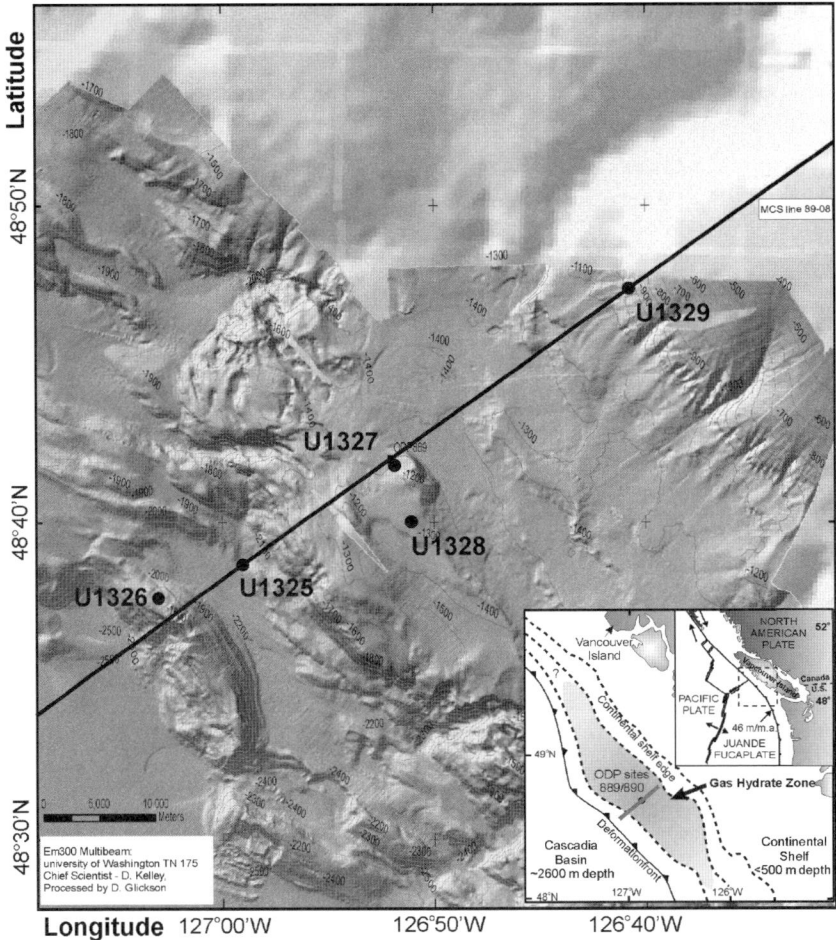

Fig. 1. Multibeam bathymetry map along transect across the accretionary prism offshore Vancouver Island established during Expedition 311 (courtesy of D. Kelly and J. Delaney, University of Washington and C. Barnes, C. Katnick, NEPTUNE Canada, University of Victoria). Inset: general tectonic location of drilling transect near previous ODP Sites 889/890. A BSR is present on c. 50% of the mid-continental slope (shaded area, from Hyndman et al. 2001).

Pleistocene turbidites for a total sediment thickness of c. 2500 m. Most of the incoming sediment is scraped off the oceanic crust and folded and thrust upward to form elongated anticlinal ridges with elevations as high as 700 m above the adjacent basins (Davis & Hyndman 1989).

Evidence for the presence of gas hydrate was first detected in 1985 and 1989 in seismic data by the occurrence of widespread bottom-simulating reflectors (BSRs). Summaries of previous gas hydrate studies including seismic methods, controlled-source electromagnetic imaging and results from drilling and coring at ODP Site 889 are given in Hyndman et al. (2001) and Spence et al. (2000).

The specific objectives of this expedition are to test gas hydrate formation models and constrain model parameters, especially those that account for the formation of concentrated gas hydrate occurrences driven by upward fluid and methane transport. This expedition concentrates on the contrast between methane transport by dispersed pervasive upward flow and focused flow in fault zones. The pervasive permeability may be on the sediment grain scale, on a centimetre scale (the scaly fabric observed in previous ODP clastic accretionary prism cores; Westbrook et al. 1994) or in closely spaced faults.

A general model for gas hydrate formation by removal of methane from upwardly expelled fluids was proposed for the Expedition 311 area (Hyndman & Davis 1992). Mainly microbial methane, inferred to be produced over a thick sediment section, migrates vertically and forms gas hydrate when it enters the stability field. The gas hydrate concentration is predicted to be greatest just above the BSR. A model has also been proposed for how free gas and the resulting BSR will be formed as the base of gas hydrate stability moves upward due to post-Pleistocene seafloor warming, uplift and sediment deposition (e.g. Paull & Ussler 1997; von Huene & Pecher 1998). In addition, physical and mathematical models have been developed for the formation of gas hydrate from upward methane advection and diffusion (e.g. Xu & Ruppel 1999).

Recently, evidence for focused fluid/gas flow and gas hydrate formation has been identified on the Vancouver Island margin. The most studied site is an active cold vent field associated with near-surface gas hydrate occurrences (e.g. Riedel et al. 2002, 2006; Schwalenberg et al. 2005). These vents are associated with fault-related conduits for focused fluid and/or gas migration associated with massive gas hydrate formation within the fault zone and represent, therefore, the opposite mechanism to the widespread fluid flow. It is so far unknown how important these cold vents are in the total budget of fluid flow in an accretionary prism. Several mechanisms have been proposed to explain the seismic blanking and nature of fluid venting in the Bullseye vent area. Riedel et al. (2002, 2006) proposed a combination of amplitude loss from diffraction at surface and subsurface heterogeneities such as carbonate outcrops and gas hydrate-filled fractures. Zühlsdorff & Spiess (2004) suggested hydraulic fracturing and associated increase in porosity, and Wood et al. (2002) suggested an up-lifted BSR and free gas as cause of seismic blanking. The objectives of coring and logging at this location were to test the different models for the cold vent structure and associated causes of seismic blanking, the rate of methane advection, and potential loss of methane into the water column.

Drilling strategy

Logging while drilling and measurement while drilling (LWD/MWD) carried out prior to coring provided a set of measurements that guided subsequent coring and special tool deployments at all five sites. Additional wireline logging at each site and two vertical seismic profiles (VSP) at Sites U1327 and U1328 were completed. A total of 1217.76 m of sediment core was recovered using the advanced piston corer (APC) and extended core barrel (XCB) systems. Standard coring was interspersed with pressure core sampler (PCS) runs for onboard degassing experiments and special Hyacinth deployments, of which four were stored under in situ pressure for subsequent shore-based studies.

Infrared (IR) imaging of the recovered core was used to quickly identify gas hydrate that may be preserved in the core for direct sub-sampling (preservation with liquid nitrogen). Special care was taken to also sample cold IR anomalies for interstitial water (IW) freshening to confirm the presence of gas hydrate as well as sampling for shore-based microbiological analyses.

At each site we tried to accomplish a three-hole conventional coring, specialized coring, and logging approach to maximize the scientific objectives. The first hole (Hole A) was always dedicated to LWD/MWD operations. All LWD/MWD holes were drilled within the first week of the expedition and were followed by coring and conventional wireline operations. The second hole (Hole B) in most cases was for continuous coring, temperature measurements, and PCS coring to establish complete downhole profiles of gas hydrate proxies (such as interstitial water chlorinity, infrared images, etc.). The third hole (Hole C) was dedicated to special tool deployments, especially the two Hyacinth pressure coring systems, the Fugro Pressure Corer (FPC) and Hyace Rotary Corer (HRC), and additional PCS deployments (see Tréhu et al. 2003 for additional information about these pressure coring systems). In all coring holes we attempted to recover a detailed temperature profile by regularly deploying the Advance Piston Corer Temperature (APCT), APC3 and the Davis Villinger Temperature Pressure (DVTP/P) tools. However, severe weather conditions during portions of the expedition resulted in degraded data quality for some deployments, resulting in less than desirable constraints on the temperature-gradient at all sites.

Transect across margin

All results from the drilling transect and the cold vent site are represented in Figures 2–4 as a collage of lithostratigraphic units and resistivity-at-the-bit (RAB) images from the LWD/MWD deployment compared with pore-fluid geochemistry (Fig. 2), sediment gas geochemistry (Fig. 3) and gas hydrate concentration estimates using the LWD/MWD resistivity data (Fig. 4). The approximate depths of the BSR as predicted from seismic site survey data are also shown.

Site U1326

Site U1326 is the westernmost transect site located on the first uplifted ridge of the accretionary wedge (Fig. 1). The sediments are heavily faulted with faults intersecting the entire sediment column from

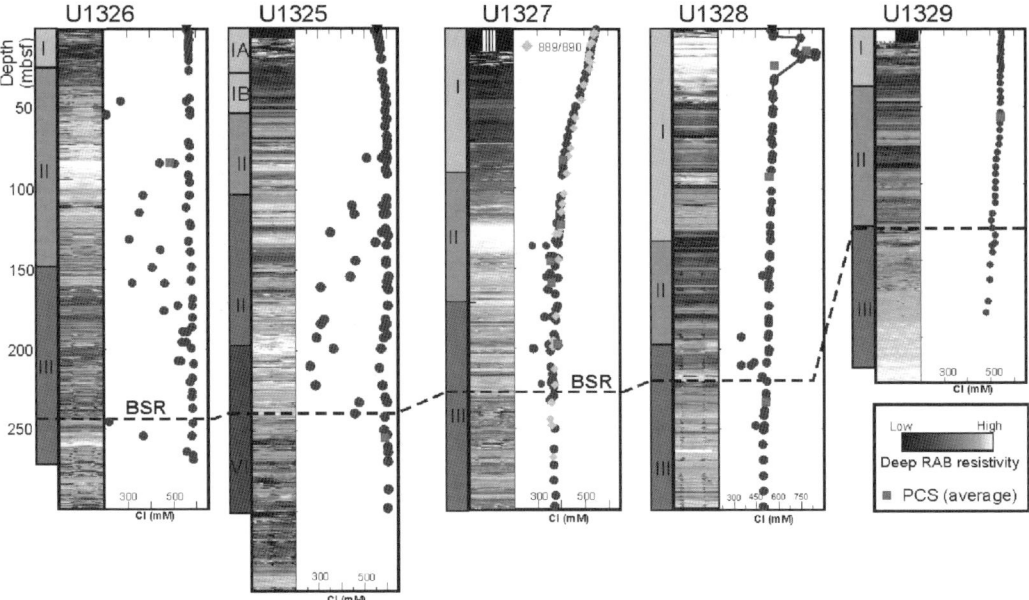

Fig. 2. Summary of Expedition 311 interstitial water (IW) chlorinity profiles, including average IW Chlorinity from PCS core. Site 889/890 data are shown for comparison. Also shown are lithostratigraphic units, LWD RAB resistivity images and location of the BSR.

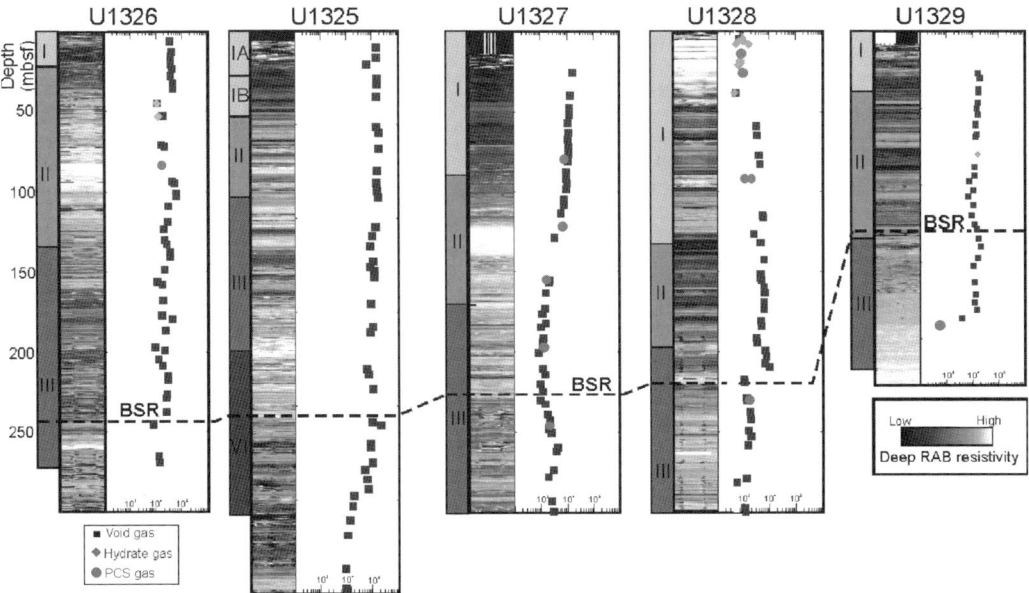

Fig. 3. Summary of Expedition 311 $C_1 : C_2$ gas profiles, including data from void, gas hydrate and PCS gas. Also shown are lithostratigraphic units and LWD RAB resistivity images and location of the BSR.

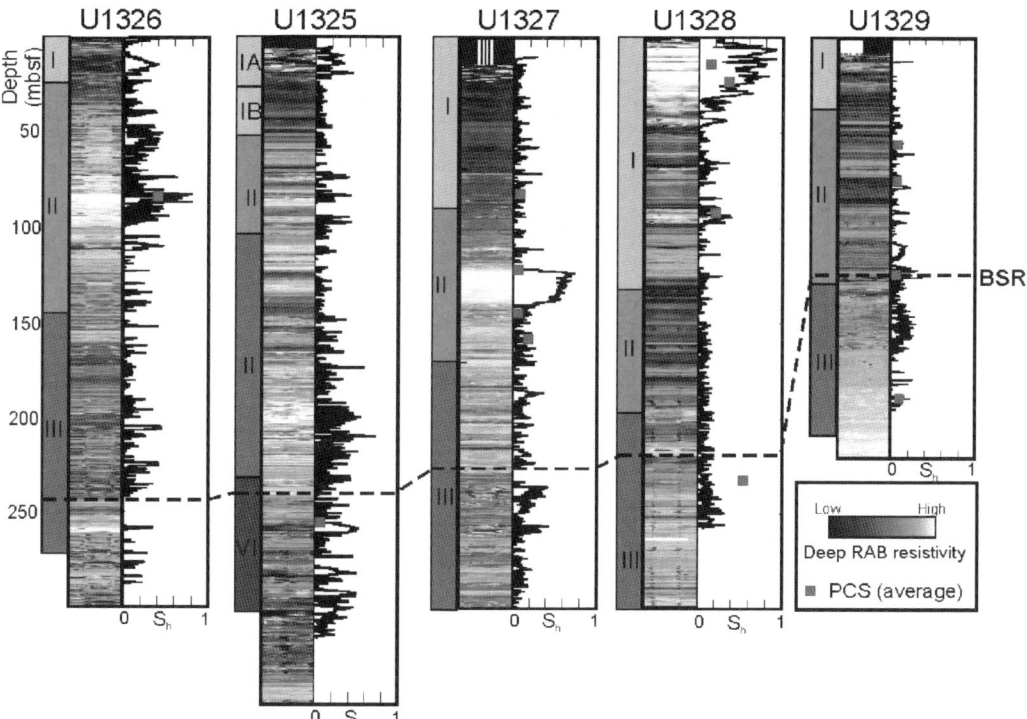

Fig. 4. Summary of Expedition 311 gas hydrate concentration (S_h) derived from LWD data using Archie's relation. Also shown are lithostratigraphic units and LWD RAB resistivity images and location of the BSR.

the seafloor to below the BSR. They also show linear outcrops on the seafloor around a prominent collapse structure near the centre of the ridge. The presence of gas hydrate was suggested by the occurrence of a widespread BSR in site-survey seismic data. LWD/MWD-derived RAB images confirmed the presence of gas hydrate, by exhibiting high electrical resistivities between 72 and 240 metres below seafloor (mbsf). The logging data yielded evidence for relatively shallow gas hydrate occurrences between $c.$ 50 and 100 mbsf. Gas hydrate concentrations as high as 80% of the pore space were inferred from the logs in a $c.$ 5 m sand-rich layer between 82 and 87 mbsf. P-wave velocities measured by wireline logging show highly elevated values between 1750 and >3000 m s^{-1} in the same high-resistivity interval (Fig. 4). The interstitial water chlorinity profile shows an almost constant, near-seawater baseline trend with depth (Fig. 2). Superimposed on this baseline, abundant low chlorinity anomalies associated with gas hydrate are present. Chlorinity anomalies associated with gas hydrate may actually extend to 270 mbsf at this site, which is deeper than the anticipated BSR depth of 230 mbsf. This apparent conflict between the predicted depth of the BSR and the observed occurrence of gas hydrate may be explained by the borehole temperature measurements that yield a methane hydrate stability zone depth of 275 ± 25 mbsf; however, for the most part, data from the temperature tool deployments were degraded and we have extrapolated the temperature data from Site U1325 for this calculation.

Hydrocarbon headspace gas measurements show that methane is the dominant hydrocarbon gas within the cored interval (Fig. 3). The C_{2+} hydrocarbon void gas concentrations were <125 ppmv for all samples. Low $C_1 : C_2$ ratios within the interval from $c.$ 35 to $c.$ 72 mbsf were associated with two recovered gas hydrate samples. With greater depth, ethane concentrations returned to the near-surface concentrations close to the predicted depth of the BSR. Isobutane concentrations were also elevated within the same interval, which may indicate the presence of Structure II gas hydrate. The occurrence of more complex hydrocarbon gases in combination with the pore-water chlorinity baseline shift with depth generally indicates the contribution of a deeper gas source.

The shallow and especially highly concentrated gas hydrate occurrence at Site U1326 combined with the lack of high gas hydrate concentrations near the BSR contradicts the former model of gas hydrate formation at an accretionary margin. Combined drilling and seismic observations suggest strong lithologic control of the gas hydrate occurrence with preferred gas hydrate formation in sandy turbidites.

Site U1325

Site U1325 is located within a major slope basin that developed eastward of the deformation front behind the first ridge of accreted sediments (Fig. 1). Most of this basin is underlain by a continuous BSR. LWD/MWD data suggest that gas hydrate is concentrated in thin sand layers within an interval between 173 and 240 mbsf close to the base of the gas hydrate stability field. The logs also show that the gas hydrate occurrence is heterogeneous, composed of alternating layers of gas hydrate–saturated sands and clay-rich layers with little to no gas hydrate (Fig. 4). This is in general agreement with the marked freshening of the interstitial waters observed in sampled sand layers. In a zone extending from c. 70 to 240 mbsf, salinity and chlorinity data show discrete excursions to fresher values, indicating that gas hydrate was present in the cores (Fig. 2). In this zone, soupy and mousse-like textures were also observed in the recovered cores. The overall increase of salinity and chlorinity with depth indicates an advective fluid transport system in the basin. The elevated salinity and chlorinity concentration at depth may be caused by low-temperature diagenetic hydration reactions, probably alteration of volcanic ash to clay minerals and/or zeolites in the deeper parts of the basin. Organic geochemical studies of headspace samples, void gas and gas samples recovered during the PCS degassing experiment indicated that the recovered gas was almost entirely methane with a small percentage of carbon dioxide (c. 0.1–0.5%). Trace quantities of C_{2+} hydrocarbons (<5 ppmv) were present above the predicted BSR depth (Fig. 3). With greater depth, the concentrations of ethane, propane and isobutane increased but did not exceed 15 ppmv. Decreasing $C_1 : C_2$ values below the BSR indicate a slight thermogenic or diagenetic contribution of ethane.

The observations at Site U1325 are opposite to those at Site U1326 in that they are more consistent with the traditional gas hydrate formation model having higher gas hydrate concentrations near the BSR. Strong lithological control of gas hydrate formation was, however, also confirmed at this site.

Site U1327

Site U1327 is located near ODP Leg 146 Sites 889/890, approximately at the mid-slope of the accretionary prism over a prominent BSR (Fig. 1). Sediments at Site U1327 are characterized by a c. 90 m thick cover of no gas hydrate-bearing slope-basin sediments underlain by a thick section of accreted sediments. Pre-coring LWD/MWD logging showed a thick section with consistently high electrical resistivities and low density values from 120–138 mbsf. The data suggest that gas hydrate is filling up to 50% of the pore volume in this interval (Fig. 4). However, the same interval was penetrated in the adjacent core holes, Hole U1327C and Hole U1327D, at much greater depths and with lower estimated gas hydrate concentrations inferred from the wireline electrical log in Hole U1327D. No evidence of this high resistivity interval was found in the wireline logs in Hole U1327E. This demonstrates large intrasite variability in gas hydrate content that is probably controlled by lithostratigraphic changes or structural complexities. This observation has important implications for the calibration of geophysical surveying techniques such as seismic methods or controlled-source electromagnetic surveys, which have detection limits on the order of several tens of metres, significantly above the size of the gas hydrate occurrences delineated during Expedition 311.

From 128 mbsf to the depth of the BSR the chlorinity and salinity profiles exhibit distinct anomalies indicating freshening from gas hydrate dissociation (Fig. 2). Chloride values as low as 70 mM correlated to individual turbidite sand layers. The strongly decreasing background chlorinity profile from the surface to below the BSR suggests mixing between in situ seawater and modified deep-sourced, relatively fresh pore water. The predominant hydrocarbon gas found in the cores from Site U1327 was methane; however, we did see an increase in ethane concentrations in the void and headspace gases collected from the stratigraphic section overlying the projected depth of the BSR (Fig. 3). The increase in ethane concentrations near the BSR is reflected in the $C_1 : C_2$ void gas ratios, which decrease with depth toward the BSR.

The combined findings at Site U1327 show a similar lithology-controlled gas hydrate setting as Site U1326 with massive shallow gas hydrate occurrences and low gas hydrate concentrations near the BSR. However, the overall underlying fluid geochemistry is completely different than at the previous two sites.

Site U1329

Site U1329 represents the eastern limit of gas hydrate occurrence on the northern Cascadia margin (Fig. 1). A faint BSR was identified in seismic data at an approximate depth of only 126 mbsf. Site U1329 is located near the foot of a relatively steep slope, and sedimentation at this site is dominated by slope processes. The stratigraphy at Site U1329 is marked by the occurrence of an unconformity at a depth of c. 130 mbsf separating younger sediments (0.3 to 2.0 Ma) from sediments of late Miocene age (>6.7 Ma). Recovered sediments did not show any sign of prominent IR anomalies or soupy or mousse-like textures indicative of the presence of gas hydrate. Similarly there are no obvious anomalies in the downhole porewater chlorinity profile (Fig. 2). The predominant hydrocarbon gas found in the cores from Site U1329 was methane; however, we did see an increase in ethane concentrations in the void gases collected from the stratigraphic section overlying the estimated depth of the BSR (Fig. 3). The $C_1 : C_2$ ratios were generally high, suggesting a microbial origin for the observed methane. Samples collected from deeper than 180 mbsf at this site, however, exhibited significant increases in C_2–C_5 concentrations, which suggests an influence from a thermogenic hydrocarbon source. The combined logging data also do not show any evidence of significant gas hydrate concentrations.

Cold vent Site U1328

Site U1328 is located within a cold vent field consisting of at least four vents associated with near-surface faults. The cold vents are characterized by seismic blank zones that are between 80 m and several 100 m wide (Riedel *et al.* 2002, 2006). The most prominent vent in the field, referred to as Bullseye vent, was the target of this site and has been the subject of intensive geophysical and geochemical studies since 1999. Site U1328 is different from all of the other sites visited during this expedition in that it represents an area of focused fluid flow.

The most striking feature in the LWD/MWD logs is the occurrence of layers with high resistivities (>25 Ωm) alternating with zones of lower resistivity (1–2 Ωm) in the upper c. 40 mbsf. The high resistivities probably indicate the presence of gas hydrate. Below this uppermost high resistivity zone, the LWD/MWD logs did not show much evidence for gas hydrate occurrence with the exception of few intervals where steeply dipping fractures with high resistivity values, probably the result of gas hydrate, were penetrated. These steeply dipping fractures may act as gas migration conduits that feed the gas hydrate accumulation observed near the seafloor. Massive gas hydrate (some methane–H_2S gas hydrate; Fig. 5b) and gas hydrate-bearing fractures (Fig. 5c) were sampled near the seafloor and evidence of gas hydrate was found in the recovered cores in the form of soupy and mousse-like textures and cold IR anomalies.

Analysis of wireline logging acoustic velocities and waveform amplitudes identified the occurrence of free gas in the interval between 210 and 220 mbsf, near the predicted depth of the BSR in Hole U1328C, where the *P*-wave velocity drops slightly and the *S*-wave velocity increases. Additional evidence for the occurrence of free gas below the BSR was noted in the analysis of the LWD/MWD acoustic coherence data, borehole fluid pressure response during drilling, and the wireline *P*-wave, resistivity, density and neutron logs. The PCS degassing experiments also indicate the presence of a free gas phase below the BSR at this site. However, it should be noted that the zero offset VSP experiment carried out in Hole U1328E did not show any sign of the presence of gas hydrate and free gas with uniform velocity

(a)

(b)

(c)
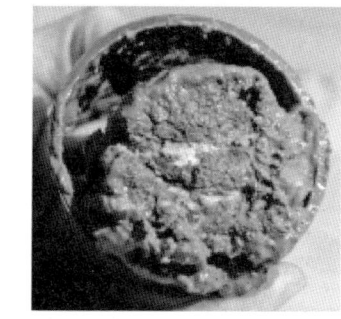

Fig. 5. Examples of gas hydrate occurrences found during IODP Expedition 311. (**a**) Typical gas-hydrate-bearing sand-dominated turbidite layer (left) separated from gas hydrate-free mud (right); (**b**) massive gas hydrate nodule found at Site U1328 (Bullseye vent); and (**c**) fracture-filling gas hydrate found Site U1328 (Bullseye vent).

near 1645 m s^{-1} was recorded for the entire depth range between 105 and 286 mbsf.

The composite chlorinity profile for this site shows a distinct increase (>30% increase relative to standard seawater) in the upper c. 40 mbsf similar to observations at the summit of southern Hydrate Ridge made during ODP Leg 204 (Fig. 2). These elevated values are interpreted to result from salt exclusion during active *in situ* gas hydrate formation. Below this depth chlorinity linearly decreases with few discrete excursions to fresher chlorinity values, indicating the presence of gas hydrate. The most prominent hydrocarbon gas at Site U1328 was methane (Fig. 3); however, ethane was also present in almost all of the headspace samples. It is also notable that the concentrations of ethane, propane and isobutane increase from the cores crossing the predicted depth of the BSR. The occurrence of propane and elevated isobutane to *n*-butane (i-C$_4$: n-C$_4$) ratio suggests that the gas hydrate near the BSR contains Structure II gas hydrate.

The combined observations at Site U1328 suggest a complex, laterally highly variable network of fractures that channel fluid and gas to the seafloor where they form a massive gas hydrate cap in the uppermost c. 40 mbsf, consistent with previous geophysical observations at Bullseye vent.

Summary

Expedition 311 established the first margin-wide transect of drill sites through an accretionary prism expressly to study gas hydrate. The sites represent various stages in the gas hydrate formation history capturing typical accretionary wedge environments such as accreted ridges and slope basins filled with either undisturbed or slightly deformed slope-sediments and accreted sediments that lack seismic coherency as a result of tectonic stresses.

Indirect evidence of the presence of gas hydrate included increased electrical resistivities and P-wave velocities on downhole logs and low-salinity interstitial water anomalies, numerous infrared cold spots and decreases in void gas C$_1$: C$_2$ ratios, as well as gas hydrate-related sedimentological mousse-like and soupy textures in recovered cores. Gas hydrate was also observed directly in recovered cores (Fig. 5), and >30 gas hydrate samples were preserved in liquid nitrogen for shore-based studies.

The combined observations show that along this transect of sites gas hydrate occurs within coarser-grained turbidite sands and silts. The occurrence of gas hydrate appears to be controlled by several key factors, including (1) local methane solubility linked with pore water salinity, (2) fluid/gas advection rates and (3) availability of suitable host material (coarse-grained sediments). The concentration of gas hydrate changes significantly as these factors vary in the sediments along the margin.

In previous published models for gas hydrate formation in an accretionary margin, the highest concentrations of gas hydrate were expected to occur near the base of the gas hydrate stability zone above the BSR, with concentrations gradually decreasing upward as a result of pervasive fluid advection from overall tectonically driven fluid expulsion. However, the results of Expedition 311 show that this model is too simple and that there are additional controlling factors. Although evidence for widespread gas hydrate-related BSRs was observed in the data, by far the largest concentrations of gas hydrate were observed well above the base of gas hydrate stability zone, at a point where the amount of methane in the pore fluid exceeded the local methane solubility threshold. This condition was most evident at Sites U1326 and U1327, where gas hydrate was observed in sections several tens of metres thick at a shallow depths of c. 100 mbsf; with concentrations exceed 80% of the pore volume. Another site of very high gas hydrate concentrations was the cold vent Site U1328, where beds containing massive forms of gas hydrate occurred within the top c. 40 mbsf with concentrations exceeding 80% of the pore space as a result of focused fluid/gas migration along imaged fault systems from underneath.

Samples and data were provided by the Integrated Ocean Drilling Program (IODP), which is funded by the US National Science Foundation and participating countries under management of the Joint Oceanographic Institutions (JOI), Inc. We thank the Captain and crew of the *Joides Resolution* and the technical staff for their support at sea. Additional thanks goes to the onboard and post-cruise scientific community, whose research results were used in this paper.

References

DAVIS, E. E. & HYNDMAN, R. D. 1989. Accretion and recent deformation of sediments along the northern Cascadia subduction zone. *Geological Society of America Bulletin*, **101**, 1465–1480.

HYNDMAN, R. D. & DAVIS, E. E. 1992. A mechanism for the formation of methane hydrate and seafloor bottom-simulating reflectors by vertical fluid expulsion. *Journal of Geophysical Research*, **97**, 7025–7041.

HYNDMAN, R. D., SPENCE, G. D., CHAPMAN, N. R., RIEDEL, M. & EDWARDS, R. N. 2001. Geophysical studies of marine gas hydrate in northern Cascadia. *In*: PAULL, C. K. & DILLON, W. P. (eds) *Natural Gas Hydrates, Occurrence, Distribution and Detection*. Geophysics Monographs, **124**, 273–295.

PAULL, C. K. & USSLER, W., III. 1997. Are low salinity anomalies below BSRs a consequence of interstitial gas bubble barriers? *Eos, Transactions of the American Geophysics Union*, **78**, F339.

RIDDIHOUGH, R. P. 1984. Recent movements of the Juan de Fuca plate system. *Journal of Geophysical Research*, **89**, 6980–6994.

RIEDEL, M., SPENCE, G. D., CHAPMAN, N. R. & HYNDMAN, R. D. 2002. Seismic investigations of a vent field associated with gas hydrates, offshore Vancouver Island. *Journal of Geophysical Research*, **107**; doi: 10.1029/2001JB000269.

RIEDEL, M., NOVOSEL, I., SPENCE, G. D., HYNDMAN, R. D., CHAPMAN, R. N., SOLEM, R. C. & LEWIS, T. 2006. Geophysical and Geochemical Signatures Associated with Gas Hydrate Related Venting at the North Cascadia Margin. *GSA Bulletin*, **118**; doi: 10.1130/B25720.1.

SCHWALENBERG, K., WILLOUGHBY, E. C., MIR, R. & EDWARDS, R. N. 2005. Marine gas hydrate signatures in Cascadia and their correlation with seismic blank zones, *First Break*, **23**, 57–63.

SPENCE, G. D., HYNDMAN, R. D., CHAPMAN, N. R., RIEDEL, M., EDWARDS, N. & YUAN, J. 2000. Cascadia Margin, Northeast Pacific Ocean: Hydrate. Distribution from geophysical observations. *In*: MAX, M. D. (ed.) *Natural Gas Hydrates*. Kluwer Academic, Dordrecht, 183–198.

TRÉHU, A. M., BOHRMANN, G. *ET AL.* 2003. *Proceedings of ODP, Initial Reports*, **204**. World Wide Web Address: http://www-odp.tamu.edu/publications/204_IR/204ir.htm assested 9 June 2006.

VON HUENE, R. & PECHER, I. A. 1998. Vertical tectonics and the origins of BSRs along the Peru margin. *Earth and Planetary Science Letters*, **166**, 47–55; doi: 10.1016/S0012-821X(98)00274-X.

WESTBROOK, G. K., CARSON, B. *ET AL.* 1994. *Proceedings of ODP, Initial Reports*, **146** (pt 1). Ocean Drilling Program, College Station, TX.

WOOD, W. T., GETTRUST, J. F., CHAPMAN, N. R., SPENCE, G. D. & HYNDMAN, R. D. 2002. Decreased stability of methane hydrates in marine sediments owing to phase-boundary roughness. *Nature (London)*, **420**, 656–660; doi: 10.1038/nature01263.

XU, W. & RUPPEL, C. 1999. Predicting the occurrence, distribution, and evolution of methane gas hydrate in porous marine sediments. *Journal of Geophysical Research*, **104**, 5081–5096; doi: 10.1029/1998JB900092.

ZÜHLSDORFF, L. & SPIESS, V. 2004. Three-dimensional seismic characterization of a venting site reveals compelling indications of natural hydraulic fracturing. *Geology*, **32**, 101–104; doi: 10.1130/G19993.1.

# Exploration strategy for economically significant accumulations of marine gas hydrate

R. L. KLEINBERG

Schlumberger-Doll Research, One Hampshire Street, Cambridge, MA 02139, USA
(e-mail: kleinberg@slb.com)

Abstract: There are at present no validated methods for reliably finding economically significant accumulations of natural gas hydrate in marine environments. The seismic bottom simulating reflector (BSR) has been regarded as a primary indicator of hydrate presence in marine environments, but the presence of a BSR conveys no information about the abundance of hydrate in the sediments above it. Seafloor features such as gas seeps, pockmarks or hydrate outcrops may be qualitative markers of deeper hydrate presence, but cannot be interpreted quantitatively. Another approach to exploration geophysics is required to find exploitable gas hydrate reservoirs with high reliability. It is known that in many cases gas is supplied to the gas hydrate stability zone primarily through faults or fractures. In a certain range of gas flux, these fissures should become mineralized with gas hydrate and form vertical or subvertical dykes. The dip and strike of these dykes are controlled by the principal stress directions, which can be predetermined. Thus multiple hydrate dykes are expected to be parallel. Even if the greatest volume of gas hydrate is to be found in sub-horizontal permeable beds, the steeply dipping mineralized conduits that fed gas to them may be the most reliable marker of substantial subsurface hydrate presence. Geological and geophysical survey methods sensitive to parallel arrays of vertical and subvertical hydrate dykes are presented.

Gas hydrate is thought to be an important component of deep water marine sediments, and is thus of interest to those looking for economically significant sources of fossil fuels. Indeed, seismic evidence of offshore gas hydrate deposits is widespread. A compilation (Kvenvolden & Lorenson 2001) documents more than one-hundred occurrences of hydrate on continental margins and in inland seas.

However, drilling campaigns in some promising offshore areas, such as Blake Ridge offshore South Carolina (Paull *et al.* 2000), and Hydrate Ridge offshore Oregon (Tréhu *et al.* 2004a), have shown gas hydrate to be generally dilute throughout the gas hydrate stability zone. Higher concentrations can be found in limited depth intervals, sometimes coincident with beds of relatively coarse permeable sand within otherwise fine-grained marine sediments.

If marine gas hydrate is abundant, the mechanisms by which it accumulates below the seafloor may make it difficult or impossible to find using the conventional techniques of exploration geophysics as presently applied. This chapter presents a new exploration paradigm based on a probable mechanism of marine hydrate deposit accumulation.

Present exploration methods

The most important technique for probing deep water geological formations is the marine seismic survey. Modern seismic surveys efficiently generate three-dimensional images of the subsurface over large areas. Gas hydrate has a distinctive seismic signature, the bottom simulating reflector (BSR) (Shipley *et al.* 1979). The BSR is seen in marine seismic images, running parallel to, and several hundred metres below, the seafloor, approximately coincident with the base of the gas hydrate stability zone (GHSZ). As such its presence has been a primary driver of many hydrate exploration campaigns.

Several major offshore drilling campaigns have been directed primarily by the presence of the BSR, including ODP Leg 164 at Blake Ridge (Paull *et al.* 2000) and the 32-well campaign in the Nankai Trough, offshore Japan, in 2004 (Takahashi & Tsuji 2005). The US Department of Energy/Chevron Joint Industry Project (JIP), which drilled at two sites in the Gulf of Mexico, selected one site based on BSR occurrence and the other based on geological evidence (Claypool *et al.* undated). BSR maps were combined with geological, geochemical and microbiological information to choose drilling sites offshore India in 2006 (Ramana *et al.* 2006).

Unfortunately, the BSR is not always a good predictor of abundant hydrate occurrence. For example, at Blake Ridge, little hydrate was found in a well drilled to a strong BSR, whereas there were hydrate shows in a well drilled in a locale where the BSR was absent (Paull *et al.* 2000).

When interpreting the significance of the BSR, three principles should be kept in mind. Firstly, very little gas is required to produce a strong seismic reflector (Domenico 1977). Secondly, very little hydrate is needed to elevate the capillary pressure of a fine-grained marine sediment to the point where it resists the entry of gas. Thirdly, an apparently continuous seismic reflector does not imply a continuous gas-saturated stratum. High-resolution seismic acquisition and processing show that a strong BSR, which appears continuous at low resolution, can be produced by small discontinuous pockets of gas (Wood et al. 2002; Dai et al. 2004).

Seismic velocity anomalies have been used to detect and/or quantify gas hydrate presence in the subsurface (Barth et al. 2004; Dai et al. 2004). Seismic and well log estimates of hydrate saturation have been compared at Hydrate Ridge (Kumar et al. 2007) and at the JIP Atwater Valley site (Dai et al. 2008). However, the ability of seismic velocity analysis to accurately predict hydrate saturation depends on an accurate transform between hydrate saturation and acoustic velocity, and a good model for the velocity profile of the sediment in the absence of hydrate.

Another characteristic of the seismic response to gas hydrate is amplitude blanking within the GHSZ, noted by early investigators of marine hydrate deposits (Shipley et al. 1979). Various theories have been proposed to explain blanking. One holds that hydrate, which increases the acoustic velocity of unconsolidated sediments, is most likely to form in high porosity, i.e. low velocity, strata, thus reducing the acoustic contrast with neighbouring strata (Lee & Dillon 2001; Hornbach et al. 2003). Blanking has also been explained by the disruption of sedimentary stratigraphy in marine environments thought to harbour hydrate (Chapman et al. 2002). A third explanation is destructive interference from vertically displaced reflectors within the Fresnel zone (Wood & Ruppel 2000). A fourth explanation attributes blanking to the presence of liquid and gas migrating upwards through conduits (Chapman et al. 2002). Still other theories remain current (Hornbach et al. 2007). Regardless of the cause, amplitude blanking has not been correlated with economically significant hydrate deposits.

Mechanisms of gas hydrate reservoir accumulation

In order to develop less ambiguous exploration methods, it is necessary to understand the mechanisms by which gas hydrate deposits are formed. Given appropriate temperature and pressure conditions, gas availability is the primary factor controlling the quantity and distribution of hydrate. The nature of a deposit depends critically on how gas is delivered to the site of hydrate production. Very low and very high fluxes of gas are less likely to create significant reservoirs than moderate fluxes. This concept has been discussed by Roberts et al. (2006). Note that the notion of 'moderate' discussed here overlaps with but is not identical to Roberts's classification.

Local production of gas

Unlike oil and gas, gas hydrate, once formed within its stability zone, will not migrate to reservoirs where it can attain a significant concentration. Thus, in the absence of external sources of gas, gas hydrate concentrations will be correlated with the quantity of organic source material originally in place. The ultimate concentration of hydrate in such deposits is limited by the concentration of total organic carbon (TOC) (Waseda 1998). A survey of marine sediments in water depths less than 3000 m revealed a maximum TOC content of 2% and an average TOC content of 0.55% (Harvey & Huang 1995), of which only a fraction is converted to methane. Thus, in deposits in which *in situ* biogenic production of methane is the only source of gas, gas hydrate concentrations are not expected to be more than a few per cent of total sediment volume (Paull et al. 1994; Milkov & Sassen 2001). Although the deposits resulting from *in situ* gas sources can be spatially extensive, they are unlikely to be concentrated enough to be economically interesting.

Low flux mechanisms

Next consider permeation of gas through the pore space of the GHSZ. The gas can be either dissolved in pore water or free in bubbles. It can originate from either microbial activity (biogenic) or high-temperature cracking of deep petroleum (thermogenic). It can also result from sedimentation, which causes pre-existing gas hydrate to decompose as it exits the GHSZ through its base, a fixed depth below the seafloor (Paull et al. 1994). A number of sophisticated models have been formulated to describe the accumulation of methane hydrate in earth formations resulting from slow permeation of methane into the gas hydrate stability zone (see e.g. Xu & Ruppel 1999). These models explain the pervasive, relatively low concentrations of gas hydrate in quiescent environments (Tréhu et al. 2006). While this class of accumulation may be widespread, and may even contain most of the Earth's endowment of gas hydrate, it is unlikely to be of economic interest.

It is instructive to contrast the problems of hydrate accumulation to the mechanism of conventional oil and gas reservoir accumulation. Oil and gas remain mobile as they move upward through a reservoir, and in fact the relative permeability of rock to hydrocarbon phases increases as those phases fill the pore space. By contrast, hydrate is immobile and blocks further influx of gas. The accumulation of hydrate from dissolved gas is self-limiting due to pore clogging and consequent reduction of hydraulic permeability (Evrenos et al. 1971; Katoh et al. 2000). Sediments that do not block water flow may block upward migration of free gas due to capillary pressure effects (Clennell et al. 2000). Thus if transport through the pore space is the only means by which gas can move up through the GHSZ, it is unlikely to create an exploitable deposit.

When free gas moves upward in disconnected bubbles, theory (Nunn 1996) and experiment (Boudreau et al. 2005) suggest that in unconsolidated muddy sediment the bubbles are thin vertical or subvertical discs. Upon reaching a pocket of trapped gas, these discs give up their gas to it and disappear. Whereas gas is known to migrate many kilometres upward to conventional hydrocarbon reservoirs, one might expect thin migrating discs of gas to form solid gas hydrate shortly after entering the GHSZ. The large surface to volume ratio of disc-like bubbles promotes efficient heat and mass transfer conducive to rapid conversion to hydrate. Thus hydrate deposits formed from isolated moving discs of gas are likely to be concentrated near the base of the GHSZ.

Transport through faults and fractures

There is a growing consensus that faults and fractures extending into and through the GHSZ are the most efficient pathways for distributing gas throughout the GHSZ. The acceptance of this idea was delayed by the knowledge that gas-phase methane cannot exist in thermodynamic equilibrium with excess water within the GHSZ. Nonetheless, the preponderance of known marine hydrate accumulations appears to be associated with fault systems (Booth et al. 1998) through which dissolved or gaseous methane can move upwards rapidly (Clennell et al. 2000). Hydrate is associated with natural gas vents and seeps in the Black Sea (Ivanov et al. 1998), off the Pacific Coast of North America (Spence et al. 2000; Riedel et al. 2002; Wood et al. 2002), in the Gulf of Mexico (Roberts 2001; Sassen et al. 2001) and elsewhere. In fact, it is estimated that most of the gas hydrate in the northwestern Gulf of Mexico is to be found in fault zones (Milkov & Sassen 2001), and hydrate deposited in the Bush Hill fault zone, in Green Canyon Block 185, contains roughly half the gas contained in the deeper, commercially exploited, Jolliet reservoir (Chen & Cathles 2003). Even the Blake Ridge deposit, situated on a passive margin, is associated with faults extending from below the bottom simulating reflector to the seafloor. These faults are believed to constitute efficient conduits for transport of methane (Rowe & Gettrust 1994; Booth et al. 1998), which may be in the gas phase (Paull et al. 1995; Gorman et al. 2002) and therefore out of thermodynamic equilibrium with surrounding sediments.

Faults and fractures in shallow marine sediments can be caused by either of two distinct mechanisms, both of which may be active in the same reservoir. Firstly, faults can form as the result of tectonics. Secondly, faults or fractures can be caused by free gas, which increases the pore pressure and thereby reduces the strength of the sediment (Grauls et al. 1998; Flemings et al. 2003; Hornbach et al. 2004; Kleinberg 2005; Liu & Flemings 2007). In either event, free or dissolved gas exploits conduits of high hydraulic conductivity to move into the GHSZ, vent to the seafloor and/or find permeable horizons within which gas can spread laterally, creating hydrate-rich layers (Milkov & Sassen 2002).

This model of gas transport is perhaps most clearly realized at Hydrate Ridge (Tréhu et al. 2004b; Weinberger & Brown 2006). Free gas is transported from a deep source through a coarse-grained, permeable layer called Horizon A. A connected path of gas in Horizon A creates an overpressure, which either fractures the overlying low-strength shallow marine sediments or remobilizes pre-existing faults. Given a sufficient flux of free gas, conduits are expected to remain open and move gas significant distances through the GHSZ. This can occur when the flux of gas is high enough to overwhelm the capacity of surrounding low permeability sediments to deliver water for complete conversion of gas to hydrate (Tréhu et al. 2004b).

In addition to the mechanisms discussed by Tréhu et al. (2004b) hydrate will form rapidly at fracture surfaces, stiffening the channel and allowing gas to flow through it without contacting liquid water. Eventually hydrate grows into the center of the fissure, creating a solid dyke. This behaviour was observed in seafloor experiments in which methane was injected into the bottom of a cylinder of alluvial mud (Brewer et al. 1997). In this experiment, the channel, from which sediment was completely displaced by the flow of gas, gradually filled with solid hydrate. Sassen et al. (2001) observed that hydrate veins could continue to expand due to the pressure associated with the crystallization of the low-density hydrate phase. Fluxes defined as 'moderate' by Roberts et al. (2006) constitute the high-flux end of this regime: in their model significant amounts of gas reach the surface, creating hydrate outcrops on the seafloor, while leaving some hydrate in the shallow subsurface.

Chen & Cathles (2003) used the differences between source gas, vent gas, and hydrate gas compositions at Green Canyon Block 185 to construct a gas hydrate accumulation model appropriate to this class of vents. The model predicts, among other things, that the accumulation of hydrate is reasonably uniform with depth over most of the gas hydrate stability zone. It also predicts that the fraction of gas precipitated with hydrate is inversely proportional to the vent rate, and that therefore accumulation rate is relatively insensitive to vent rate over a substantial range of rates. A more recent model predicts that hydrate saturation is greatest near the seafloor (Liu & Flemings 2007).

At mass and energy fluxes defined as 'high' by Roberts et al. (2006), gas vents copiously to the seafloor, creating such features as mud volcanoes. The thermal regime of the sediments is disturbed to the extent that hydrate cannot form in the gas conduit, but can accumulate at the periphery (Milkov 2000).

Morphology of faulted or fractured hydrate reservoirs

In the Earth, faults are likely to be planar, with strike parallel to the direction of maximum horizontal stress. According to Coulomb theory their dip is $45° + \varphi/2$, where φ is the friction angle of the marine sediment in the absence of hydrate, which is typically equal to $20°$ (Kleinberg 2005; Kleinberg & Dai 2005). This may explain the discontinuities with dip angles of $50-60°$ that are commonly observed in shallow marine sediments (Rowe & Gettrust 1993; Gorman et al. 2002; Holbrook et al. 2002; Hornbach et al. 2004, 2007). Alternatively, fractures open when the pore pressure exceeds the minimum formation stress. Then the fracture plane is normal to the minimum stress direction, which is not necessarily horizontal in shallow marine sediments. Near the seafloor, the minimum principal stress can be vertical, which may be a factor in the development of mushroom-like vent structures observed near the seafloor at Hydrate Ridge (Milkov et al. 2005) and Blake Ridge (Hornbach et al. 2007).

Hydrate dyke systems will in general cut across stratigraphic boundaries. When a fault or fracture intersects a layer of coarse sand, gas can spread horizontally to produce a hydrate horizon that coincides with local stratigraphy; this is called a 'combination accumulation' by Milkov & Sassen (2002), and may constitute a promising hydrate prospect.

Borehole image logs provide the most direct evidence for the importance of steeply dipping conduits in generating hydrate reservoirs. Weinberger & Brown (2006) and Cook et al. (2008) used resistivity-at-bit (RAB) borehole images to show that hydrate-rich intervals are often coincident with steeply dipping fractures. Hydrate in a subhorizontal bed is sometimes associated with a nearby steeply dipping mineralized fracture, which presumably served as a gas conduit, consistent with the combination accumulation model.

Natural gas vents on the seafloor appear to move. In Gulf of Mexico Green Canyon Block 204, dead chemosynthetic communities are found where vent activity is diminishing (Roberts et al. 2006), suggesting that the energy source has decayed or changed its location. Geochemical evidence suggests that vent rates at Green Canyon Block 185 vary by a factor of three in time scales of a few years (Chen & Cathles 2003).

There are a number of reasons why the subsurface gas plumbing should change over time (Hovland 2002); filling and plugging of gas conduits with hydrate is surely one of these. Once a conduit is filled with hydrate, it is the least permeable and mechanically strongest feature in the sediment, and will not be fractured again. Free gas will find another path having the same dip and strike as previous fractures. The surface expression of the seep is likely to change. Thus one expects a series of parallel hydrate-filled dykes in the GHSZ, possibly marked on the seafloor by dead or abandoned chemosynthetic communities. If the sediment faults by the Mohr–Coulomb mechanism, chevrons of hydrate can open upward with fracture dips symmetrical about the axis of maximum horizontal stress. This pattern has been observed in a hydrate deposit in the Gulf of Mexico (Cook et al. 2008).

Detection of hydrate reservoirs

Evolution of vents

Even if a hydrate reservoir's main accumulation is substantially below the seafloor, its most obvious expression may be seafloor gas vents or seeps. These are manifested directly by abnormally high methane concentrations in seawater or bubbles issuing from the seafloor, and indirectly by chemosynthetic communities and carbonate mounds or pavements (Sassen et al. 2001; Whelan et al. 2005). Vents and seeps are localized, and appear and disappear on short time scales.

From the standpoint of hydrate exploration, dead vents are at least as interesting as active vents. When conduits become clogged with hydrate, the locus of gas transport changes from one fissure to another. The result is areas of inactive chemosynthetic communities and authigenic carbonate pavements on the

Seismic surveys

Hydrate-filled vertical or subvertical dykes are attractive targets for geophysical prospecting. They have relatively uniform strike and dip, which are predictable based on reservoir stresses. They are likely to extend from the base of the GHSZ to a depth below the seafloor defined by the sulfate-methane interface (Borowski *et al.* 1996), which is shallow if the gas flux during hydrate formation has been high. Finally, they cut across prevailing sedimentary structures.

Conventional marine seismic surveys use sources and streamers of hydrophones towed near the sea surface. This geometry is optimal for detecting horizontal or near-horizontal acoustic anomalies. Steeply dipping faults can be detected by imaging the vertical displacement of adjacent horizontal reflectors. However, fractures produced by gas overpressure, unaccompanied by faulting, are not detected, even if the fracture is filled with high-velocity hydrate. Ray path analysis shows that little if any energy is reflected back to sea surface receivers from steeply dipping dykes. Moreover, large lateral changes of velocity are not acknowledged by conventional seismic processing algorithms.

Hydrates can be detected in conventional marine surveys by their influence on seismic wave velocities. This is most clearly illustrated by *velocity and amplitude anomalies* (VAMPs), which are velocity pull-up pseudostructures above a BSR, sometimes paired with velocity push-down bright spots below the BSR, a combination suggestive of hydrate above free gas (Barth *et al.* 2004). Velocity analysis has been used to estimate hydrate saturation from conventional seismic surveys (Ecker *et al.* 2000; Dai *et al.* 2004; Kumar *et al.* 2007; Dai *et al.* 2008); it has been universally assumed that hydrate partially fills the pore space of sediments in laterally extended horizontal or sub-horizontal strata. However, these velocity analyses will be misleading if, instead of partially filling sediment pore space, hydrate is present mostly in the form of vertical or subvertical dykes.

When a borehole is available, two-dimensional walkaway vertical seismic profiling (VSP) can be used to detect the presence of parallel arrays of subvertical structures. At Hydrate Ridge, sound propagating horizontally along the east–west axis travelled significantly slower than sound propagating horizontally along the north–south axis, or vertically. This was presented as evidence for an array of subvertical structures with a common north–south strike (Kumar *et al.* 2006).

Hornbach *et al.* (2007) used a high-frequency chirp echosounder to image the gas vent structure of Blake Ridge down to about 40 m below seafloor. A complex three-dimensional vent structure was found, which became more planar with increasing depth below the seafloor. A number of such structures were found within a kilometre of the main Blake Ridge vent site.

Electromagnetic methods

Gas hydrate is similar to ice, and is therefore an electrical insulator. The electrical conductivity of a dyke depends on the geometrical relationship between insulating hydrate and co-existing conducting saline water. (1) Sediment-excluding solid hydrate lenses have been observed in drill core. If it blocks all conduction paths, gas hydrate is a perfect insulator at radio frequencies and below. (2) It is possible for liquid water to survive the growth of hydrate lenses, leading to low, but non-zero, conductivity. (3) If hydrate grows in the pore space of sediment, mineral grains remain liquid-water-wet, Archie's Law applies and the conductivity depends on hydrate saturation (Murray *et al.* 2006).

Hydrate accumulations are more resistive than marine sediments, which are normally saturated with salt water and have conductivities around 1 S m^{-1}. Therefore controlled source electromagnetic (CSEM) techniques can be used to find and characterize gas hydrate deposits. Conductivity contrast, near-seafloor occurrence, vertical orientation, and uniform dip and strike, make arrays of gas hydrate dykes ideal targets for marine CSEM surveys.

A number of theoretical studies have described the principles of seafloor electromagnetic surveys (Chave *et al.* 1991; Edwards 1997). The emphasis of these works has been on one-dimensional Earth models, in which conductivity changes only with depth. Field studies have been carried out offshore Vancouver (Yuan & Edwards 2000; Willoughby *et al.* 2005) and offshore Oregon (Weitemeyer *et al.* 2005, 2006a) where seismic and drilling programs indicated that gas hydrate was present. In some cases, data processing has been used to parameterize a horizontally stratified Earth, in which the electrical conductivity is isotropic within each horizontal layer (Yuan & Edwards 2000; Willoughby *et al.* 2005). In other cases, data has been displayed as a pseudo-section, in which each conductivity measurement is associated with a point inside the sediment that lies at the intersection of perpendicular lines drawn from the position of the transmitter and receiver respectively (Weitemeyer *et al.* 2006b). Neither method is particularly sensitive to the presence of arrays of vertical conductivity anomalies.

Insulating or low-conductivity dykes can be detected individually, or by measurement of the seafloor conductivity anisotropy. Ideally, the transmitter should be towed perpendicular to strike, i.e. in the direction of the minimum horizontal stress. This can be determined either by geological considerations, or by borehole stress measurements (Murray *et al.* 2006).

Conclusions

Seismic and electromagnetic exploration programs, as currently implemented, may be missing economically significant deposits of gas hydrate on continental slopes. An exploration paradigm based on the common observation that gas is supplied to the gas hydrate stability zone through faults and fractures may be more reliable. Whether or not hydrate dykes are the principal repositories of gas hydrate deposits in a reservoir, they offer a specific, near seafloor, readily detectable and relatively unambiguous target for geological and geophysical exploration. Because dykes are likely to form parallel arrays in accord with geomechanical principles, survey design can be optimized by employing these principles. For example, controlled source electromagnetic surveys should be towed across the probable strike of dyke arrays.

Inactive seafloor vents and dead chemosynthetic communities may be markers for hydrate-clogged vents and associated permeable horizons. If so, they may be exploration targets at least as significant as active gas vents feeding thriving biological assemblages.

I would like to thank J. Haldorsen, B. Hardage, N. Bangs, M. Hornbach, R. Sassen, A. Orange, R. Matsumoto and G. Barth for helpful discussions. C. Ruppel, K. Yamamoto and a referee offered critical reviews of earlier versions of this work.

References

BARTH, G. A., SCHOLL, D. W. & CHILDS, J. R. 2004. Quantifying the methane content of natural gas and gas hydrate accumulations in the deep water basins of the Bering Sea. Extended abstract, *AAPG Hedberg Conference*, Vancouver, 12–16 September 2004.

BOOTH, J. S., WINTERS, W. J., DILLON, W. P., CLENNELL, M. B. & ROWE, M. M. 1998. Major occurrences and reservoir concepts of marine clathrate hydrates: Implications of field evidence. *In*: HENRIET, J.-P. & MIENERT, J. (eds) *Gas Hydrates, Relevance to World Margin Stability and Climatic Change*. Geological Society, London, Special Publications, **137**, 113–127.

BOROWSKI, W. S., PAULL, C. K. & USSLER, W. 1996. Marine pore-water sulfate profiles indicate *in situ* methane flux from underlying gas hydrate. *Geology*, **24**, 655–658.

BOUDREAU, B. P., ALGER, C. *ET AL.* 2005. Bubble growth and rise in soft sediments. *Geology*, **33**, 517–520.

BREWER, P. G., ORR, F. M. *ET AL.* 1997. Deep ocean field test of methane hydrate formation from a remotely operated vehicle. *Geology*, **25**, 407–410.

CHAPMAN, N. R., GETTRUST, J. F. *ET AL.* 2002. High resolution deep towed multichannel seismic survey of deep sea gas hydrates off western Canada. *Geophysics*, **67**, 1038–1047.

CHAVE, A. D., CONSTABLE, S. C. & EDWARDS, R. N. 1991. Electrical exploration methods for the seafloor. *In*: NABIGHIAN, M. N. (ed.) *Electromagnetic Methods in Applied Geophysics – Applications, Part B*. Society of Exploration Geophysicists, Tulsa, OK, 931–966.

CHEN, D. F. & CATHLES, L. M. 2003. A kinetic model for the pattern and amounts of hydrate precipitated from a gas stream: Application to the Bush Hill vent site, Green Canyon Block 185, Gulf of Mexico. *Journal of Geophysical Research*, **108**, 2058; doi: 10.1029/2001JB001597.

CLAYPOOL, G. E., JONES, E., CONTE, A., BLOYS, B., SCHULTHEISS, P. J. & DUGAN, B. undated. Gulf of Mexico Gas Hydrates JIP Cruise Prospectus. World Wide Web Address: http://www.netl.doe.gov/technologies/oil-gas/FutureSupply/MethaneHydrates/rd-program/GOM_JIP/GOM_SciencePlan.html.

CLENNELL, M. B., JUDD, A. & HOVLAND, M. 2000. Movement and accumulation of methane in marine sediments: Relation to gas hydrate systems. *In*: MAX, M. D. (ed.) *Natural Gas Hydrate in Oceanic and Permafrost Environments*. Kluwer, Dordrecht, 105–122.

COOK, A. E., GOLDBERG, D. & KLEINBERG, R. L. 2008. Fracture-controlled gas hydrate systems in the Gulf of Mexico. *Marine and Petroleum Geology*, **25**, 932–941.

DAI, J., BANIK, N., GILLESPIE, D. & DUTTA, N. 2008. Exploration for gas hydrates in the deep water northern Gulf of Mexico, Part II: Model validation by drilling. *Marine and Petroleum Geology*, **25**, 845–859.

DAI, J., XU, H., SNYDER, F. & DUTTA, N. 2004. Detection and estimation of gas hydrates using rock physics and seismic inversion: Examples from the northern deepwater Gulf of Mexico. *The Leading Edge*, **23**, 60–66.

DOMENICO, S. N. 1977. Elastic properties of unconsolidated porous sand reservoirs. *Geophysics*, **42**, 1339–1368.

ECKER, C., DVORKIN, J. & NUR, A. M. 2000. Estimating the amount of gas hydrate and free gas from marine seismic data. *Geophysics*, **65**, 565–573.

EDWARDS, R. N. 1997. On the resource evaluation of marine gas hydrate deposits using sea floor transient electric dipole-dipole methods. *Geophysics*, **62**, 63–74.

EVRENOS, A. I., HEATHMAN, J. & RALSTIN, J. 1971. Impermeation of porous media by forming hydrates in situ. *Journal of Petroleum Technology*, **23**, 1059–1066.

FLEMINGS, P. B., LIU, X. & WINTERS, W. J. 2003. Critical pressure and multiphase flow in Blake Ridge gas hydrates. *Geology*, **31**, 1057–1060.

GORMAN, A. R., HOLBROOK, W. S., HORNBACH, M. J., HACKWITH, K. L., LIZARRALDE, D. & PECHER, I. 2002. Migration of methane gas through the hydrate stability zone in a low-flux hydrate province. *Geology*, **30**, 327–330.

GRAULS, D., BLANCHE, J.-P. & POUDRE, J.-L. 1998. Hydrate sealing efficiency from seismic AVO and hydromechanical approaches. In: *Proceedings of the International Symposium on Methane Hydrates: Resources in the Near Future?* Japan National Oil Corporation, Chiba, 81–86.

HARVEY, L. D. D. & HUANG, Z. 1995. Evaluation of the potential impact of methane clathrate destabilization on future global warming. *Journal of Geophysical Research*, **100**, 2905–2926.

HOLBROOK, W. S., LIZARRALDE, D. ET AL. 2002. Escape of methane gas through sediment waves in a large methane hydrate province. *Geology*, **30**, 467–470.

HORNBACH, M. J., HOLBROOK, W. S., GORMAN, A. R., HACKWITH, K. L., LIZARRALDE, D. & PECHER, I. 2003. Direct seismic detection of methane hydrate on Blake Ridge. *Geophysics*, **68**, 92–100.

HORNBACH, M. J., RUPPEL, C. D. & VAN DOVER, C. L. 2007. Three-dimensional structure of fluid conduits sustaining an active deep marine cold seep. *Geophysical Research Letters*, **34**, L05601; doi: 10.1029/2006GL028859.

HORNBACH, M. J., SAFFER, D. M. & HOLBROOK, W. S. 2004. Critically pressured free gas reservoirs below gas hydrate provinces. *Nature*, **427**, 142–144.

HOVLAND, M. 2002. On the self-sealing nature of marine seeps. *Continental Shelf Research*, **22**, 2387–2394.

IVANOV, M. K., LIMONOV, A. F. & WOODSIDE, J. M. 1998. Extensive deep fluid flux through the sea floor on the Crimean continental margin (Black Sea). In: HENRIET, J.-P. & MIENERT, J. (eds) *Gas Hydrates, Relevance to World Margin Stability and Climatic Change*. Geological Society, London, Special Publications, **137**, 195–213.

KATOH, A., NAKAYAMA, K., BABA, K. & UCHIDA, T. 2000. Model simulation for generation and migration of methane hydrate. *Energy Exploration and Exploitation*, **18**, 401–422.

KLEINBERG, R. L. 2005. Mechanical stability of seafloor sediments with application to gas hydrate deposits. *Proceedings of the Fifth International Conference on Gas Hydrates*, **3**, 736–748.

KLEINBERG, R. L. & DAI, J. 2005. Estimation of mechanical properties of natural gas hydrate deposits from petrophysical measurements. *OTC 17205, Offshore Technology Conference*, Houston, TX, 2–6 May 2005.

KUMAR, D., SEN, M. K. & BANGS, N. L. 2007. Gas hydrate concentration and characteristics within Hydrate Ridge inferred from multi-component seismic reflection data. *Journal of Geophysical Research*, **112**, B12306, doi: 10.1029/2007JB004993.

KUMAR, D., SEN, M. K., BANGS, N. L., WANG, C. & PECHER, I. 2006. Seismic anisotropy at Hydrate Ridge. *Geophysical Research Letters*, **33**, L01306; doi: 10.1029/2005GL023945.

KVENVOLDEN, K. A. & LORENSON, T. D. 2001. The global occurrence of natural gas hydrate. In: PAULL, C. K. & DILLON, W. P. (eds) *Natural Gas Hydrates: Occurrence, Distribution, and Detection*. Geophysical Monographs **124**. American Geophysical Union, Washington, DC, 3–18.

LEE, M. W. & DILLON, W. P. 2001. Amplitude blanking related to the pore-filling of gas hydrate in sediments. *Marine Geophysical Researches*, **22**, 101–109.

LIU, X. & FLEMINGS, P. B. 2007. Dynamic multiphase flow model of hydrate formation in marine sediments. *Journal of Geophysical Research*, **112**, B03101; doi: 10.1029/2005JB004227.

MILKOV, A. V. 2000. Worldwide distribution of submarine mud volcanoes and associated gas hydrates. *Marine Geology*, **167**, 29–42.

MILKOV, A. V. & SASSEN, R. 2001. Estimate of gas hydrate resource, northwestern Gulf of Mexico continental slope. *Marine Geology*, **179**, 71–83.

MILKOV, A. V. & SASSEN, R. 2002. Economic geology of offshore gas hydrate accumulations and provinces. *Marine and Petroleum Geology*, **19**, 1–11.

MILKOV, A. V., CLAYPOOL, G. E., LEE, Y.-J. & SASSEN, R. 2005. Gas hydrate systems at Hydrate Ridge offshore Oregon inferred from molecular and isotopic properties of hydrate-bound and void gases. *Geochimica et Cosmochimica Acta*, **69**, 1007–1026.

MURRAY, D., KLEINBERG, R. L. ET AL. 2006. Formation evaluation of gas hydrate reservoirs. *Petrophysics*, **47**, 129–137.

NUNN, J. A. 1996. Buoyancy-driven propagation of isolated fluid-filled fractures: Implications for fluid transport in Gulf of Mexico geopressured sediments. *Journal of Geophysical Research*, **101**, 2963–2970.

PAULL, C. K., MATSUMOTO, R., WALLACE, P. J. & DILLON, W. P. 2000. *Proceedings of the Ocean Drilling Program, Scientific Results, Leg 164*. Ocean Drilling Program, College Station, TX.

PAULL, C. K., USSLER, W. & BOROWSKI, W. S. 1994. Sources of biogenic methane to form marine gas hydrates: In situ production or upward migration? In: SLOAN, E. D., HAPPEL, J. & HNATOW, M. A. (eds) *International Conference on Gas Hydrates*. Annals of the New York Academy of Sciences, **715**, 392–409.

PAULL, C. K., USSLER III, W., BOROWSKI, W. S. & SPIESS, F. N. 1995. Methane-rich plumes on the Carolina continental rise: Associations with gas hydrates. *Geology*, **23**, 89–92.

RAMANA, M. V., RAMPRASAD, T., DESA, M., SATHE, A. V. & SETHI, A. K. 2006. Gas hydrate related proxies inferred from multidisciplinary investigations in the Indian offshore areas. *Current Science*, **91**, 183–189.

RIEDEL, M., SPENCE, G. D., CHAPMAN, N. R. & HYNDMAN, R. D. 2002. Seismic investigations of a vent field associated with gas hydrates, offshore Vancouver Island. *Journal of Geophysical Research*, **107**, 2200; doi: 10.1029/2001JB000269.

ROBERTS, H. H. 2001. Fluid and gas expulsion on the northern Gulf of Mexico continental slope: Mud-prone to mineral-prone responses. In: PAULL, C. K. & DILLON, W. P. (eds) *Natural Gas Hydrates: Occurrence, Distribution and Detection*. Geophysical Monographs **124**. American Geophysical Union, Washington, DC, 145–161.

ROBERTS, H. H., HARDAGE, B. A., SHEDD, W. W. & HUNT, J. 2006. Seafloor reflectivity – An important seismic property for interpreting fluid/gas expulsion geology and the presence of gas hydrate. *The Leading Edge*, **25**, 620–628.

ROWE, M. M. & GETTRUST, J. F. 1993. Fine structure of methane hydrate bearing sediments on the Blake outer ridge as determined from deep-tow multichannel seismic data. *Journal of Geophysical Research*, **98**, 463–473.

ROWE, M. M. & GETTRUST, J. F. 1994. Methane hydrate content of Blake Outer Ridge sediments. *In*: SLOAN, E. D., HAPPEL, J. & HNATOW, M. A. (eds) *International Conference on Natural Gas Hydrates*. Annals of the New York Academy of Sciences, **715**, 492–494.

SASSEN, R., LOSH, S. L. *ET AL*. 2001. Massive vein-filling gas hydrate: Relation to ongoing gas migration from the deep subsurface in the Gulf of Mexico. *Marine and Petroleum Geology*, **18**, 551–560.

SHIPLEY, T. H., HOUSTON, M. H. *ET AL*. 1979. Seismic evidence for widespread possible gas hydrate horizons on continental slopes and rises. *American Association of Petroleum Geologists Bulletin*, **63**, 2204–2213.

SPENCE, G. D., HYNDMAN, R. D., CHAPMAN, N. R., RIEDEL, M., EDWARDS, N. & YUAN, J. 2000. Cascadia Margin, Northeast Pacific Ocean: Hydrate distribution from geophysical investigations. *In*: MAX, M. D. (ed.) *Natural Gas Hydrate in Oceanic and Permafrost Environments*. Kluwer, Dordrecht, 183–198.

TAKAHASHI, H. & TSUJI, Y. 2005. Multi-well exploration program in 2004 for natural hydrate in the Nankai Trough. *OTC-17162, Offshore Technology Conference*, Houston, TX, 2–5 May 2005.

TRÉHU, A. M., LONG, P. E. *ET AL*. 2004a. Three dimensional distribution of gas hydrate beneath southern Hydrate Ridge: Constraints from ODP Leg 204. *Earth and Planetary Science Letters*, **222**, 845–862; erratum: *Earth and Planetary Science Letters*, **227**, 557–558.

TRÉHU, A. M., FLEMINGS, P. B. *ET AL*. 2004b. Feeding methane vents and gas hydrate deposits at south Hydrate Ridge. *Geophysical Research Letters*, **31**, L23310; doi: 10.1029/2004gl021286.

TRÉHU, A. M., RUPPEL, C. *ET AL*. 2006. Gas hydrates in marine sediments. *Oceanography*, **19**, 124–142.

WASEDA, A. 1998. Organic carbon content, bacterial methanogenesis, and accumulation processes of gas hydrates in marine sediments. *Geochemical Journal*, **32**, 143–157.

WEINBERGER, J. L. & BROWN, K. M. 2006. Fracture networks and hydrate distribution at Hydrate Ridge, Oregon. *Earth and Planetary Science Letters*, **245**, 123–136.

WEITEMEYER, K., CONSTABLE, S., KEY, K. & BEHRENS, J. 2005. The use of marine EM methods for mapping gas hydrates. *OTC 17170, Offshore Technology Conference*, Houston, TX, 2–5 May 2005.

WEITEMEYER, K., CONSTABLE, S. & KEY, K. 2006a. Marine EM techniques for gas-hydrate detection and hazard mitigation. *The Leading Edge*, **25**, 629–632.

WEITEMEYER, K. A., CONSTABLE, S. C., KEY, K. W. & BEHRENS, J. P. 2006b. First results from a marine controlled source electromagnetic survey to detect gas hydrates offshore Oregon. *Geophysical Research Letters*, **33**, L03304; doi: 10.1029/2005GL024896.

WHELAN, J., EGLINTON, L., CATHLES, L., LOSH, S. & ROBERTS, H. H. 2005. Surface and subsurface manifestations of gas movement through a N-S transect of the Gulf of Mexico. *Marine and Petroleum Geology*, **22**, 479–497.

WILLOUGHBY, E. C., SCHWALENBERG, K., EDWARDS, R. N., SPENCE, G. D. & HYNDMAN, R. D. 2005. Assessment of marine gas hydrate deposits: A comparative study of seismic, electromagnetic and seafloor compliance methods. *Proceedings of the Fifth International Conference on Gas Hydrates*, **3**, 802–811.

WOOD, W. T. & RUPPEL, C. 2000. Seismic and thermal investigations of the Blake Ridge gas hydrate area: A synthesis. *In*: PAULL, C. K., MATSUMOTO, R., WALLACE, P. J. & DILLON, W. P. (eds) *Proceedings of the Ocean Drilling Program, Scientific Results, Leg 164*. Ocean Drilling Program, College Station, TX.

WOOD, W. T., GETTRUST, J. F., CHAPMAN, N. R., SPENCE, G. D. & HYNDMAN, R. D. 2002. Decreased stability of methane hydrates in marine sediments owing to phase-boundary roughness. *Nature*, **420**, 656–660.

XU, W. & RUPPEL, C. 1999. Predicting the occurrence, distribution, and evolution of methane gas hydrate in porous marine sediments. *Journal of Geophysical Research*, **104**, 5081–5095.

YUAN, J. & EDWARDS, R. N. 2000. The assessment of marine gas hydrates through electrical remote sounding: Hydrates without a BSR? *Geophysical Research Letters*, **27**, 2397–2400.

Hydrocarbon gas hydrates in sediments of the Mississippi Canyon area, Northern Gulf of Mexico

T. MCGEE[1]*, L. MACELLONI[1,2], C. LUTKEN[1], A. BOSMAN[2], C. BRUNNER[3], R. ROGERS[4], J. DEARMAN[4], K. SLEEPER[1] & J. R. WOOLSEY[1]

[1]*Center for Marine Resources and Environmental Technology, University of Mississippi, Hattiesburg, Mississippi, USA*

[2]*Department of Earth Science, University of Rome 'La Sapienza', Rome, Italy*

[3]*Department of Marine Sciences, University of Southern Mississippi, Starkville, Mississippi, USA*

[4]*Department of Chemical Engineering, Mississippi State University, Starkville, Mississippi, USA*

Corresponding author (e-mail: tmm@olemiss.edu)

Abstract: The Gulf of Mexico Hydrates Research Consortium has begun installing a seafloor observatory to monitor gas hydrate outcrops and the hydrate stability zone in Mississippi Canyon Area Lease Block 118. Relevant background information concerning the Mississippi Canyon Area and gas hydrate occurrences in the northern Gulf of Mexico is presented. Microbial influences and possible scenarios of hydrate accumulation are considered. The design of the observatory was based on field data recorded in the Mississippi Canyon Area, principally lease block 118 (MC118) and the vicinity of lease block 798 (MC798). Swath bathymetry by autonomous underwater vehicle played a large part, as did seismic imaging within the hydrate stability zone and core sampling. These data and the results of their analyses are discussed in detail. Discussion and interim conclusions are presented.

In an effort to expand the information base on how natural gas hydrates form in sediments and how they affect seafloor stability, the University of Mississippi organized the Gulf of Mexico Hydrates Research Consortium in 1999. Its principal goal is to develop and install a seafloor observatory capable of monitoring a hydrate outcrop and its environment more-or-less continuously for a period of 5–10 years. In 2006, the Consortium included researchers from more than 30 universities and corporations. A Board of Scientific Advisers oversees project management, site selection, geochemical aspects, microbial aspects, seismo-acoustic aspects, data quality control and data archiving. A summary of project history, an overview of observatory design and a list of organizations that participate in the scientific Board is given by McGee (2006).

A number of possible sites for the observatory were considered by the Site Selection Committee. Data presented here were collected in the vicinity of two sites of known hydrate occurrence in the Mississippi Canyon Area sites; lease block 118 (MC118) and lease block 798 (MC798). A lease block in the northern Gulf of Mexico comprises a square of sides 3 statute miles (5 km).

Consortium members voted in October 2004 to select a prominent seafloor mound located in MC118 as the observatory site. The Minerals Management Service (MMS) of the US Department of Interior reserved the vicinity of the mound exclusively for Consortium activities. Pre-installation studies began in January 2005 and the first monitoring components were installed during May of the same year. Additional components were scheduled for deployment during the following September but the aftermath of hurricane Katrina forced a delay until June 2006. In addition to information on regional contexts, this contribution includes early results and tentative conclusions produced by Consortium activities prior to the resumption of the installation.

Quaternary geology of the Mississippi Canyon area

The Quaternary geology of the Mississippi Canyon region is dominated by climate-controlled depositional cycles during which progradation and aggradation across the continental shelf were controlled by sea level and complicated by salt tectonics and slope failure (Berryhill *et al.* 1986). During each cycle, fluvial material was deposited across the shelf and directly onto the upper continental slope as the sea receded. Each cycle was closed by rapid

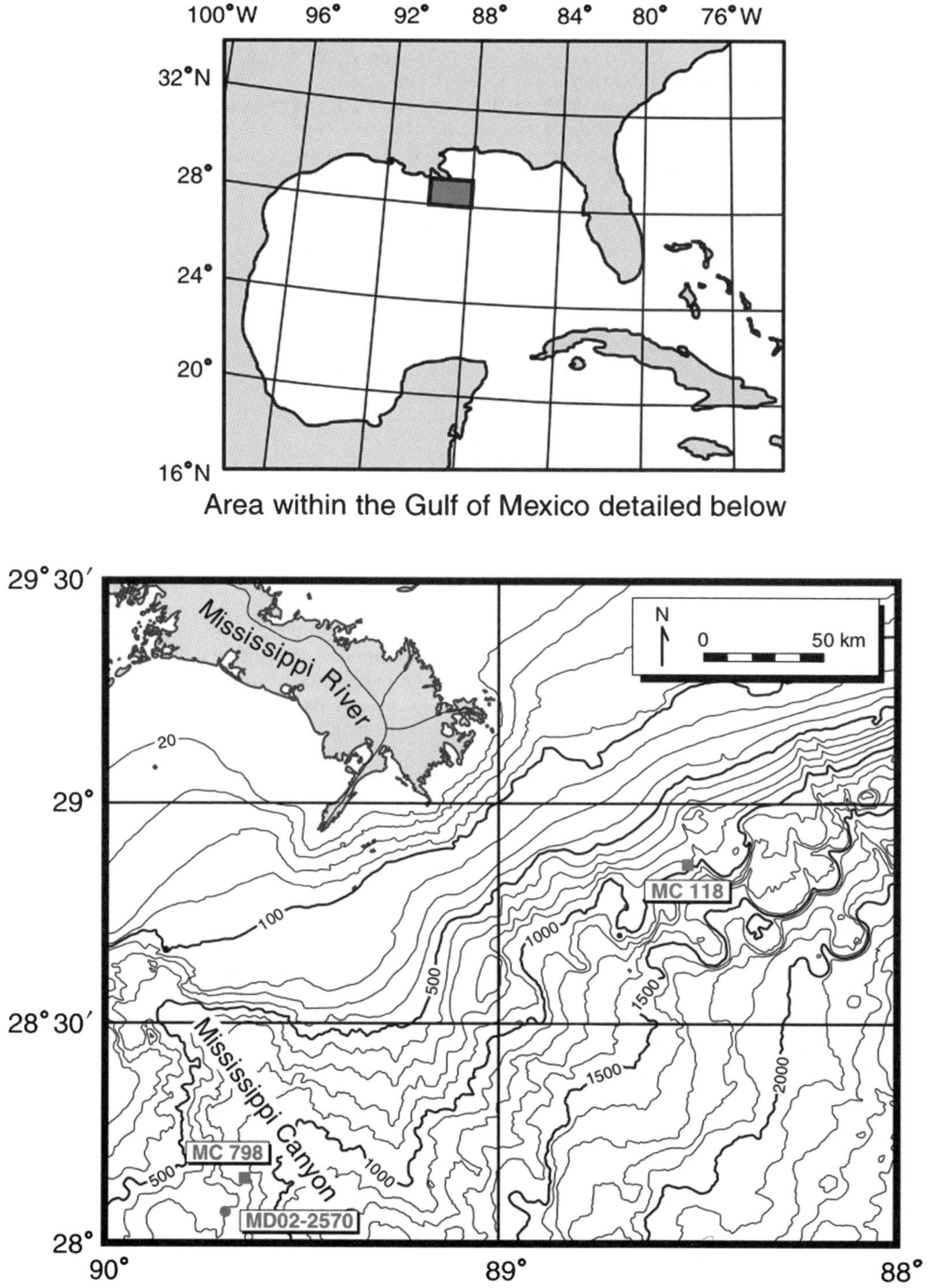

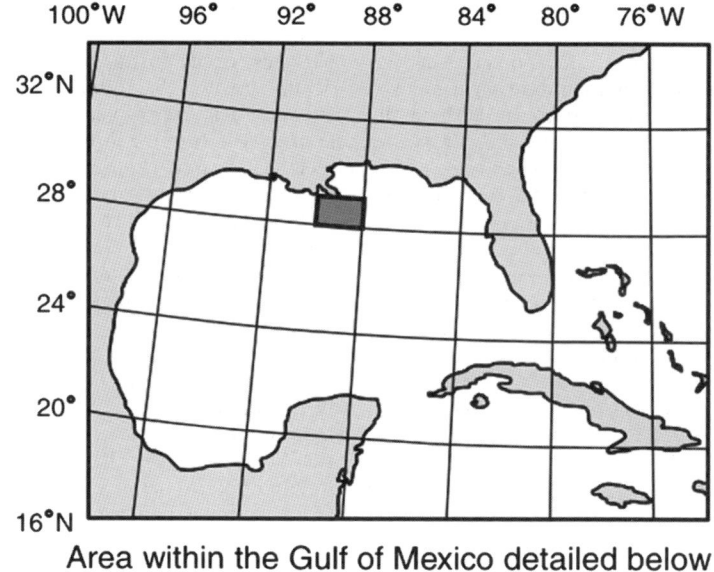

Fig. 1. Map showing the locations of Mississippi Canyon, MC118, MC798 and MD02-2570 (the 28.5 m core collected by R/V *Marion Dufresne*).

marine transgression onto the shelf which caused the coastline to retreat and depocentres to move to the inner shelf. Subsequent effects of salt tectonics and resultant faulting/slumping were superimposed on this cyclic pattern. On the continental slope, the weight of prograding shelf material drove up diapirs and curvilinear pressure ridges from underlying salt deposits, most likely the Jurassic Louann Salt (Halbouty 1979). Mini-basins formed by salt withdrawal between the salt structures were filled with slumped, turbidic and hemipelagic sediment in various quantities depending on local conditions.

The formation of the bathymetric feature known as Mississippi Canyon (Fig. 1) was influenced greatly by a large-scale slope failure. Several climate-controlled depositional cycles that predate the Mississippi Canyon, including the last Wisconsin low stand (c. 24–14 ka), are visible on seismic profiles (Coleman et al. 1983). The final Wisconsin depositional cycle began during 25–27 ka with shallow-water deposition which was interrupted by a mega-slump event (Coleman et al. 1983). All of the Wisconsin and some older units were truncated by a trough-shaped surface that subsequently comprised the floor of a gigantic incised canyon. Incising ceased about 20 ka when infilling began. Thus, initial opening of the Canyon was catastrophic and its complete excavation was accomplished in less than 7 ka. The final Wisconsin cycle closed when the last rise in sea level initiated fill and transgressive deposits that are now capped by Holocene high-stand drape.

At present, the flanks of the Canyon are subject to climate-driven depositional cycles similar to those described above and continue to be geologically active. Low-stand loading of the shelf edge and oversteepening of the upper slope accelerate salt movement, which activates faults and initiates slumps into the Canyon and minibasins (Berryhill et al. 1986). Hydrocarbon fluid flow along faults promotes active hydrate formation at the seafloor and, perhaps, within the conduits (Roberts & Carney 1997). Slumps that are currently creeping downslope along decollements may be related to hydrate and gas deposits (Cooper & Hart 2003). Mud volcanoes that pierce the seafloor spew fluidized sediment and reworked fossils into flanking basins (Kohl & Roberts 1994, 1995; Roberts & Carney, 1997). These processes aid hydrocarbon migration and entrapment.

Gas hydrates in the northern Gulf of Mexico

Hydrates can form with many kinds of gas but those that form with hydrocarbon gases are of greatest interest to the energy community (Milkov & Sassen 2001). In the northern Gulf of Mexico, hydrocarbon gas hydrates, also called natural gas hydrates, occur on the continental slope in water depths greater than 500 m where bottom-water temperatures are typically about 7 °C (Sassen et al. 1999). In some instances the gas is microbial methane (Sassen et al. 2003) but more commonly it is a mixture of thermogenic gases that have migrated up faults from deep petroleum reservoirs (Sassen et al. 1999). These hydrates are stable in the sediments to a depth below the seafloor that depends on the composition of the gas as well as on the pressure and temperature (Milkov & Sassen 2003).

Information concerning the hydrate stability zone in the Gulf of Mexico is sparse. The petroleum industry has drilled many boreholes that pass through the stability zone but there has been little incentive to log information in the relatively shallow sediments where hydrates might exist. Although local thicknesses of the stability zone are not well documented, it is estimated that the zone generally varies from zero thickness in water depths less than about 450 m to more than 1100 m thick in water 2000 m deep and perhaps thicker in deeper water (Milkov & Sassen 2000). Certainly, a more detailed knowledge of the thickness of the hydrate stability zone would improve the understanding of hydrate occurrence in the Gulf of Mexico.

Conventional wisdom expects the base of the hydrate stability zone (BHSZ) to comprise a contrast in seismic impedance with the impedance above the BHSZ being greater than that below it. This leads to the usual expectation that the BHSZ produces a seismic reflection of negative polarity. When the hydrate-forming gas is a mixture of gases with different hydrate stability conditions, the BHSZ may produce a sequence of reflections with slightly different travel times (Wenyue Xu, pers. com.). Since the BHSZ results from a phase change rather than a change in geologic properties, the reflection(s) from it can cut across reflections associated with geologic structure and/or layering. Once a reflection from the BHSZ is identified, the travel time between it and the seafloor is used to calculate the thickness of the associated stability zone. If the speed of propagation through the zone is not known, it is common to assume a constant speed of 1500 m s^{-1}. In most instances this assumption yields an estimate of minimal thickness.

In regions of the world where the geothermal gradient does not vary from place to place, the BHSZ occurs at a constant depth below the seafloor and a reflection from it is called a bottom-simulating reflection (BSR). In such cases the seismic characteristics of negative polarity, crosscutting other reflections and bottom simulation, are useful for identifying the BSR (Dillon et al. 1994). The geothermal gradient in the northern Gulf of Mexico can vary significantly, however. The variation is due largely to the region being underlain by a salt formation which exhibits high heat conductivity and irregular relief that pierces overlying sedimentary structure. In such a scenario it is common for the location of the BHSZ to be related more to proximity to the salt than to depth below the seafloor and, therefore, for bottom simulation not to occur. This can make identification of the BHSZ difficult, reducing it to a search for negative reflections that cut across geologic reflections. Since Gulf sediments are typically alternating sand–shale sequences, several negative reflections can be present and an interpretation based on whether or not they crosscut other reflections can become rather subjective. For this reason identifying the BHSZ using conventional seismic information often fails in the Gulf of Mexico.

Other information concerning naturally occurring hydrates in the Gulf of Mexico comes from observations of outcropping hydrates and chemical analyses of samples. Many outcrops of hydrates have been observed briefly by manned and unmanned submersibles and over longer periods of time by autonomous devices such as recording thermometers and remote low-light cameras (e.g. MacDonald *et al.* 2005; Vardaro *et al.* 2005; Roberts & Carney 1997). At some localities massive hydrate layers as much as 1 m thick are seen at and immediately below the seafloor. In cores of shallow sediment, it is more common to observe small nodules or grains of hydrate dispersed in fine-grained sediments. Deeper information is derived from samples of hydrates, gases and sediments collected from seafloor vents (e.g. Sassen & MacDonald 1997; Sassen *et al.* 1998). Chemical analyses of such samples have provided the most direct information on conditions within and below the hydrate stability zone.

Microbial influences

Microbes proliferate around gas hydrate outcrops because a synergistic relationship is involved. The carbon-rich gases that form natural gas hydrates provide sustenance for microbes and biosurfactants produced by microbes enhance the formation of hydrates (Lee 2001). Laboratory experiments show that small concentrations of water-borne biosurfactants wetting porous mineral surfaces have the effect of substantially decreasing the induction time (the time it takes hydrates to begin forming after the appropriate combination of pressure and temperature is reached) and increasing the formation rate (how rapidly hydrate volume increases after the induction time) of gas hydrates (Rogers *et al.* 2002, 2003). Moreover, biosurfactants exhibit surface specificities for particular mineral surfaces; that is, biosurfactant adsorption on a specific mineral surface results in hydrates nucleating on that surface (Rogers *et al.* 2003). Surfaces of smectite clay, common in the fine-grained sediments of the northern Gulf, serve this purpose particularly well.

Sulfate from the seawater permeates seafloor sediments. In the presence of methane, archaea work in consortia with sulfate-reducing bacteria to reduce the sulfate to H_2S while concurrently oxidizing methane to form hydrogencarbonate (HCO_3), also known as bicarbonate (Boetius *et al.* 2000). The HCO_3 precipitates as carbonate (methane-derived authigenic carbonate; MDAC) at the sulfate–methane transition zone (SMTZ). The presence of MDAC is therefore an indicator of the SMTZ (Rogers, pers. com.). As the upward flux of methane through the sediments increases, the SMTZ ascends toward the sediment–sea interface. Near gas hydrate outcrops and methane gas vents, the bottom of the SMTZ has been reported to be only a few centimetres below the seafloor (Boetius *et al.* 2000).

Possible scenarios of hydrate accumulation in the northern Gulf of Mexico

There are two types of sediment in the northern Gulf of Mexico that may have sufficient permeability to host significant accumulations of gas hydrate: sandy sediments such as channel-overbank deposits and fine-grained sediments that have acquired fracture porosity. Since the sandy sediments contain a significant fine-grained fraction, it is possible for smectite to be present in both these sediment types.

Visual observations, vent-gas chemistry and laboratory experiments suggest different styles of hydrate accumulation depending on whether or not the faults that provide pathways for fluid migration from depth intersect the seafloor. Where the pathways extend into the stability zone but do not reach the seafloor, it is thought that the ascending gas percolates into sediments which are sufficiently permeable to allow the circulation of gas and water necessary for hydrate formation. Where the pathways extend through the stability zone and intersect the seafloor, a portion of the hydrocarbon fluids is vented directly into the water column. Microbial action near the seafloor then precipitates minerals, most commonly carbonates, which accumulate to form mounds near the vents. If gas flow is sufficient and the pressure and temperature are conducive to the formation of hydrates, masses of pure hydrate can be incorporated into the mounds. A photograph of such a mound is shown in Figure 2. The vicinity is littered with mussel shells (mainly dead and

Fig. 2. A seafloor mound in the northern Gulf of Mexico (© Ian MacDonald). The bright white material in the foreground is outcropping gas hydrate. The vicinity of the mound is littered with mainly dead, disarticulated shells of chemosynthetic mussels and portions of its surface are covered by (orange-coloured) microbial mats. A thermistor is inserted into a borehole on top of the mound to record the temperature inside the mound. Lights of the Johnson Sea Link manned submersible are visible in the background.

disarticulated) and other chemosynthetic detritus. Parts of the mound's surface are covered by orange-coloured microbial mats. The bright white material seen outcropping from the lower part of the mound is natural gas hydrate.

Imaging by autonomous underwater vehicle

Very high resolution images of the seafloor and shallow sub-bottom were obtained in both MC118 and MC789 using a commercially available autonomous underwater vehicle (AUV) that operates in a 'mow the grass' survey mode. The surveys consisted of running parallel tracks 200 m apart. The acquired data included swath bathymetry, chirp-sonar sub-bottom profiles and side-scan-sonar mosaics.

The swath-bathymetry and chirp-sonar data was further processed. Careful attention to small navigation inconsistencies allowed the spatial resolution of the swath images to be improved by reducing the bin size from 5 to 1 m. Back-scatter images were calculated from the improved swath data. The chirp profiles generally provided about 50 m sub-bottom penetration. Their scale and vertical exaggeration were manipulated during further processing to resolve layer thicknesses and fault displacements to within a metre or so.

Seismic reflection imaging within the hydrate stability zone

As discussed above, observations of outcrops indicate that individual layers of massive hydrate are usually less than a metre thick and theoretical calculations indicate that the BHSZ in deep water may be many hundreds of metres below the seafloor. This makes it difficult to acquire seismic reflection data capable of resolving sufficiently fine detail throughout the hydrate stability zone. Experience is that conventional exploration methods do not resolve fine detail above the BHSZ and conventional high-resolution methods are not powerful enough to reach the BHSZ in deep water. What is required is a technique which provides hundreds of metres of penetration while maintaining sub-metre resolution of layer thicknesses.

A technique developed to provide such a capability uses an energy source at the water surface and a single receiver deployed at depth. It is called the surface-source-deep-receiver (SSDR) technique (McGee 2000). The source is located at the surface to minimize the ambient pressure that must be

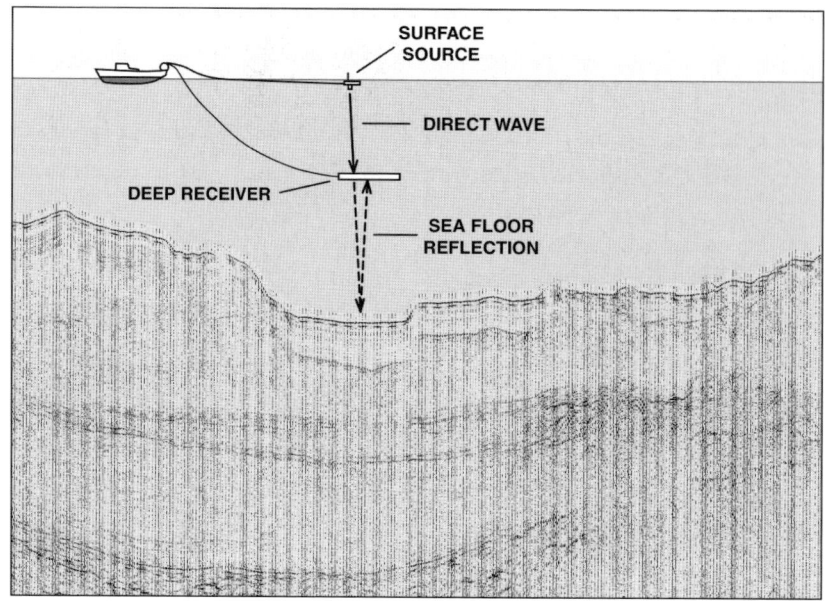

Fig. 3. Recording geometry for the surface-source-deep-receiver (SSDR) technique.

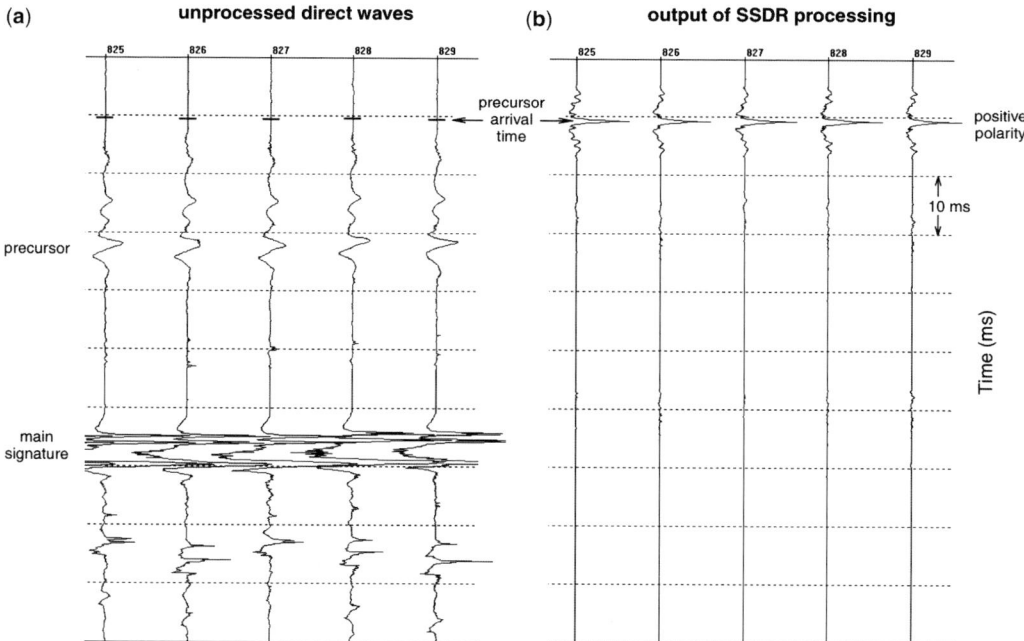

Fig. 4. Example of SSDR processing applied to direct waves generated by an 80 in^3 watergun: (**a**) unprocessed direct waves and (**b**) output of SSDR processing of direct waves. The arrival of the direct wave is taken to be the onset of the watergun precursor. The central peak of the processed output occurs at that time. The direction of the central peak excursion defines positive polarity. The amplitude of the central peak defines unit amplitude for the trace. Note that the duration of the direct waveform is 90$^+$ ms. The processing collapses it into a symmetrical waveform about 10 ms long and most of the energy is concentrated within ±1 ms of the central peak.

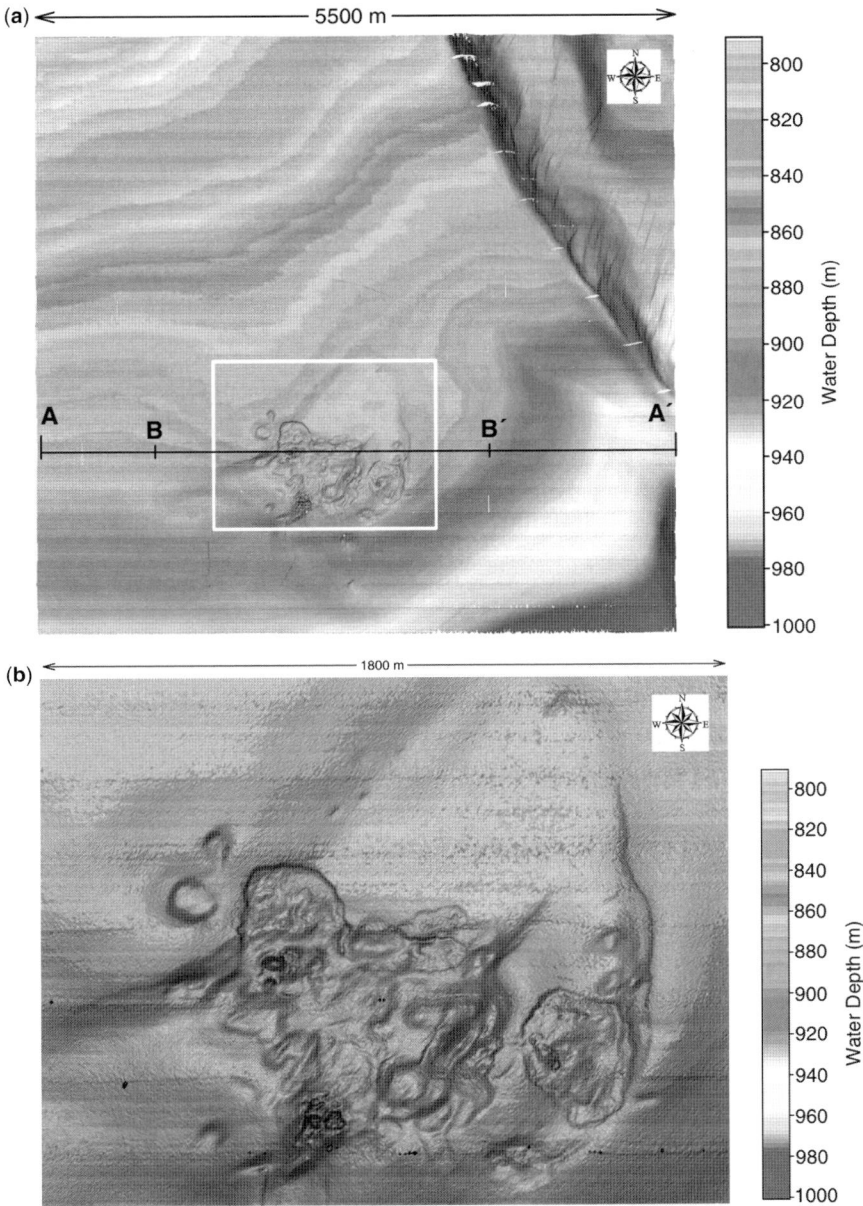

Fig. 5. Swath bathymetry acquired by AUV in MC118: (**a**) image of the entire block. Note that the seafloor is rather featureless except for a submarine canyon crossing the northeastern portion of the block and an area of irregular relief in the south-central portion. The irregular relief marks the mound being monitored by the seafloor observatory. Seismo-acoustic profiles along sections A–A′ and B–B′ are presented in Figure 6. (**b**) Enlargement of the outlined portion of (a). Note the sharply defined, nearly circular depressions. One of these depressions was visited by manned submersible and hydrate was observed to outcrop within it (Sassen *et al.* 2006).

exceeded and still have enough excess energy to penetrate the entire hydrate stability zone. The recording geometry, illustrated in Figure 3, is such that the source is located directly above the receiver and the receiver depth is great enough for the direct wave from source to receiver to approximate a vertically travelling plane wave. The receiver is deployed far enough above the water bottom that, for each shot, a complete source signature is recorded without interference from the water-bottom reflection. Post-acquisition digital processing removes the phase of the source-signature from the direct wave and the sequence of sub-bottom reflections. As shown in Figure 4, the direct wave is collapsed into a symmetrical wavelet whose central lobe defines positive polarity. If the central lobe is sufficiently narrow, its amplitude can be considered unity. Similarly, sub-bottom reflections are collapsed into nearly symmetrical wavelets whose main lobes are calibrated in polarity and amplitude. Resolution is usually improved significantly.

The SSDR technique has been tested at a well site in 1500 m of water. A suite of logs had been recorded in the shallow portion of the well because a shallow-flow hazard was suspected. The technique resolved layers less than 10 cm thick near the seafloor and about 60 cm thick more than 300 m below the seafloor (McGee 2001).

SSDR profiling was done in both MC118 and MC798. Data acquisition in MC798 occurred some years before that in MC118, while the field instrumentation was still in an early stage of development. Thus SSDR data quality in MC118 is better than that in MC798.

Mississippi Canyon 118

MC118 is located 200 km northeast of the bathymetric feature called Mississippi Canyon (Fig. 1). The improved swath bathymetry image in MC118 (Fig. 5a) shows a seafloor that is smooth and nearly featureless except for a submarine canyon crossing its northeastern portion and an area of irregular relief in its south-central portion. The irregular relief is related to a mound composed largely of authigenic carbonates (mainly $CaCO_3$) in which hydrates have accumulated and from which gas is being vented (Sassen & Roberts 2004). This is the mound being monitored by the seafloor observatory. It obviously belongs to the scenario in which migration pathways extend through the hydrate stability zone and vent hydrocarbon fluids directly into the sea water.

The area outlined in the south-central portion of Figure 5a is enlarged in Figure 5b to illustrate details of the mound's surface. Sharply defined, nearly-circular features are visible at several locations. One of these was inspected by a manned submersible and found to be a deep, well-like depression within which hydrates could be seen projecting from the walls (Sassen et al. 2006). Less sharply defined depressions appear to be vent sites which have become inactive and are covered by a thin deposit of sediment.

The compositions of three samples of vent gas and one of gas from intact hydrate are given in Table 1 (Sassen et al. 2006). The hydrate is Structure II with lower methane content and higher

Table 1. *Composition of gas samples collected in mc118*

								Gas composition					
	C_1 (%)	$\delta^{13}C$ C_1 (‰)	δD C_1 (‰)	C_2 (%)	$\delta^{13}C$ C_2 (‰)	C_3 (%)	$\delta^{13}C$ C_3 (‰)	$i\text{-}C_4$ (%)	$\delta i\text{-}C_4$ (‰)	$n\text{-}\delta C_4$ (%)	$n\text{-}C_4$ (‰)	neo-C_5 (%)	$i\text{-}C_5$ (%)	$n\text{-}C_5$ (%)	$\delta^{13}C$ CO_2 (‰)
Vent gas samples														
Dive 4414	94.6	−45.8	−164	3.1	−27.1	1.4	−24.5	0.3		0.4		0.0	0.1	0.1	+21.2
Dive 4414	94.4	−45.7	−162	3.1	−26.8	1.4	−24.1	0.4		0.4		0.0	0.2	0.1	+21.5
Dive 4414	96.5	−45.7	−163	2.4	−26.8	0.5	−24.1	0.1		0.3		0.0	0.1	0.1	+17.8
Gas hydrate sample	70.0	−46.7	−169	7.5	−29.2	15.9	−26.5	4.4	−27.5	1.1	−25.5	0.0	1.0	0.1	−15.3

Adapted from Sassen et al. (2006). Three vent gas samples and one gas sample from intact hydrate are represented.

propane content than those found in most hydrate samples previously collected in the Gulf of Mexico (Sassen et al. 2006).

A total of 27 chirp sonar profiles were recorded by AUV in MC118. The chirp profile shown in Figure 6a is located along section A–A' in Figure 5a. It provides about 50 m sub-bottom penetration at most places except near the carbonate/hydrate mound where there is little penetration. Faults visible to the east and west of the mound are displaced downward toward the mound. The syncline east of the mound appears to predate the small fault there (Sleeper et al. 2006).

An orthogonal grid of 76 SSDR profiles were recorded over the mound in MC118 using an 80 in^3 watergun. The SSDR profile in Figure 6b is located along section B–B' in Figure 5b. Some distance from the mound sub-bottom penetration is more than 500 m (>400 m). Directly over the mound, chaotic seismic energy extends as much as 200 ms below the seafloor reflection. It appears to be backscatter from seafloor roughness rather than reflections from sub-bottom interfaces.

Examples of SSDR wiggle traces are shown in Figure 7. They illustrate reflection polarities within the rectangular area outlined in Figure 6b. The seafloor reflection is seen to be of positive polarity (as defined in Fig. 4). The designation 'negative/positive' marks a two-reflection sequence located 80 m below the seafloor; the upper reflection is of negative polarity and is followed closely by a reflection of positive polarity. Such a doublet is consistent with reflection from a thin layer whose interior acoustic impedance is lower than that of

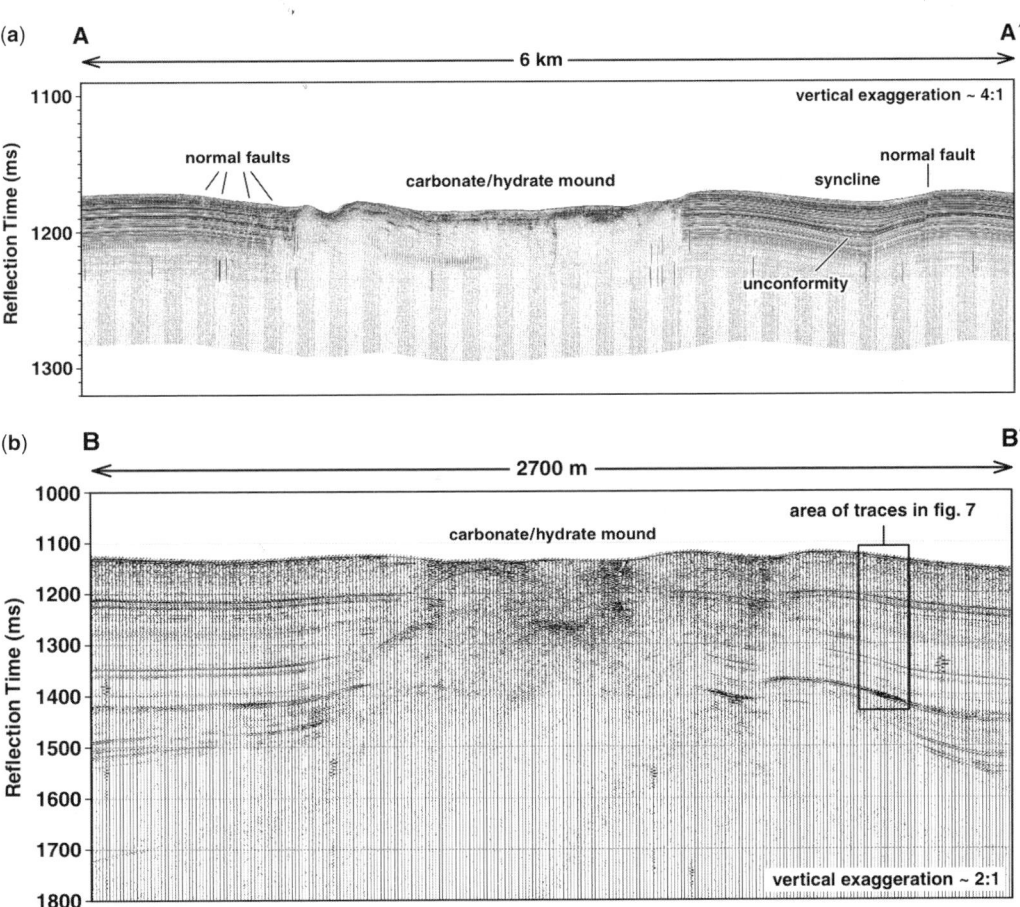

Fig. 6. Seismo-acoustic profiles in MC118 along sections plotted in Figure 5a: (**a**) chirp sonar profile along section A–A'. Note that the mound exhibits little positive relief and that faults east and west of the mound are displaced downward toward it. The syncline east of the mound apparently predates the normal fault there. (**b**) SSDR profile along section B–B'. Note penetration exceeds 400 ms (>300 m) except on the mound where there is little coherent penetration.

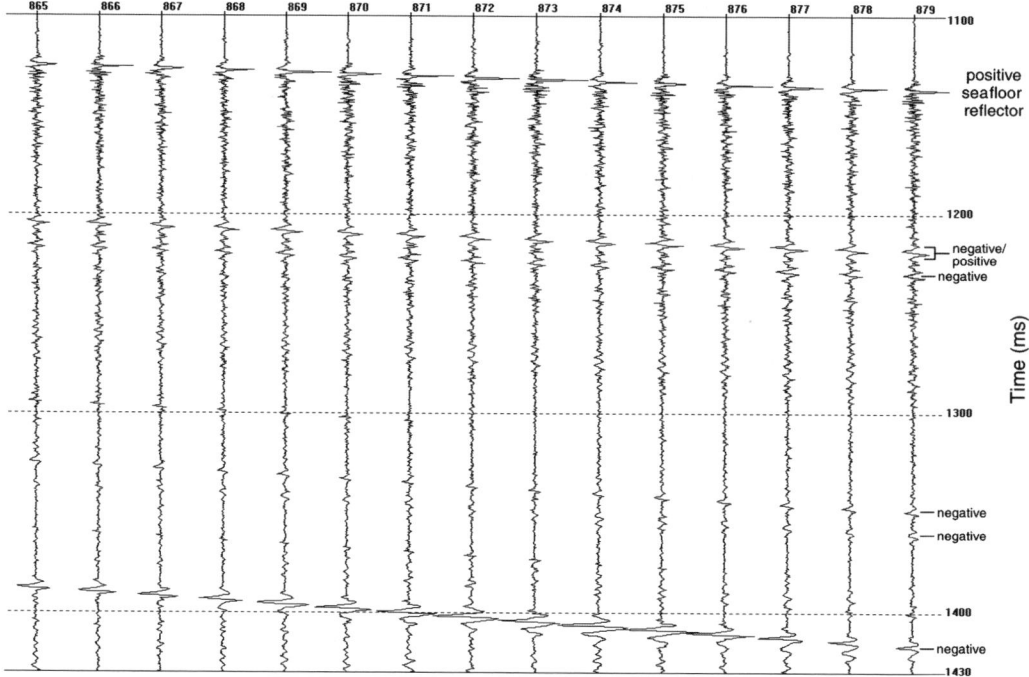

Fig. 7. Wiggle traces illustrating reflection polarity within the area outlined in Figure 6b. Note that the seafloor reflection is of positive polarity as defined in Figure 4. The designation 'negative/positive' 80 ms below the seafloor indicates a two-reflection sequence; the first of negative polarity followed closely by the second of positive polarity. Such a doublet is consistent with a thin layer whose interior has acoustic impedance lower than that of the material above and below it. Deeper reflections are generally of negative polarity, the strongest being near 1400 ms.

the sediment above and below it. Deeper reflections are seen to be generally of negative polarity, the strongest occurring near 1400 ms.

Mississippi Canyon 798

MC798 is located in a salt-withdrawal minibasin near the crest of the southwestern flank of Mississippi Canyon (Fig. 1). The improved swath bathymetry in MC798 is shown in Figure 8a. It depicts a seafloor of variable relief that has been subjected to repeated episodes of slumping. The shallower northwestern and southwestern portions of the area exhibit irregular remnants of seafloor that predate the slumping. The deeper northeastern half of the block exhibits smoother surfaces of sediment that has slumped into Mississippi Canyon, the centre of which is located about 10 km to the northeast. There are no obvious vent sites, thus much of the area seems to belong to the scenario in which migration pathways do not reach the seafloor.

The outlined area in the northwestern portion of Figure 8a is enlarged in Figure 8b to more clearly show the polygonal pattern that exists on the pre-slump seafloor. Note that the pattern is absent on the slumped seafloor.

The outlined area in the southeastern portion of Figure 8a is enlarged in Figure 8c to show more clearly the mud volcano there. Neurauter & Bryant (1989) reported sampling hydrates from its summit. A survey by TDI-Brooks (2001) found heat-flow values of $6-14$ mW m^{-2} in the vicinity of the mud volcano and an anomalously high value of almost 25 mW m^{-2} at the base of the volcano's northeastern flank. Based on near-bottom temperatures measured during the heat-flow survey, Lewis (2001) suggests that a stream of warm brine issues from the seafloor at that location. Evidence of a fluid flow denser than seawater is provided by the gully eroded into the adjacent seafloor.

Figure 9 is a contour map of the mud volcano in Figure 8c. Conical promontories are visible on and near its summit. These may be mineral accumulations around vent sites and thus be the source of the gas that generated the hydrate sample. The eroded gully is seen to be 2–3 m deep.

A total of 15 chirp-sonar profiles were recorded by AUV in MC798. Figure 10a shows the chirp profile recorded along section A–A′ in Figure 8a.

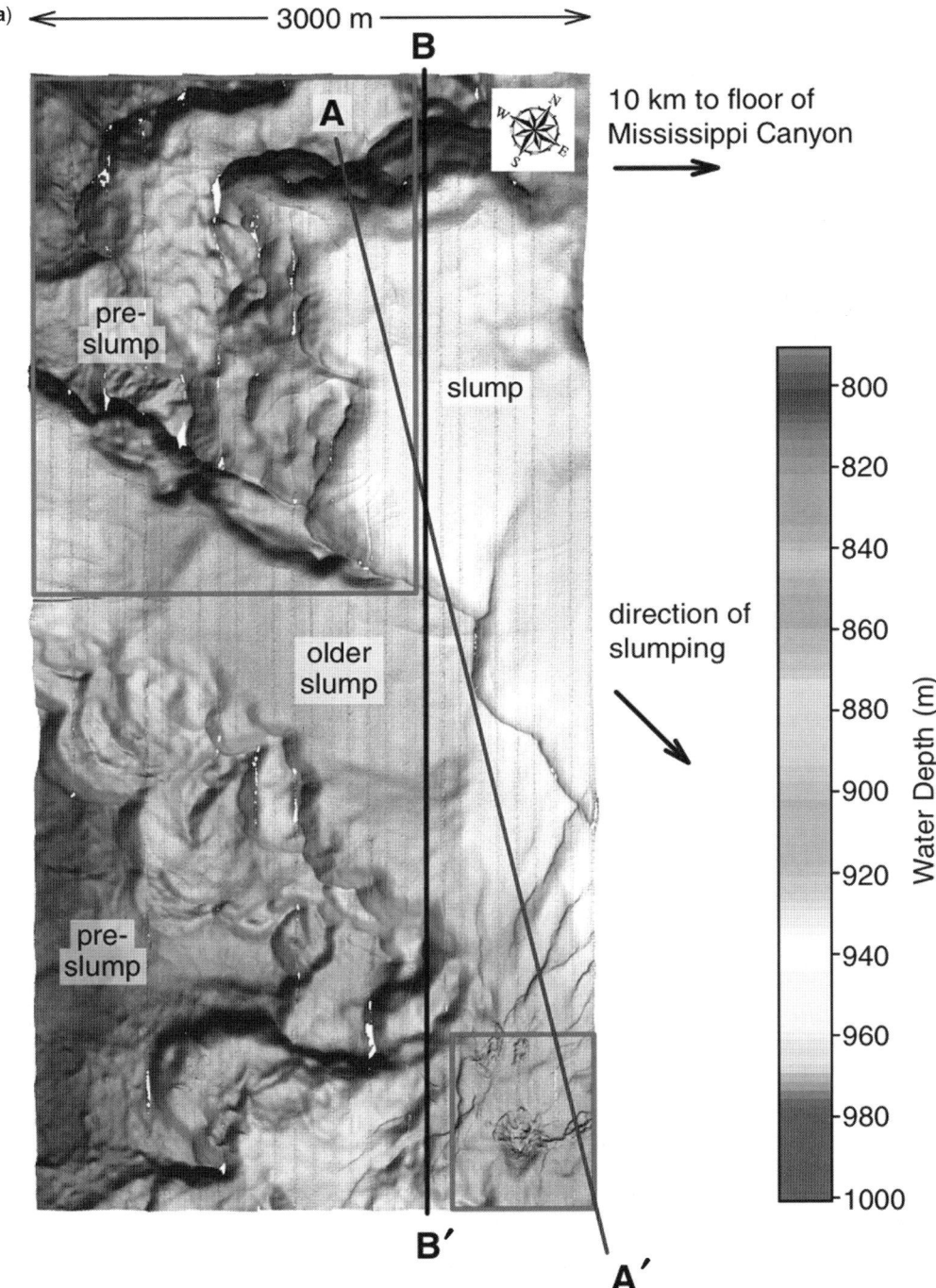

Fig. 8. Swath bathymetry acquired by AUV in MC798: (**a**) image of the entire survey. Note areas of pre-slump seafloor and two generations of slumped seafloor. The direction of slumping is SE into Mississippi Canyon. Seismo-acoustic profiles along sections A–A′ and B–B′ are presented in Figure 10. (**b**) Enlargement of the northwestern portion outlined in (a). Note that the polygonal pattern of seafloor relief present in pre-slump areas is absent in slump areas. (**c**) Enlargement of the southeastern portion outlined in (a). Note the mud volcano and the gully northeast of it. The rectangle outlines the area shown in Figure 9.

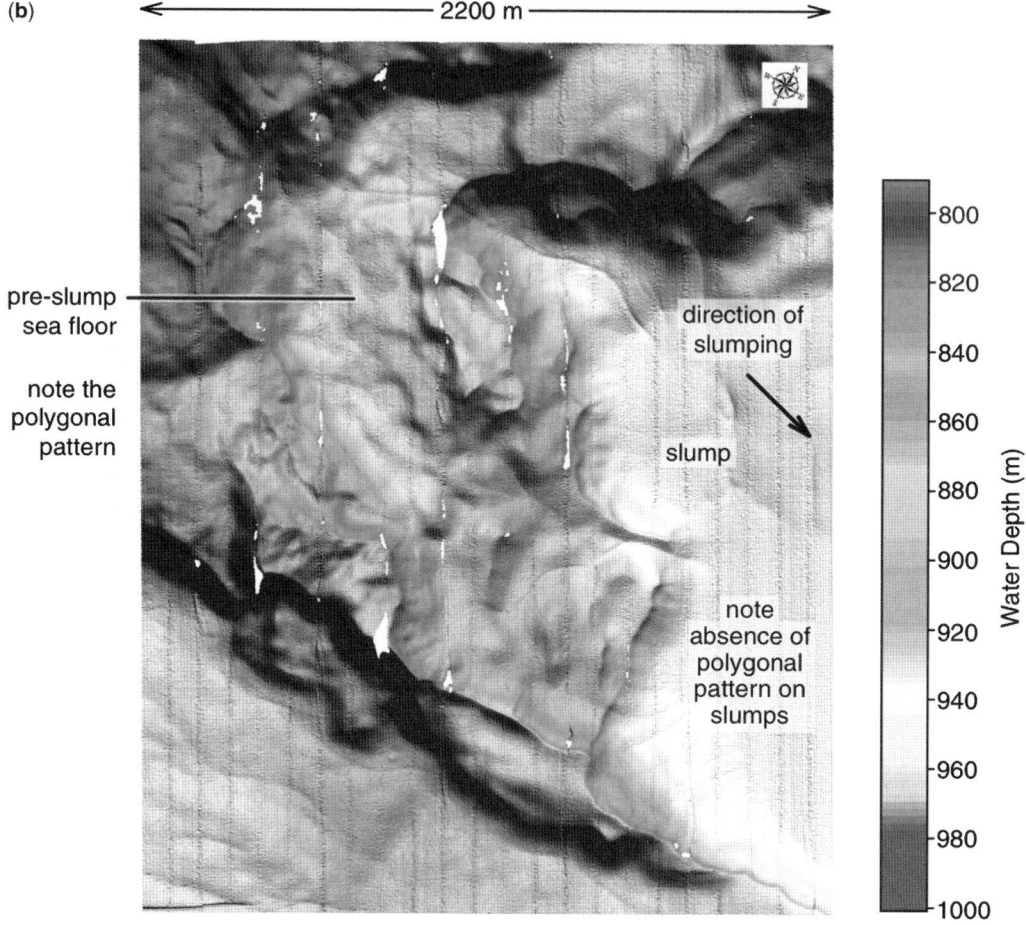

Fig. 8. (*Continued*).

The bathymetric depression on the left marks a slump block that has moved into the plane of the profile.

Several types of seismic reflection profiles have been collected in MC798 including eight SSDR profiles acquired early in the development of the deep-tow hydrophone. Figure 10b shows the SSDR profile recorded along section B–B' in Figure 8a. Subsequently collected core samples confirm that the sediments near the seafloor are fine grained. The wedge-shaped body whose upper boundary occurs between 1300 and 1350 ms is interpreted to be a sandy over-bank deposit and, on the basis of nearby multi-channel profiles, the chaotic reflections to the right of the sandy deposit are interpreted as marking the salt diapir that bounds the salt-withdrawal minibasin (Lutken *et al.* 2003). Note the sequence of reflections about 20 ms long near 1600 ms. On the basis of heat flow, Lewis (2001) suggests that it marks the BHSZ. The fact that its amplitude decreases as it approaches the salt supports that suggestion.

Figure 11 shows SSDR trace segments within the rectangular areas outlined in Figure 10a to illustrate reflection polarities. Figure 11a shows traces in the upper outlined area. The seafloor reflection is seen to be of positive polarity. Little coherent reflected energy is visible until 55 m below the seafloor when the first reflection of a two-reflection sequence labeled 'positive/negative' arrives. It is of positive polarity and is followed immediately by a reflection of negative polarity. Such a sequence is consistent with a thin layer of acoustic impedance higher than that of the material above and below it. A few milliseconds later, a second two-layer sequence labelled 'negative/positive' arrives. Its first reflection is of negative polarity and is followed by a reflection of positive polarity. The second

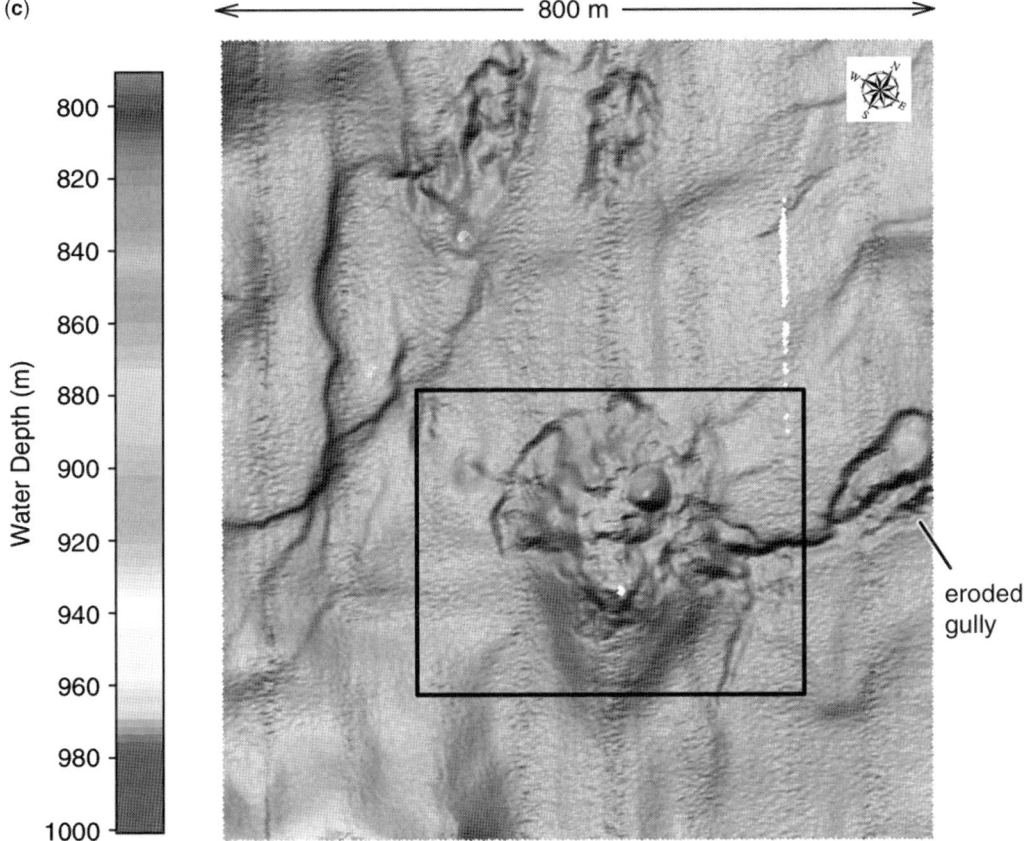

Fig. 8. (*Continued*).

sequence is consistent with a thin layer of acoustic impedance lower than that of the material above and below.

Figure 11b shows traces in the lower outlined area. They span a portion of the reflection sequence that Lewis (2001) suggests may mark the BHSZ. The sequence appears as a series of alternating negative and positive polarities. Although the sequence appears well defined on the variable density plot of Figure 10b, the energy on the wiggle traces is of such low frequency that it is not possible to delineate individual reflection waveforms. The loss of higher frequencies is apparently due to absorption during transmission through the sub-bottom.

Analyses of core samples

Core samples of sub-bottom sediments were taken at several sites in and near MC798 and in MC118. MC798 cores were taken in both pre-slumping and slumped areas and are typically 4–5 m long. At MC118 sites distant from the mound, typical cores are 2–3 m long. Cores on the mound are very short due to the hard ground encountered there.

Laboratory experiments were carried out to form natural gas hydrates in sediment sub-sampled from the cores. Sub-samples were taken at intervals of 3 m or less. The sub-samples were placed immediately into Zip-lock plastic bags and remained sealed in air-conditioned rooms until tested. To prepare a sub-sample for testing, 20 g of sediment was dispersed in 60 g of clean sand. This provided sufficient permeability and porosity to maximize the surface area of sediment that would be exposed to the hydrocarbon gas pressurizing the test cell. Original pore waters were preserved in the sediments and no other water was added. The mixture was placed into a 60 ml Teflon container with twelve 32 mm holes drilled into its sides for gas access. The sample container was placed into a 400 ml stainless steel test cell from Parr

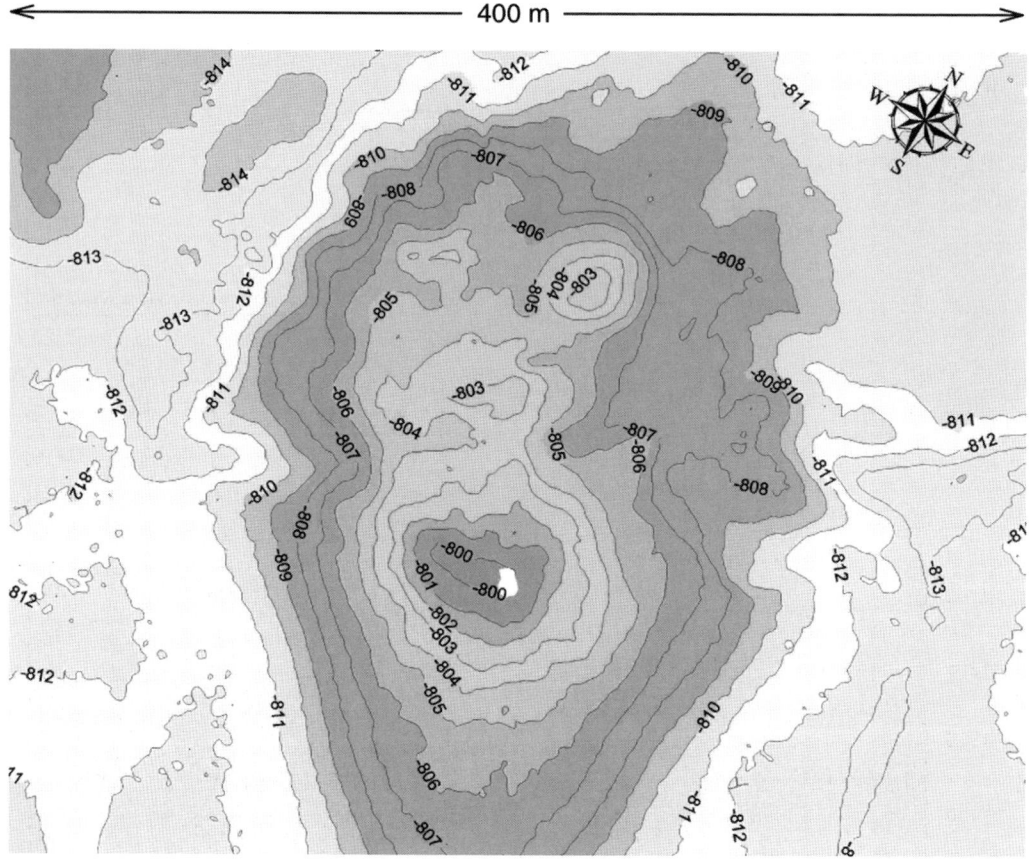

Fig. 9. Contour map of the rectangular area outlined in Figure 8c. The contour interval is 1 m. Conical promontories near the summit of the mud volcano may mark sites of fluid venting. The gully at the base of the northeastern flank is seen to be 2–3 m deep.

Instrument Company. An RTD resistance temperature detector was placed just below the surface of the sand–sediment mixture. A pressure transducer measured internal pressure of the reaction vessel. The cell was purged of air and pressurized to 2280 kPa with natural gas (90% methane, 6% ethane, 4% propane). After the system had equilibrated for 2 h at 21 °C in a constant-temperature bath, pressure was adjusted to 2210 kPa and the system allowed to re-equilibrate for another hour. The test cell was immersed in a constant-temperature bath at 0.5 °C. Pressure and temperature were measured every 2 min as the test cell cooled. Data collection was by an Omega Daqbook 120 equipped with DBK9 Data Acquisition System and DasyLab Software. A plot of pressure v. temperature was made.

Induction time was calculated from the elapsed time between the equilibrium pressure–temperature and hydrate formation as the sub-sample cooled. Normalized induction time was calculated by dividing the induction time of the specific sub-sample by the induction time for the sub-sample taken from 1.0 m depth in the core. Formation rate was determined for each 2 min interval as the number of moles of gas going into solid solution from the gas phase per unit time; the Peng–Robinson equation of state was used. The maximum formation rate was divided by the maximum formation rate for the sub-sample from 1.0 m depth in the core to obtain a normalized maximum formation rate. Duplicate runs were made for each sub-sample. Representative results are shown in Figure 12. They generally exhibit a pattern of induction time decreasing and formation rate increasing with depth in the core.

The cores were analysed to determine planktonic foraminiferal biostratigraphy, lithology and distribution of sand. Figure 13 compares representative results in each block to illustrate the main

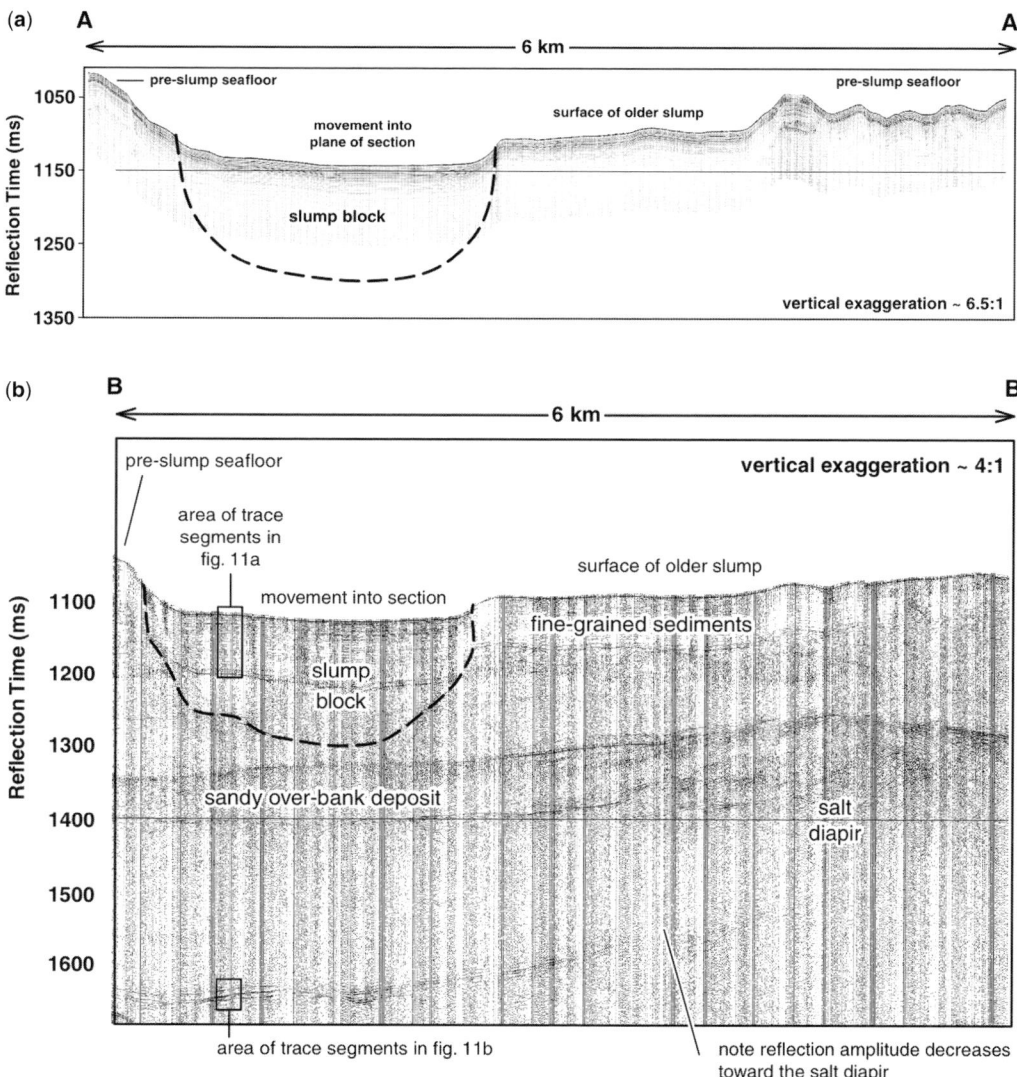

Fig. 10. Seismo-acoustic profiles in MC798 along sections plotted in Figure 8a: (**a**) chirp sonar profile along section A–A′. Note that both pre-slump and slumped seafloors are represented. There are two generations of slumping. (**b**) SSDR profile along section B–B′. It is interpreted by Lutken *et al.* (2004) to represent fine-grained sediments over a sandy over-bank deposit and the top of a salt diapir. The slump block on the left of the figure cuts through the fine-grained material but does not disturb the sandy material. Heat flow measurements suggest that the sequence of reflectors in the vicinity of 1600 ms marks the base of the hydrate stability zone (Lewis 2001).

sedimentary differences between the blocks. The MC798 core is 5 m long from a site near the centre of the block. The MC118 core is 3 m long from a site about 2 km north of the mound.

In both blocks the lithology is dominated by mud. The lithology in MC798 consists of a massive, olive grey mud with minor foraminiferal sand above and sparse sand below. In MC118, a massive light olive grey mud with minor foraminiferal sand lies above a distinctly layered, finer-grained and darker olive-grey mud which has notable mottles in its upper part and a distinctive reddish layer near its base. This unit is underlain by a massive mud, which lies above another, layered, unit. The MC118 section contains 5 times more sand than that in MC798.

Biostratigraphic results show that planktonic foraminiferal zones from the Holocene and latest

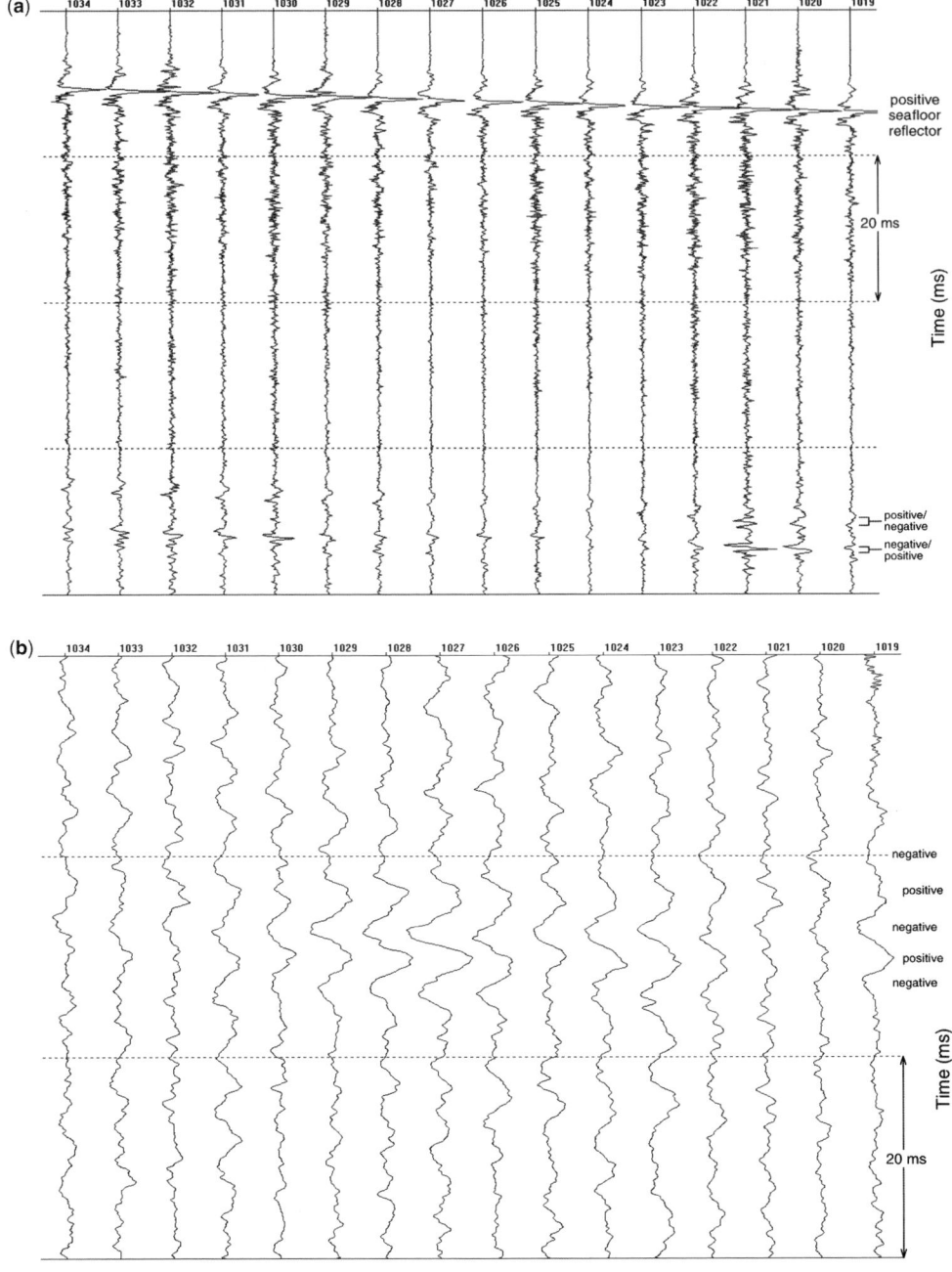

Fig. 11. Wiggle-trace segments illustrating polarity of reflections within the rectangular areas outlined in Figure 10b: (**a**) the upper outlined area. The seafloor reflection is seen to be of positive polarity (as defined in Fig. 4). There is little coherent reflected energy until the arrival of a two-reflection sequence labelled 'positive/negative'. It is composed of an upper reflection of positive polarity and a lower reflection of negative polarity. It is consistent with a thin layer of acoustic impedance higher than that of the material above and below it. A few milliseconds later, a second two-layer sequence labelled 'negative/positive' arrives. Its upper reflection is of negative polarity and is followed by the lower reflection of positive polarity. The second sequence is consistent with a thin layer of acoustic impedance lower than that of the material above and below. (**b**) The lower outlined area. The reflection sequence near 1600 ms is seen to be of alternating positive and negative polarities but individual reflections are not clearly defined.

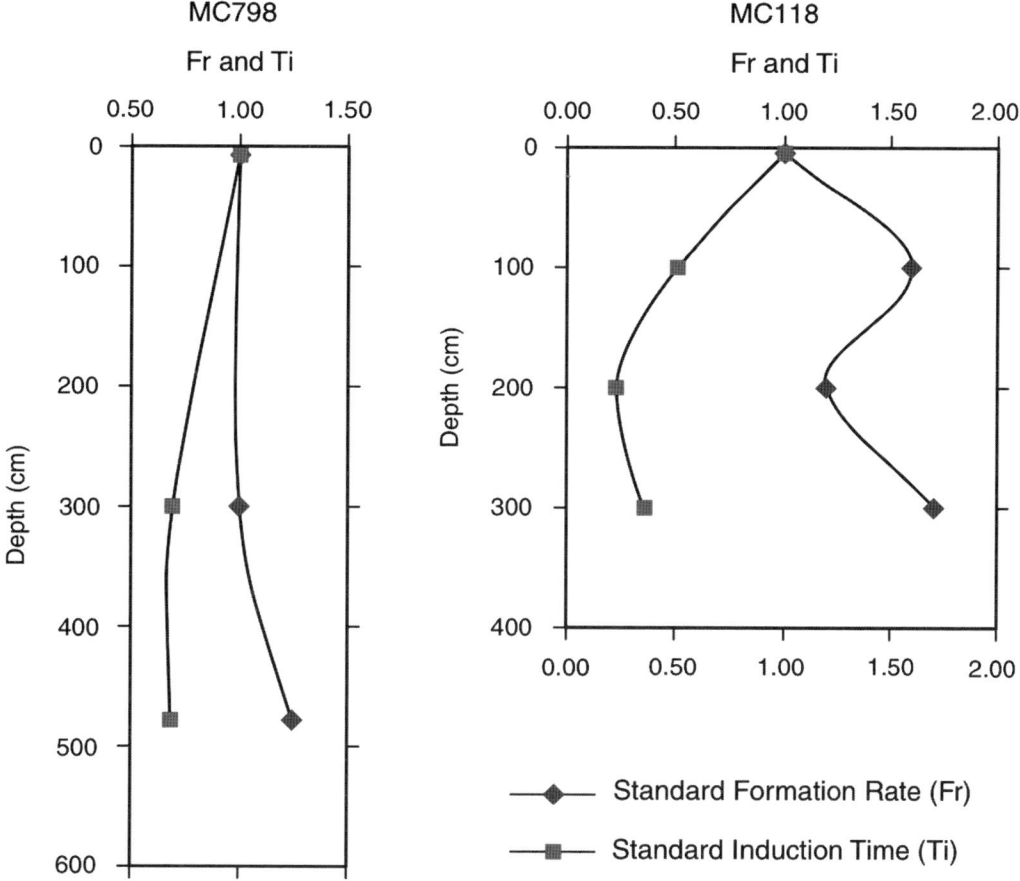

Fig. 12. Comparison of normalized hydrate induction times and formation rates determined by forming natural gas hydrates in sediment sub-sampled from cores. Normalization is accomplished by dividing each measurement by the corresponding measurement at a 1 m depth in the core.

Pleistocene (Ericson & Wollin 1968, as modified by Flower & Kennett 1990) were recovered in or near both blocks. Cores from MC798 only recovered the interglacial (Holocene) Z Zone whereas cores in MC118 recovered not only the Z Zone but also part of the Y Zone (all of sub-zone Y1 and at least part of sub-zone Y2). In the Gulf of Mexico, the base of the Z Zone dates to about 9.8 ka (Flower & Kennett 1990; Poore et al. 2003) and the base of the Y1 sub-zone to about 14 ka (Flower & Kennett 1990). The distinctive reddish layer mentioned above consistently lies near the Y1–Y2 boundary.

In addition to the cores in MC798 and MC118, other cores were collected a few kilometres southwest of MC798. The longest (28.5 m) was collected by the R/V *Marion Dufresne* at the location in Figure 1 labelled 'MD02-2570'. It samples the entire Holocene interglaciation and all of the last deglaciation, extending into and perhaps a little below the glacial maximum. Sedimentation rates determined from the biostratigraphy of all cores are presented in Figure 14. Rates in MC118 are $9-11$ cm ka^{-1} during the Holocene and $24-35$ cm ka^{-1} during the interval of time 9.8–14 ka, which spans the earliest Holocene and the latest Pleistocene. Holocene rates in and near MC798 are in the range $25-54$ cm ka^{-1}. Most cores in the vicinity of MC798 provide no information on sedimentation rates prior to the Holocene because they are too short to sample the Y1–Y2 boundary. The 28.5 m core does sample that boundary and indicates a sedimentation rate of 214 cm ka^{-1} during 9.8–14 ka. Thus sedimentation rates are greater in the vicinity of MC798 than in MC118. The quantitative difference is a factor

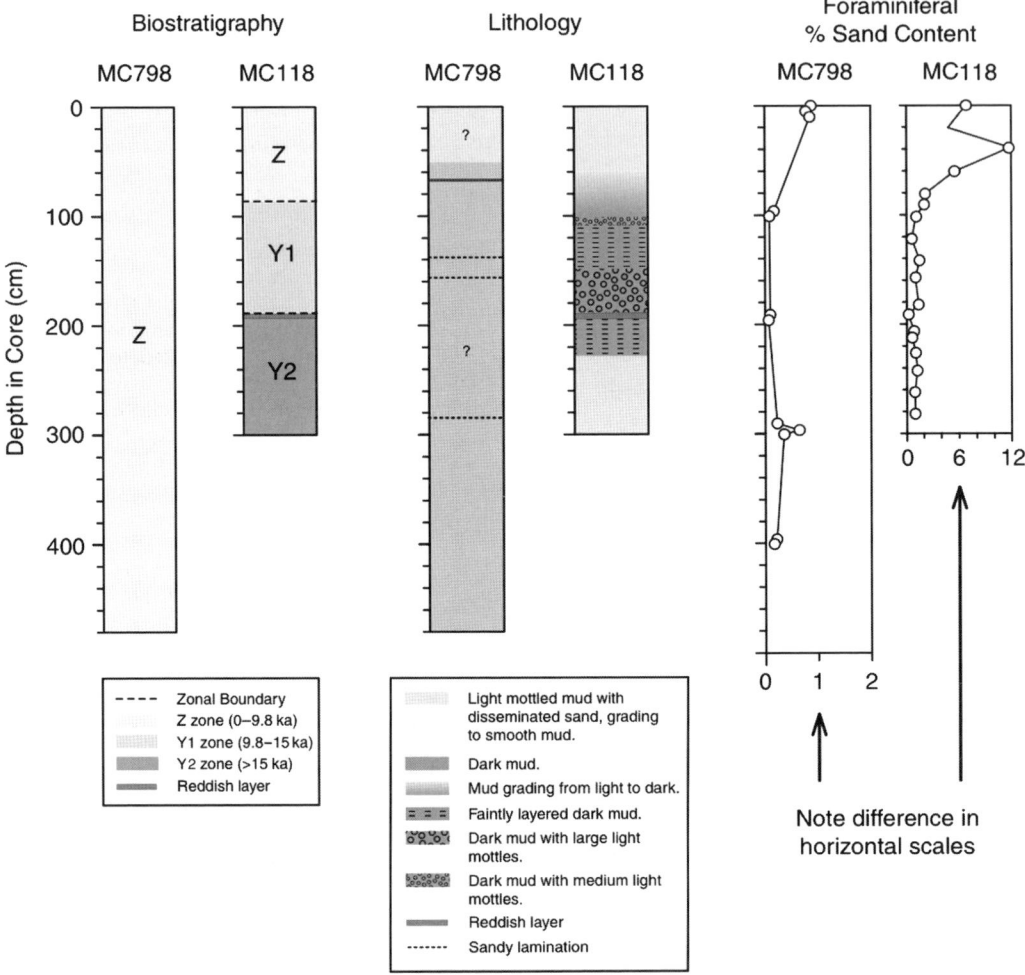

Fig. 13. Comparison of lithology, percentage sand and planktonic foraminiferal biostratigraphy in representative cores. The lithology in both MC118 and MC798 is dominated by mud. The section in MC118 contains five times more sand than that in MC798. The sand is foraminiferal. Cores in MC798 are entirely Z zone (Holocene less than 9.8 ka). Cores in MC118 contain sediments from the Z zone and two sub-zones of the Y zone: Y1 (9.8–14 ka) and Y2 (greater than 14 ka). A distinctive reddish layer occurs near the Y1–Y2 boundary.

of 2–6 during the Holocene and a factor of 6–9 during and immediately after the last deglaciation.

Discussion and conclusions

Some of the earliest activities supported by the Gulf of Mexico Hydrates Research Consortium were experiments that used a mixture of hydrocarbon gases and sediment cored in the Mississippi Canyon Area rather than the 'pure' methane and artificial sand–clay 'sediment' often used when forming hydrates in the laboratory. It was observed that the change reduced the induction time and increased the formation rate of laboratory hydrates. Lee (2001) discovered that biosurfactants associated with microbes in the natural materials were responsible for the observed effects. This revelation caused the study of microbial influences to become a Consortium priority. It was learned that laboratory hydrates form preferentially on surfaces of smectite rather than on kaolinite or sand surfaces (Rogers et al. 2003). This is significant when considering possible scenarios of hydrate accumulation in the northern Gulf of Mexico where smectite is ubiquitous. In fact, it is becoming accepted that

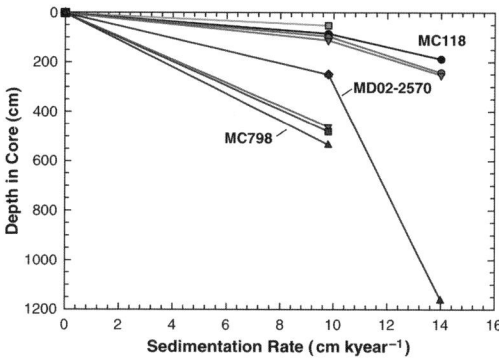

Fig. 14. Sedimentation rates determined from the biostratigraphy in cores. Holocene rates are 9–11 cm ka^{-1} in MC118 and 25–54 cm ka^{-1} in the vicinity of MC798. The latest Pleistocene-to-earliest Holocene rates are 24–35 cm ka^{-1} in MC118 and 214 cm ka^{-1} in the 28.5 m core collected by the R/V *Marion Dufresne*.

microbial studies are basic to understanding geochemical and geophysical aspects of sediment-hosted hydrates everywhere.

The use of surface vessels and lengthy cables to collect data near the seafloor in deep water is time-consuming and subject to inaccuracies in sensor locations. The Consortium used an AUV to survey MC798 in 2002 and to survey MC118 in 2006. Experience has shown that a well-operated AUV produces data quality superior to that produced by a surface-based operation and does so at less cost. It is concluded that AUVs have become essential to accurate, economical seafloor studies in water more than about 800 m deep.

It was realized that the chirp-sonar data suffer from a deficiency that is not immediately obvious. The reflected chirp signal is rectified during the processing that produces the graphic display. Unfortunately, this eliminates any possibility of using reflection polarity to distinguish between hard ground and free gas. This is important when studying hydrate accumulations and severely limits the value of the chirp data.

The spatial resolution of the swath bathymetry data recorded by AUV in MC118 and MC798 was improved from 5 m to about 1 m by careful reprocessing. Back-scatter images were derived from the improved swath data and compared to the side-scan-sonar images that had been recorded. The two were generally comparable but the back-scatter had better resolution. It may develop that AUVs will no longer be equipped with side-scan sonar, relying on back-scatter imagery instead.

The SSDR seismic profiling technique has proved to be effective for imaging the hydrate stability zone. In spite of strong noise due to the cable strumming deep-towing, Figure 4 illustrates that the technique can compress the principal energy content of an 80 in^3 watergun source by about a factor of 50. This is sufficient for the amplitude of the central lobe of the processed direct wave to approximate unity. It then becomes possible to calculate normal-incidence reflection and transmission coefficients in the sub-bottom and, thereby, to estimate acoustic impedance contrasts across sub-bottom reflectors.

Provided that the seismic source provides sufficient energy, the amount of penetration provided by the SSDR technique is limited to the travel time of the surface ghost, i.e. the time it takes the upward-travelling sub-bottom reflections to be reflected at the water surface and travel back downward to the hydrophone. Thus the maximum usable penetration on a SSDR profile is approximately equal to the depth of the receiver.

Comparison of wiggle traces in Figure 7 with those in Figure 11a, b illustrates how much the quality of SSDR data improved during the four years that passed between surveying MC789 and MC118. The next planned field improvement will be to mount the hydrophone on an AUV, thereby eliminating the deep-tow cable and its strumming. Since the AUV will be able to operate much deeper than the towing depth, the usable sub-bottom penetration in deep water will increase substantially.

The swath bathymetry in Figure 8a shows that some areas of MC798 have slumped and others have not. The slump block in Figure 10b is about 200 ms (*c*. 150 m) thick. The slump movement is confined to the shallower fine-grained sediments and does not disturb the wedge-shaped sediment body interpreted by Lutken *et al.* (2003) to be a sandy over-bank deposit. A polygonal pattern is visible on the seafloor that has not slumped but is not seen on the surface of slumps (Fig. 8b). The pattern may indicate that polygonal faulting was present prior to the episodes of slumping. Polygonal faulting provides the fracture porosity that allows conventional oil and gas to be produced from fine-grained sediments in many parts of the world (Cartwright *et al.* 2003). If pre-slump fracture porosity in the fine-grained sediments of MC798 permitted sufficient circulation of gas and water, hydrate formation enhanced by the presence of smectite would have occurred (McGee *et al.* 2004). Subsequent slump movement could have exposed the fracture-filling hydrates directly to seawater, causing them to dissociate and liberate water which would have fluidized the slumping material and eliminated all traces of the polygonal pattern.

The 28.5 m core collected near MC798 (Fig. 1) by the R/V *Marion Dufresne* provided convincing evidence that long cores are necessary for understanding Holocene events in the Mississippi Canyon Area. It provided sedimentation rates for the entire Holocene interglaciation and all of the last deglaciation, extending into and perhaps a little below the glacial maximum. A potential example of useful information that could be provided by a long core is the unconformity within the syncline east of the mound in MC118 (Fig. 6a). That unconformity could be penetrated by a 20^+ m core and, perhaps, provide evidence concerning the age of the synclinal deformation.

The AUV surveys were carried out by C&C Technologies of Lafayette, Louisiana. Post-processing of the multi-beam data was done at Department of Earth Science of the University of Rome 'La Sapienza'. Software for processing SSDR profiles was written by Lookout Geophysical Company of Palisade, Colorado. Figure 10a was provided by Laura Giacomini. All figures were prepared for publication by Paul Mitchell. Funding for the Gulf of Mexico Hydrates Research Consortium is provided by the Minerals Management Service (MMS) of the Department of the Interior, the National Energy Technology Laboratory (NETL) of the Department of Energy and the National Institute for Undersea Science and Technology (NIUST) of the National Oceanographic and Atmospheric Administration (NOAA) of the Department of Commerce.

References

BERRYHILL, J. L. JR., SUTER, J. R. & HARDIN, N. S. 1986. Late Quaternary facies and structure, northern Gulf of Mexico: Interpretations from seismic data. *American Association of Petroleum Geologists Studies in Geology*, **23**.

BOETIUS, A., RAVENSCHIAG, K. ET AL. 2000. A marine microbial consortium apparently mediating anaerobic oxidation of methane. *Nature*, **407**, 623–626.

CARTWRIGHT, J., JAMES, D. & BOLTON, A. 2003. The genesis of polygonal fault systems: A review. *In*: VAN RENSBERGEN, P., HILLIS, R. R., MALTMAN, A. J. & MORLEY, C. K. (eds) *Subsurface Sediment Mobilization*. Geological Society, London, Special Publications, **216**, 223–243.

COLEMAN, J. M., PRIOR, D. B. & LINDSAY, J. F. 1983. *Deltaic Influences on Shelfedge Instability Processes*. SEPM Special Publications, **33**, 121–137.

COOPER, A. & HART, P. 2003. High-resolution seismic-reflection investigation of the northern Gulf of Mexico gas-hydrate-stability zone. *Marine Petroleum Geology*, **19**, 1275–1293.

DILLON, W. P., LEE, M. Y. & COLEMAN, D. F. 1994. Identification of marine hydrates in situ and their distribution off the Atlantic coast of the United States. *In*: SLOAN, E. D., HAPPEL, J. & HNATOW, M. A. (eds) *International Conference on Natural Gas Hydrates*, Annals of the New York Academy of Sciences, **715**, 364–380.

ERICSON, D. B. & WOLLIN, G. 1968. Pleistocene climates and chronology in deep-sea sediments. *Science*, **162**, 1227–1234.

FLOWER, B. P. & KENNETT, J. P. 1990. The Younger Dryas cool episode in the Gulf of Mexico. *Paleoceanography*, **5**, 949–961.

HALBOUTY, M. T. 1979. *Salt Domes: Gulf Region, United States and Mexico*, 2nd edition. Gulf, Houston, TX.

KOHL, B. & ROBERTS, H. H. 1994. Fossil foraminifera from four active mud volcanoes in the Gulf of Mexico. *Geo-Marine Letters*, **14**, 126–134.

KOHL, B. & ROBERTS, H. H. 1995. Mud volcanoes in the Gulf of Mexico: A mechanism for mixing sediments of different ages in slope environments. *Transactions Gulf Coast Association of Geological Societies*, **45**, 351–359.

LEE, M.S. 2001. *The effects of biosurfactants on gas hydrate formation in ocean sediments*. Thesis submitted for M.S. degree in Chemical Engineering, Mississippi State University.

LEWIS, T. 2001. *Analysis of Recent Heat-flow Measurements Made by TDI-Brooks*. Report to the Gulf of Mexico Gas Hydrates Research Consortium, unpublished.

LUTKEN, C. B., GERESI, E., MCGEE, T., WOOD, W. T. & LOWRIE, A. 2004. Complex geology creates difficult interpretational scenarios in hydrated areas along western flank of Mississippi Canyon, northern Gulf of Mexico. *Abstracts Volume of the AAPG Annual Convention*, Dallas, TX, A89.

LUTKEN, C., DOURIE, A. ET AL. 2003. Interpretation of high resolution seismic data from a geologically complex continental margin, northern Gulf of Mexico. *GCAGS/GCSSEPM Transactions*, **53**, 504–516.

MACDONALD, I. R., BENDER, L. C., VARDARO, M., BERNARD, B. & BROOKS, J. R. 2005. Thermal and visual time-series at a seafloor gas hydrate deposit on the Gulf of Mexico slope. *Earth and Planetary Science Letters*, **233**, 45–59.

MCGEE, T. M. 2000. Pushing the limits of high-resolution in marine seismic profiling. *Journal of Environmental and Engineering Geophysics*, **5**, 43–53.

MCGEE, T. M. 2001. Pushing the limits of seismic resolution in deep water. *Proceedings of the Eleventh International Offshore and Polar Engineering Conference*, Stavanger, Norway, 42–45.

MCGEE, T. M. 2006. A sea-floor observatory to monitor gas hydrates in the Gulf of Mexico. *SEG Leading Edge*, **25**, 644–647.

MCGEE, T. M., LUTKEN, C. B. ET AL. 2004. Can fractures in soft sediments host significant quantities of gas hydrates? *AAPG Hedberg Conference*, Vancouver.

MILKOV, A. & SASSEN, R. 2000. Thickness of the gas hydrate stability zone, Gulf of Mexico continental slope. *Marine and Petroleum Geology*, **17**, 981–991.

MILKOV, A. V. & SASSEN, R. 2001. Estimate of gas hydrate resource, northwestern Gulf of Mexico continental slope. *Marine and Petroleum Geology*, **179**, 71–83.

MILKOV, A. V. & SASSEN, R. 2003. Preliminary assessment of resources and economic potential of individual gas hydrate accumulations in the Gulf of Mexico continental slope. *Marine and Petroleum Geology*, **20**, 111–128.

NEURAUTER, T. W. & BRYANT, W. R. 1989. Gas hydrates and their association with mud diapir/mud volcanos on the Louisiana continental slope.

Proceedings of 21st Offshore Technology Conference, **5944**, 599–607.

POORE, R. Z., DOWSETT, J. J. & VERARDO, S. 2003. Millennial- to century-scale variability in Gulf of Mexico Holocene climate records. *Paleoceanography*, **18**, 1048.

ROBERTS, H. H. & CARNEY, R. S. 1997. Evidence of episodic fluid, gas, and sediment venting on the northern Gulf of Mexico continental slope. *Economic Geology*, **92**, 863–879.

ROGERS, R. E., KOTHAPALLI, C. K. & LEE, M. S. 2002. Influence of microbes on gas hydrate formation in porous media. *Proceedings of the Fourth International Conference on Gas Hydrates*, Yokohama, 4175–4187.

ROGERS, R. E., KOTHAPALLI, C. K., LEE, M. S. & WOOLSEY, J. R. 2003. Catalysis of gas hydrates by biosurfactants in seawater-saturated sand/clay. *Canadian Journal of Chemical Engineering*, **81**, 1–8.

SASSEN, R. & MACDONALD, I. R. 1997. Hydrocarbons of experimental and natural gas hydrates, Gulf of Mexico continental slope. *Organic Geochemistry*, **26**, 289–293.

SASSEN, R. & ROBERTS, H. H. 2004. Site selection and characterization of vent gas, gas hydrate and associated sediments. *Final Technical Progress Report of the Gulf of Mexico Seafloor Stability and Gas Hydrate Monitoring Station Project*, 28–227. World Wide Web Address: http://www.osti.gov/bridge/purl.cover.jsp?purl=/838881-59VaGw/native/.

SASSEN, R., MACDONALD, I. R. ET AL. 1998. Bacterial methane oxidation in sea-floor gas hydrate: Significance to life in extreme environments. *Geology*, **26**, 851–854.

SASSEN, R., MILKOV, A. V. ET AL. 2003. Gas venting and subsurface charge in the Green Canyon area, Gulf of Mexico continental slope: Evidence of a deep bacterial methane source? *Organic Geochemistry*, **34**, 1555–1564.

SASSEN, R., ROBERTS, H. H. ET AL. 2006. The Mississippi Canyon 118 gas hydrate site: A complex natural system. *Proceedings of 38th Offshore Technology Conference*, **18132**.

SASSEN, R., SWEET, S. T. ET AL. 1999. Geology and Geochemistry of gas hydrates, central Gulf of Mexico continental slope. *Transactions of the Gulf Coast Association of Geological Societies*, **49**, 462–468.

SLEEPER, K., LOWRIE, A., BOSMAN, A., MACELLONI, L. & SWANN, C. T. 2006. Bathymetric mapping and high resolution seismic profiling by AUV in MC118 (Gulf of Mexico). *Proceedings of 38th Offshore Technology Conference*, **18133**.

TDI-Brooks. 2001. *The 2000 University of Mississippi Mississippi Canyon Heat Flow Program*. TDI-Brooks International Inc., Technical Report, **01–697**.

VARDARO, M., MACDONALD, I. R., BENDER, L. C. & GUINASSO, N. L. JR. 2005. Dynamic biological and physical processes observed at a gas hydrate outcropping on the continental slope of the Gulf of Mexico. *Geo-Marine Letters*, **26**, 6–15.

Gas hydrate forming fluids on the NE Sakhalin slope, Sea of Okhotsk

L. L. MAZURENKO[1†], T. V. MATVEEVA[1]*, E. M. PRASOLOV[1,2], H. SHOJI[3],
A. I. OBZHIROV[4], Y. K. JIN[5], J. POORT[6], E. A. LOGVINA[1], H. MINAMI[3], H. SAKAGAMI[3],
A. HACHIKUBO[3], A. S. SALOMATIN[4], A. N. SALYUK[4], E. B. PRILEPSKIY[2] & CHAOS
2003 SCIENTIFIC TEAM

[1]*All-Russia Research Institute for Geology and Mineral Resources of the World Ocean,
1, Angliyskiy prospect, 190121 St Petersburg, Russia*

[2]*Centre of Isotopic Research of VSEGEI, 190 136, 74, Srednii prospect V.O.,
St Petersburg, Russia*

[3]*Kitami Institute of Technology, New Energy Research Center, 165 Koen-cho, 090-8507
Kitami, Japan*

[4]*PV.Ilyichev's Pacific Oceanographical Institute, Baltiyskaya Street, 690041,
Vladivostok, Russia*

[5]*Korea Polar Research Institute (KOPRI), 1903 Get-Pearl Tower, Songdo Technopark, 7-50
Songdo-dong, Yunsu-ku Incheon, 406-840, Korea*

[6]*Renard Centre of Marine Geology, Ghent University, Krijgslaan 281, S8, B-9000
Ghent, Belgium*

[†]*Leonid Mazurenko died in July 2007*

*Corresponding author (e-mail: tv_matveeva@mail.ru)

Abstract: An area of focused fluid venting off NE Sakhalin, Sea of Okhotsk, was investigated in 2003 during the 31st and 32nd international expeditions of R/V *Akademik M. A. Lavrentyev* within the framework of the CHAOS Project. More than 40 structures related to seafloor gas venting were discovered and gas hydrates were sampled from three of these: CHAOS, Hieroglyph and Kitami. Geochemical analyses were used to define the mechanisms of gas hydrate accumulation and the sources of fluids involved. Chemical and isotopic analyses of the interstitial and hydrate waters suggest that hydrates were formed from seawater (or *in-situ* pore water) and an ascending fluid enriched in salts. Hydrate formation occurs at locations of the most intensive saline water upflow, and this is probably a function of the gas solubility in water in equilibrium with hydrate. The water involved in gas hydrate formation consists of about 70% pore water derived from the host sediment and 30% from the ascending fluid. The overall isotopic composition of the 'fluid' taking part in hydrate formation was calculated as $\delta^2H \approx -11‰$ and $\delta^{18}O \approx -1.5‰$.

Gas hydrates in marine sediments are considered as a potential hydrocarbon energy resource and as an important matter of concern regarding the impact of their dissociation upon global climate change. In both cases, it is essential to estimate the quantities of gas hydrates present in marine sediments and to understand the processes leading to gas hydrate accumulation and decomposition. Of particular interest are shallow gas hydrates associated with fluid venting structures. These gas hydrates are capable of forming geochemical barriers to the gas flux, and therefore reducing the amount of gas entering the hydrosphere from the seafloor. To calculate the global methane budget in marine sediments, an evaluation has to be made of the different processes involved, such as gas diffusion, oxidation and formation of methane-derived carbonates and hydrates. The composition of the original gas and of the gas locked up in carbonates or hydrates is often rather different, and understanding the gas origin is important in evaluating the role of methane hydrate in the carbon cycle (Dickens 1999).

Although the origin and composition of fluids discharging at the venting locations have been the subject of numerous studies (Luff & Wallmann 2003; Aloisi *et al.* 2004; Hensen *et al.* 2007), the influence of fluid composition on hydrate formation is often not taken into account. In this study we made an attempt to consider the sources of fluids involved in the formation of hydrates associated with the gas vents located off NE Sakhalin Island in the Sea of Okhotsk (Fig. 1). The Sakhalin shelf and continental slope towards the Derugin Basin form a remarkable active hydrocarbon seep region supplied by deeply sourced hydrocarbons through active fault systems. There are a large number of already identified and potential oil and gas reserves, with about 70 deposits estimated on the island and adjacent shelf (Dmitrievsky et al. 2004).

Methane hydrate occurrences related to active gas vents in the Sea of Okhotsk are known from offshore Paramushir Island and offshore NE Sakhalin Island (Zonenshayn et al. 1987; Ginsburg et al. 1993; Soloviev et al. 1994). Off Sakhalin Island, gas vents were identified on the continental slope at a water depth of 620–1040 m using echo sounding and by means of methane concentration measurements (Obzhirov 1992; Ginsburg et al. 1993). More than 100 additional locations with features typical of gas emission were discovered during repeated expeditions in the framework of the German–Russian KOMEX Project (Biebow & Hutten 1999; Biebow et al. 2003) and the Russian–Japanese–Korean CHAOS Project (Shoji et al. 2005). Here we report on new samples and analysis obtained during the CHAOS Project from a gas venting zone that is about 13 km long in a SW to NE direction, and about 11.5 km wide. Chemical and isotope analyses of interstitial and hydrate waters from three hydrate-rich venting structures are presented and interpreted in relation to the hydrate-forming fluid.

Geological setting

The Sea of Okhotsk is a large marginal sea at the northwestern rim of the Pacific located on the Okhotsk plate. The western boundary of the Okhotsk Plate follows the Hokkaido–Sakhalin dextral shear zone, which can be traced over a distance of 2000 km from Hokkaido Island up to northern tip of Sakhalin Island (Fig. 1a). Numerous compression and extension structures owe their origin to the complex structure of this plate boundary (Savostin et al. 1983; Rozhdestvensky 1986; Fournier et al. 1994). The northeastern slope of Sakhalin Island has strong tectonic activity (seismicity above 5) and delimits the Derugin Basin, which has more than 2 km of sedimentary infill (Gnibidenko & Khvedchuk 1982; Fournier et al. 1994). The Derugin Basin is outlined by the 1500 m isobath. It has a gentle but distinct bottom relief, which is largely a refection of rotated block structures in the basement. Gas accumulation in the sediments of the Sea of Okhotsk is made favourable by the subarctic climate and the prevailing hydrologic regime, which result in high primary productivity, low bottom water temperatures,

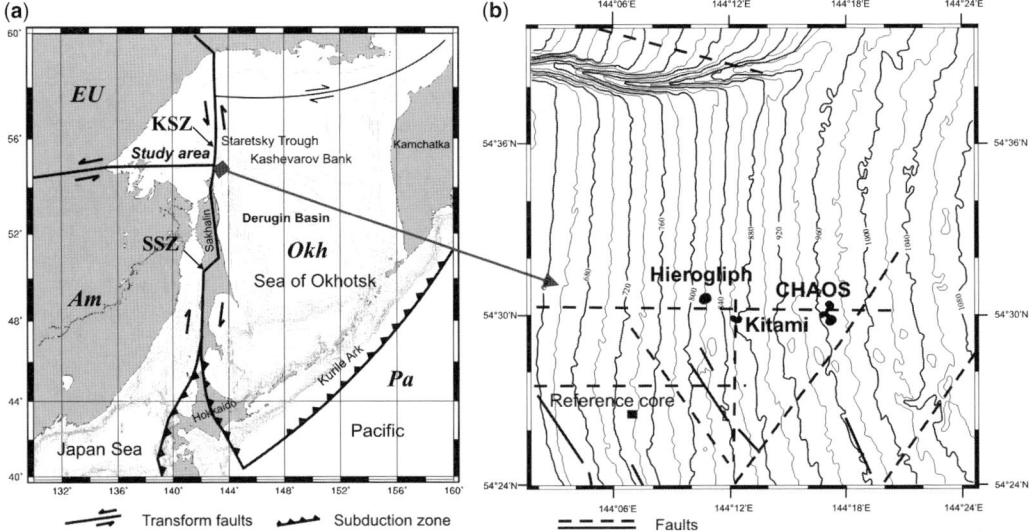

Fig. 1. (a) Main morphological features of the Okhotsk Sea and plate geometry of the northwestern Pacific Shear zones: SSZ, Sakhalin; KSZ, Kashevarov. Plates: Pa, Pacific; EU, Eurasian; Okh, Okhotsk; Am, Amur. (b) Bathymetric map of the study area off NE Sakhalin Island (Dullo et al. 2004) based upon side-scan sonar data obtained during the 31st and 32nd cruises of the R/V Akademik M. A. Lavrentyev. The positions of venting structures are marked by filled circles.

high sedimentation rates (3.8–100 cm per ka; Gorbarenko et al. 1990; Botsul et al. 1999; Astakhov et al. 2000), and a large flux of organic carbon to the seafloor (Koblents-Mishke 1967; Bogorov 1974). Sediments with an organic carbon content of more than 1.5% extend from the continental rise into the adjacent basins, including the Derugin Basin (Bezrukov 1960; Lisitsyn 1978). The sediments in the Derugin Basin are dominated by ice-rafted detritus, mass wasting deposits and hemipelagics of high organic content (Sancetta 1981; Gorbarenko et al. 1990; Gorbarenko 1991; Botsul et al. 1999; Astakhov et al. 2000; Shiga & Koizumi 2000; Wong et al. 2003).

The study area (Fig. 1b) is also characterized by the presence of numerous sub-meridional fault zones and transverse faults within the sediments. Dislocation structures several hundred metres wide are formed at fault junctions and have created pathways for gas migration (Matveeva et al. 2005; Shoji et al. 2005).

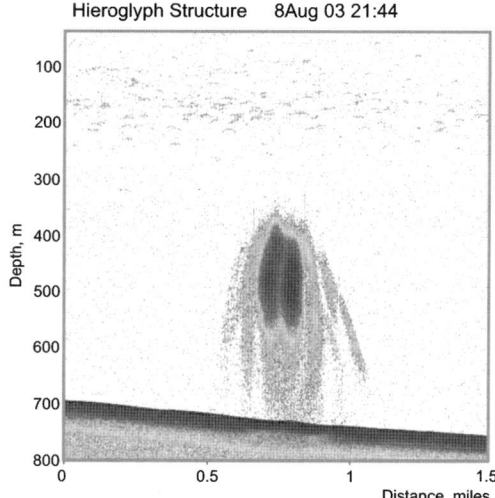

Fig. 2. Echosounder record across the Hieroglyph gas venting structure with a 'flare'-type hydroacoustic anomaly related to gas bubbles in the overlying seawater.

New gas hydrate-related venting structures

During the 31st and 32nd cruises of the R/V *Akademik M. A. Lavrentyev* carried out in the framework of the CHAOS (hydro-Carbon Hydrate Accumulations in the Okhotsk Sea) International Research Project (Matveeva et al. 2005; Shoji et al. 2005), about 40 new structures related to fluid venting were discovered. This involved using the SONIC-3 deep-towed system (side-scan sonar combined with 3.5 kHz sub-bottom profiler), together with seismoacoustic and echosounder profiling. Gas plumes in the water column above vent structures were observed on echosounder records as 'flare-type' anomalies (Fig. 2). On side-scan records these flare locations correspond to structures of enhanced backscatter with an isometric (Fig. 3a) and concentric (Fig. 3b) form. The sub-bottom profiler records of these gas seepage structures display extensive sediment blanking (Fig. 3c) characterized by a loss of correlation of the reflection signals due to the presence of free gas (Baranov et al. 2005; Shoji et al. 2005). All the fluid vents were considered to be at the locations of potential gas hydrate accumulations (Mazurenko et al. 2006).

Coring using a gravity-corer (GC) and a hydrocorer (HC) was carried out within three of the newly observed structures, named CHAOS (11 coring stations), Hieroglyph (three coring stations) and Kitami (two coring stations), and also at one reference location away of the venting structures (Fig. 1b). Gas hydrates were sampled from all of the cored structures (Mazurenko et al. 2005). The largest and complex CHAOS structure was chosen for a detailed study. It comprises a few small vents with different intensities of fluid expulsion and creates at least three gas plumes in the water column. The latter are visible both on the echo sounder and side-scan sonar records.

Five of the cores recovered gas hydrates at different sub-bottom intervals (Fig. 4, Tables 1–3). The hydrate-bearing sediments are homogenous, olive-green, silty-clay, diatomic mud with numerous inclusions of authigenic carbonates of varied shapes and sizes and buried shells and debris of *Calyptogena* bivalves. Most of the recovered cores were considerably enriched in gas and smelt of H_2S. It should be noted that the reference core also had traces of gas present, a characteristic of the high productivity sediments of the Derugin Basin. Hydrate-bearing cores were often fluidized due to hydrate decomposition during recovery. Cores were described and photographed; and the gas hydrate aggregates were decomposed for water and gas analyses. Hydrate gases were obtained from the gas hydrate aggregates during spontaneous degassing in salt-saturated water. Sediment samples for pore water squeezing were collected within half an hour of retrieval. Waters were squeezed using a pressure filtration system and analysed in onshore laboratories. Authigenic carbonate inclusions were collected for chemical and isotope studies. Bottom water samples were taken from the head of the corer.

Structures of gas hydrate-bearing sediment

The formation of gas hydrates significantly alters the structure and physical properties of the

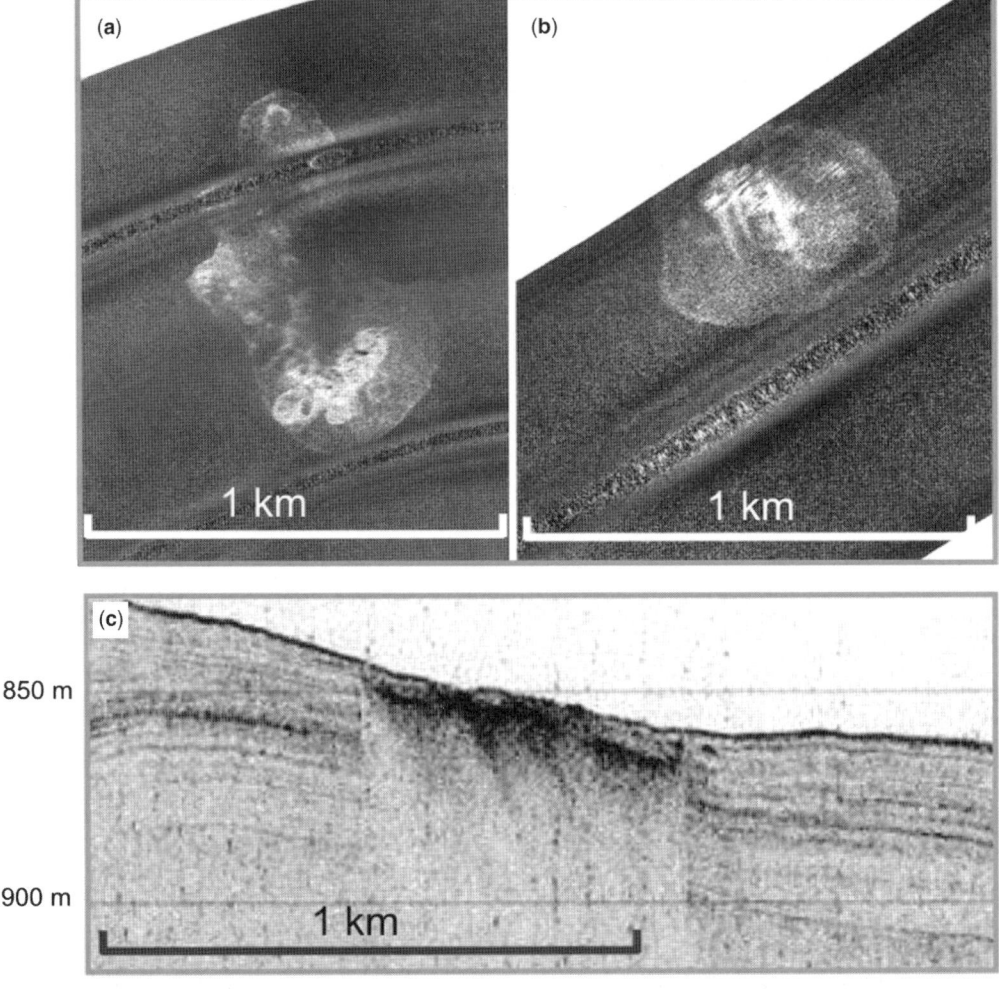

Fig. 3. Side-scan sonar records of the CHAOS (**a**) and the Hieroglyph (**b**) fluid venting structures; and seismic profile through the Kitami fluid venting structure (**c**).

sediments. Based on macroscopic observations and structural similarities with cryogenic types of sedimentary arrangements, Ginsburg & Soloviev (1998) introduced a classification of gas hydrate-bearing sedimentary structures based on the interrelation between gas hydrates and the hosting sediments due to different modes of fluid infiltration and sedimentary lithologies (Fig. 4). We applied this classification in this study and characterized four specific types of sediments possessing gas hydrates as detailed below.

All gas hydrate-containing cores were characterized by lenticular-bedded structures similar to cryogenic structures (Fig. 4a). The lenticular subparallel hydrate schlieren alternating with hydrate laminae and lenses vary from 0.5 to 3 cm thick. Often two nearby hydrate layers are connected by inclined hydrate layers. This structure is typically formed along gas venting fronts due to the processes of water segregation by diffusing gas (Ginsburg & Soloviev 1998). During this process water segregates from the host sediments and migrates to the front of hydrate formation.

Other hydrate structures include micro-aggregates, vein and void-filling hydrates, and massive hydrates. Hydrate micro-aggregates as well as isometric inclusions occur in various sizes, from micro-inclusions to aggregates up to 3 cm in diameter, and rarely as large lumps of hydrate (Fig. 4b). These hydrate manifestations cause a porphyraceous structure in the sediments. The mechanism of formation of this structure is similar to the formation of the lenticular-bedded one, but differs due to factors controlling the fields of chemical potentials of water and

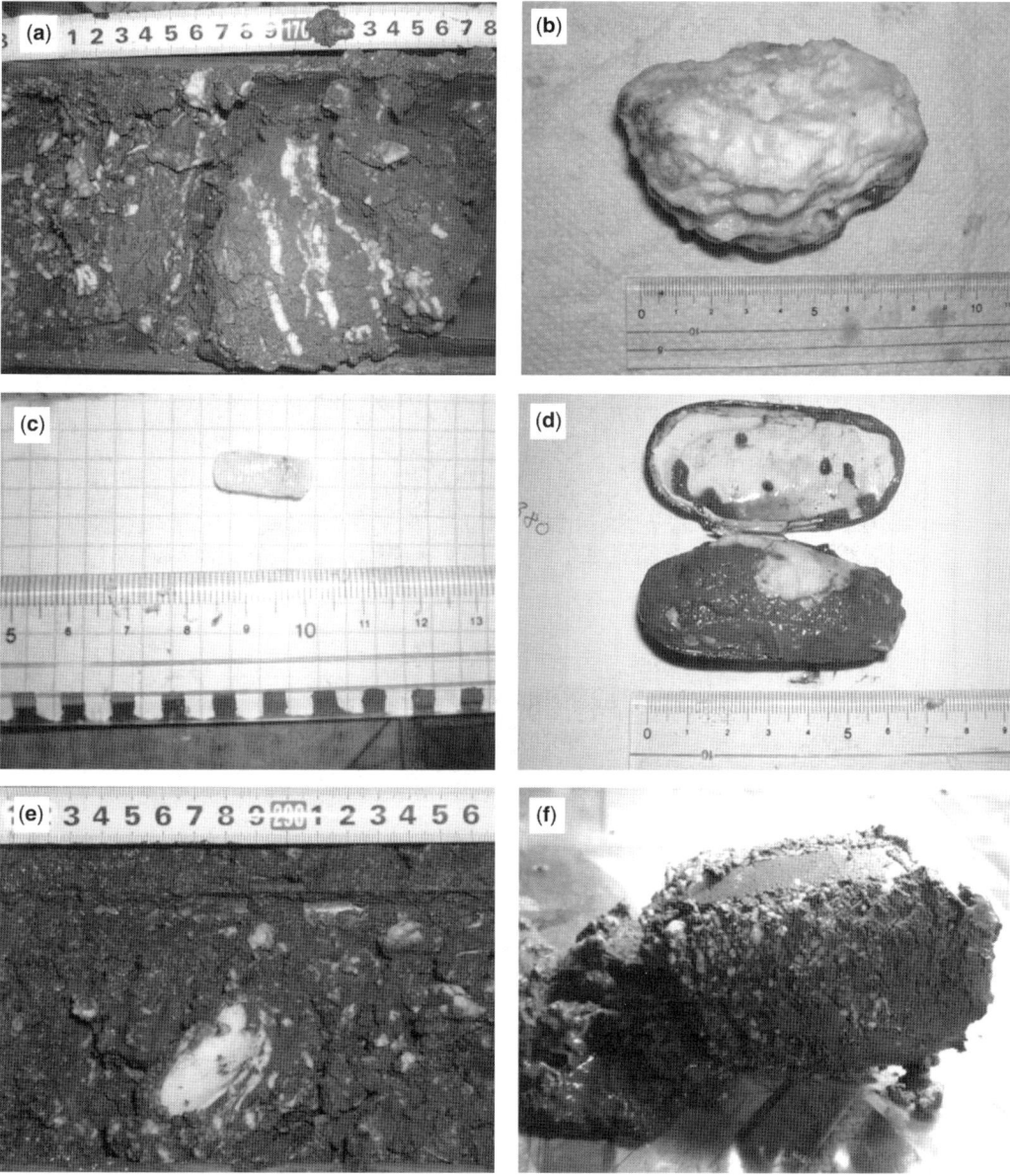

Fig. 4. Gas hydrate-induced structures of the sediments, recovered by gravity coring during the CHAOS-2003 Cruise: (**a**) lenticular-bedded hydrate-forming structure of sediments, core LV32-13GC, the Hieroglyph structure; (**b**) massive hydrate lump, core LV32-11HC, the Kitami structure; (**c**) subvertical cylindrical inclusion of gas hydrates, core LV31-34GC, the CHAOS structure; (**d**) *Calyptogena* bivalve shell filled by hydrate-bearing sediments (lower fragment), core LV31-34GC, the CHAOS structure; (**e**) gas hydrates observed around bivalve shell, core LV31-34GC, the CHAOS structure; (**f**) hydrate-cement structure of the sediments, core LV32-16GC, the CHAOS structure.

gas in the hydrate formation zone (Ginsburg & Soloviev 1998).

Subvertical cylindrical inclusions with diameters up to 0.5 cm are linked to vein-type structures (Fig. 4c). Vein hydrates result from precipitation by an infiltrating fluid (gas-rich water) in a fractured media. The variety of vein structures in hydrate-bearing sediments depends on the type of fracturing or void generation. In particular, gas hydrates fill voids in bivalve shells, around pebbles and

Table 1. *Results of chemical and isotopic measurements of the water samples obtained off NE Sakhalin during cruises 31 and 32 of R/V Akademik M. A. Lavrentyev*

Coring station	Sub-bottom depth (cm)	Concentration, mM							δD, ‰ SMOW	$\delta^{18}O$, ‰ SMOW
		K^+	Na^+	Mg^{2+}	Ca^{2+}	Cl^-	SO_4^{2-}	HCO_3^-		
					Chaos site					
LV31-24GC	25	10.9	427.4	51.4	3.6	529.0	8.6	17.2	−3.0	0
	65	10.6	427.4	52.0	3.6	529.0	9.5	14.4	−3.8	0
	95	10.8	419.3	52.3	4.6	513.2	5.2	20.7	−2.7	0
	115	10.6	419.3	53.4	3.3	513.2	9.0	21.7	−2.6	−0.1
	143	11.5	438.7	48.6	7.9	510.8	16.7	12.4	−3.1	−0.1
LV31-27HC	5	12.2	444.2	65.9	8.3	545.8	28.6	1.0	−1.2	0.1
	35	11.7	446.4	60.6	9.0	545.8	22.6	4.0	−1.6	−0.1
	55	11.7	441.1	56.5	3.9	553.8	8.5	14.0	−1.8	−0.1
	75	11.9	433.4	55.7	3.6	547.8	5.2	11.5	−0.9	0.3
	95	12.0	452.1	51.8	2.0	557.8	5.0	19.6	−1.1	0.3
	115	11.7	446.1	53.7	4.3	553.8	4.0	23.3	−0.4	0.5
	135	10.7	418.7	47.5	2.4	513.9	2.9	20.1	1.3	1.1
	165	11.4	445.1	53.9	2.0	545.8	1.3	33.8	0.4	1.1
	185	11.6	434.5	53.9	2.0	525.9	1.2	28.3	0.2	1.0
	205	12.0	464.1	58.0	4.9	565.7	3.1	34.4	−1.1	0.7
GHW	207	*3.3*	*113.9*	*12.9*	*5.1*	*138.6*	*5.2*	*5.6*	*11.2*	*1.0*
	225	12.0	459.0	55.3	2.0	557.8	2.2	36.5	0	1.0
	245	12.2	451.3	58.4	3.0	553.8	2.6	40.5	−0.6	1.1
	250	12.2	433.0	58.9	2.3	533.9	2.4	39.0	−1.5	0.8
	255								4.7	1.5
LV31-34GC	5	12.0	425.8	52.9	24.3	561.7	2.2	1.4	−0.3	0.6
	35	12.0	425.8	53.1	15.3	541.8	19.9	2.0	−0.5	0.6
	60	12.0	425.8	51.0	8.6	547.8	7.5	10.2	−1.0	0.5
	85	12.3	425.8	52.9	7.4	547.8	7.5	10.2	0.2	0.5
GHW	100	*4.3*	*161.3*	*17.1*	*7.1*	*180.2*	*3.2*	*18.0*	*8.0*	*0.9*
	105	10.5	353.9	42.6	5.9	452.2	4.7	15.8	2.9	0.9
	120	12.3	429.3	51.2	7.1	533.9	6.5	22.6	0	0.6
	145	12.0	429.3	52.0	8.2	533.9	4.1	31.2	−2.1	0.9
	175	12.5	445.8	53.5	6.7	533.9	3.2	34.2	−2.2	0.7
GHW	190								12.9	1.5
	195	12.3	445.4	52.7	8.2	535.8	2.8	37.6	−2.0	0.6
	215	12.3	446.5	52.2	7.1	539.8	3.0	36.6	−2.2	0.6
	240	12.0	420.0	51.0	3.9	504.0	1.7	37.1	0.3	0.7
GHW	247	*5.3*	*171.6*	*22.0*	*3.1*	*205.9*	*2.8*	*11.0*	*9.6*	*2.4*
	265	13.1	445.8	55.5	5.5	537.9	1.0	39.4	−0.2	0.7
	285	12.9	434.5	54.9	3.9	537.9	2.8	34.2	0.1	0.7
GHW	285								5.3	*0.2*
	305	12.6	420.0	52.0	3.9	517.9	3.0	29.8	0.4	0.7
LV32-16GC	10	9.1	367.7	47.1	5.9	460.9	11.6	18.2	−1.3	0.2
	65	9.1	362.9	49.0	6.8	460.9	12.0	19.2	−1.6	0.2
	115	10.3	399.9	47.2	6.3	497.0	5.6	17.2	−1.1	0.6
GHW[a]	130/1	*1.2*	*44.6*	*14.1*	*1.7*	*70.6*	*1.1*	*0.6*	*14.2*	*2.1*
GHW[a]	130/2	*4.8*	*226.0*	*25.5*	*1.6*	*257.0*	*2.1*	*17.2*		
	165	9.4	370.9	44.5	4.9	460.9	13.5	6.2	−1.1	0.4
	195	9.3	372.5	46.3	2.8	476.2	10.6	10.2	−1.0	0.6
	245	10.6	357.3	46.1	3.6	478.2	7.1	12.6	−4.7	−0.1
	265	12.6	323.3	45.7	4.3	474.2	9.3	14.3	−1.4	0.2
	305	13.5	347.7	50.5	5.1	486.1	5.8	7.2	−1.1	0.4
	355	10.4	421.1	47.6	1.5	525.6	0.3	8.8	−1.5	0.4
GHW[a]	370/1	*2.3*	*86.1*	*12.7*	*2.6*	*102.8*	*1.5*	*1.2*	*10.6*	*1.7*
GHW[a]	370/2	*2.0*	*78.3*	*10.6*	*0.8*	*90.9*	*1.4*	*4.2*		
GHW[a]	370/3	*2.2*	*86.1*	*9.4*	*1.2*	*10.8*	*1.3*	*2.4*		
GHW	380	*4.2*	*168.2*	*19.3*	*5.227*	*205.5*	*0*	*2.0*	*8.4*	*1.8*
	405	11.3	390.4	79.1	3.92	545.4	0	10.6	−2.2	−0.1

(Continued)

Table 1. *Continued*

Coring station	Sub-bottom depth (cm)	Concentration, mM							δD, ‰ SMOW	δ^{18}O, ‰ SMOW
		K^+	Na^+	Mg^{2+}	Ca^{2+}	Cl^-	SO_4^{2-}	HCO_3^-		
LV31-19HC	5	10.8	445.3	58.1	4.9	533.1	20.3	2.8		
	35	11.4	436.3	57.3	4.9	529.2	14.5	2.8		
	75	10.7	437.3	56.9	4.9	529.2	14.1	3.2		
	105	11.4	440.5	54.9	4.5	519.3	13.6	5.0		
	135	11.4	438.9	53.0	4.5	523.3	11.3	11.0		
	175	11.5	435.7	53.0	4.5	523.3	10.3	12.4		
	205	11.5	435.7	53.0	3.6	519.3	7.3	21.8		
	255	11.6	435.7	53.0	3.6	523.3	8.5	18.5		
	295	11.4	438.9	54.5	2.6	523.3	3.0	31.8		
	335	11.7	432.4	53.7	2.4	521.3	4.5	25.6		
	375	11.5	434.8	54.5	2.0	521.3	4.1	31.2		
	414	11.4	438.9	53.0	3.9	521.3	6.2	21.6		
	422	11.4	438.9	54.9	2.0	521.3	2.0	39.2		
LV31-30GC	5	12.0	467.7	57.5	23.4	552.7	28.6	1.0	−0.4	0.5
	45	12.0	467.7	55.5	12.8	554.7	29.6	1.4	−0.7	0.3
	75	12.0	464.5	59.8	12.6	556.6	25.8	1.8	−0.4	0.2
	115	11.8	464.5	58.4	10.6	552.7	32.4	3.3	−1.0	0
	155	11.7	464.5	59.2	12.6	554.7	19.7	8.4	−1.0	−0.1
	205	12.0	464.5	57.5	9.2	560.0	18.3	4.9	−0.6	−0.2
	245	12.0	464.5	53.1	8.6	552.7	17.9	13.0	−0.4	−0.2
	285	12.1	464.5	58.0	7.1	554.7	18.2	4.8	−0.7	0.2
	315	12.0	464.5	54.9	6.3	553.7	15.1	4.2	−0.3	0.1
	355	12.0	464.5	54.9	4.3	562.5	7.1	23.6	−0.2	0.1
	385	12.0	467.7	57.6	5.1	556.6	8.9	27.5	−0.1	0.1
	415	12.0	451.6	58.0	6.3	558.6	18.0	8.0	−0.3	0.2
LV31-31HC	5	11.7	464.7	56.9	10.2	549.0	27.8	1.4		
	45	11.9	451.8	57.1	10.8	549.0	26.9	3.2		
	95	11.7	451.8	50.4	4.9	549.0	9.1	13.8		
	135	11.6	456.6	56.9	3.9	549.0	8.7	17.6		
	175	12.1	462.7	52.2	3.5	549.0	8.1	14.4		
	225	12.1	463.1	53.0	3.1	549.0	8.9	17.2		
	265	11.9	464.7	54.1	3.2	549.0	9.4	24.5		
	290	12.1	464.7	57.5	2.9	551.0	7.3	20.2		
LV31-36HC	5	11.5	458.2	56.9	10.2	548.9	28.3	2.6		
	45	12.1	448.6	56.3	9.4	541.0	25.9	6.2		
	95	11.7	435.7	54.1	5.9	533.1	15.4	11.2		
	145	11.7	435.7	53.0	5.5	525.2	8.1	22.2		
	195	12.0	435.7	54.1	3.5	525.2	6.6	25.8		
	255	12.0	445.3	53.7	3.1	525.2	5.8	28.2		
LV32-01GC	25	11.2	440.5	55.4	9.8	529.2	31.3	2.8		
	105	12.2	456.6	57.3	10.2	550.9	25.9	10.8		
	105	10.6	394.5	52.6	9.8	483.8	21.6	9.4		
	205	10.8	445.3	55.7	4.3	533.1	16.5	23.0		
	305	10.8	438.1	53.1	5.1	523.3	7.3	38.6		
	505	10.9	438.1	54.5	5.5	513.4	6.6	45.2		
	605	11.1	435.7	53.0	5.9	509.4	6.4	44.6		
LV32-03GC	5	11.5	475.2	53.0	12.9	552.9	34.9	2.8		
	45	10.8	435.7	53.0	10.6	517.3	31.7	2.8		
	85	13.1	438.1	53.0	9.8	517.3	25.3	6.2		
	125	10.6	430.8	52.6	8.2	513.4	15.6	15.0		
	165	10.7	435.7	51.8	7.8	513.4	16.3	20.0		
	205	10.4	423.6	50.6	3.1	505.5	8.8	22.8		
	245	10.7	434.8	51.0	5.1	533.1	2.4	26.6		
	285	10.9	427.6	51.0	4.9	509.4	5.6	32.0		
	325	10.4	429.2	49.0	4.7	509.4	3.6	28.2		
	365	10.8	427.6	51.0	3.1	517.3	3.9	34.6		

(*Continued*)

Table 1. Continued

Coring station	Sub-bottom depth (cm)	Concentration, mM							δD, ‰ SMOW	$\delta^{18}O$, ‰ SMOW
		K^+	Na^+	Mg^{2+}	Ca^{2+}	Cl^-	SO_4^{2-}	HCO_3^-		
	405	10.8	429.2	50.6	3.5	517.3	2.1	34.2		
	445	10.5	427.6	53.3	3.5	517.3	2.6	34.0		
	485	11.0	447.0	53.4	3.9	529.2	2.4	36.6		
	525	11.0	446.9	53.4	5.5	525.2	6.9	34.2		
	565	12.3	478.6	62.0	2.0	577.2	3.2	34.4		
LV32-06GC	15	12.0	447.0	55.8	7.4	535.7	26.7	2.4	−1.4	0.1
	45	12.2	476.8	56.3	5.1	581.7	17.3	8.1	−1.4	−0.2
	75	12.2	444.4	50.2	4.7	559.5	2.9	8.2	−0.5	0.8
	95	11.0	427.6	65.0	5.9	525.8	0	21.3	−8.2	−1.2
	125	12.1	451.3	53.3	2.0	563.5	2.3	18.4	−1.9	0.2
	165	11.0	374.2	52.1	2.4	484.1	3.5	18.3	−4.5	0
	205	12.8	443.7	54.1	1.2	527.7	3.9	15.4	−0.1	0.5
	245	11.4	418.3	49.2	1.7	531.7	5.0	12.2	−11.4	−1.1
	275	11.1	439.6	51.0	2.4	555.5	9.5	4.0	−2.9	−0.6
	315	11.4	426.0	52.5	2.8	541.8	12.6	4.6	−1.7	0.4
	355	13.7	504.1	59.0	0.8	635.4	8.1	13.2	−1.6	1.0
	405	11.9	444.4	56.0	2.0	549.8	5.8	59.2	−7.5	−0.3
	455	12.5	449.4	55.3	2.4	551.6	1.6	30.4	−2.3	0.1
	480	12.7	452.0	53.9	3.5	559.5	2.3	21.6	1.3	2.0

Gas hydrate-bearing intervals and bottom water samples containing gas hydrate aggregates are set in bold type.
GHW, gas hydrate water.
[a]Subsamples taken from the same hydrate aggregate.

inside/outside authigenic carbonates (Fig. 4d and e). Formation of the structures can be explained by differing stabilities and rates of gas hydrate formation, giving an overall propensity to form in heterogeneities within the sediments. The porous medium may influence the equilibrium conditions of hydrate formation (thermodynamic effect) and its rate of growth (kinetic effect) (Ginsburg & Soloviev 1998). The overall effect is that the hydrophilic surface of sediment particles lowers the chemical potential of pore water. As a result, a higher concentration of hydrate-forming gas is needed for hydrate formation at a given temperature, or a lower temperature is required at a given gas concentration (Yousif & Sloan 1991; Melnikov & Nesterov 1996). The influence of the porous medium upon hydrate formation manifests itself through a variation in hydrate formation conditions for pores of different sizes – the chemical potential of water in large pores being rather higher than those in small pores, thus favouring hydrate formation in larger pores (Chersky & Mikhailov 1990). Observations of subvertical cylindrical inclusions and tubes of gas hydrate result from the same process: gas hydrates forming inside worm burrows and in other free pore spaces formed by bioturbation.

Several intervals of the sediment cores were represented by massive structures of hydrate-cemented sediments (Fig. 4f). This structure forms as a result of hydrate precipitation during fluid infiltration through a granulated reservoir and requires the presence of relatively coarser-grained layers within finely-grained sediments. The hydrate-cemented interval of core LV32-16GC (280–380 cm subbottom depths) has a particularly high content of carbonate inclusions of various sizes and shapes. The authigenic carbonate content in this interval was visually estimated to be as much as 25% (by sediment volume). These carbonates form coarser-grained sediments and are probably related to the formation of hydrate cement.

Methods of geochemical analysis

Geochemical analysis performed on pore waters, hydrate fluids and carbonate samples include major element, gas and isotopic compositions. Major element geochemistry of 206 water samples was determined in VNIIOkeangeologia using a method described by Reznikov & Mulikovskaya (1956). Cl^-, Ca^{2+} and Mg^{2+} were determined by titration (argento-, acide- and complexometry, respectively), SO_4^{2-} species was determined by weight, and Na^+ and K^+ using flame-photometry. Gas chromatography of four hydrate gas samples was performed in the laboratory of VNIGRI in St Petersburg. O_2, N_2, CO_2 and CH_4 were determined using a thermal conductivity detector

Table 2. *Results of chemical and isotopic measurements of the water samples obtained off NE Sakhalin during cruises 31 and 32 of R/V Akademik M. A. Lavrentyev*

Coring station	Sub-bottom depth (cm)	Concentration, mM							δD, ‰ SMOW	δ^{18}O, ‰ SMOW
		K$^+$	Na$^+$	Mg^{2+}	Ca^{2+}	Cl$^-$	SO$_4^{2-}$	HCO$_3^-$		
					Hieroglyph site					
LV32-13GC	15	12.3	425.8	53.9	10.2	513.2	30.8	1.4	−1.6	0
	35	11.5	406.4	51.7	7.8	505.2	21.0	1.9	−8.8	−0.9
	75	11.1	396.8	47.4	3.6	497.1	5.2	3.6	−5.4	−0.3
	95	11.2	406.4	47.8	3.6	501.1	4.3	13.4	−10.0	−1.2
GHW	*107*	*2.7*	*87.1*	*9.6*	*4.7*	*108.2*	*2.7*	*7.2*	*12.9*	*1.4*
	115	9.6	319.3	39.2	4.7	396.8	8.7	4.0	−0.3	0.3
GHW	*127*	*2.6*	*90.3*	*11.0*	*1.9*	*108.2*	*2.6*	*5.4*	*12.2*	*1.4*
	135	9.3	322.6	38.2	4.7	396.8	7.9	6.0	−0.3	0.6
	150	3.5	125.8	13.3	3.9	148.3	3.8	8.8	10.0	1.1
	155	8.3	290.3	36.5	3.9	364.7	4.7	18.4	2.0	0.3
GHW	*158*	*3.8*	*135.5*	*18.0*	*3.3*	*168.3*	*1.5*	*10.4*	*9.3*	*1.0*
	160	4.3	148.4	18.0	2.8	180.4	1.1	11.8	9.4	1.0
GHW	*170*	*3.2*	*108.1*	*4.9*	*1.9*	*116.2*	*2.2*	*12.0*	*11.0*	*1.5*
	175	8.2	277.4	35.3	3.2	344.7	7.9	14.2	−3.5	0.6
LV31-28GC	5	12.2	443.6	66.7	10.2	553.8	29.1	2.4	−0.3	0.5
	55	12.0	431.2	62.0	10.6	553.8	18.2	1.9	−0.2	0.3
	105	12.5	436.3	62.7	10.6	502.0	28.2	2.6	−0.4	0.3
	155	12.3	439.2	62.7	8.3	539.8	23.5	4.2	−1.2	0.4
	205	12.1	439.2	64.7	6.7	541.8	18.6	15.0	−0.6	0.4
	255	12.3	439.1	63.5	6.3	537.9	12.6	26.5	−0.4	0.7
	305	11.9	430.9	63.9	5.1	537.8	7.1	38.2	−0.9	0.3
	355	12.1	430.1	61.1	6.3	537.8	3.2	38.0	−1.2	0.2
	405	12.3	430.9	57.7	7.1	537.8	4.1	45.8	−0.9	0.3
	455	12.3	435.8	57.5	4.9	535.4	2.4	41.8	−1.2	0.2
	485	12.9	435.8	57.4	4.7	535.1	3.2	42.4	−0.2	0.6
	505	12.9	446.0	57.4	3.0	537.8	2.2	46.7	−1.0	0.2
	525	12.9	446.0	55.7	4.1	537.8	2.4	46.4	−0.5	0.4
LV32-07GC	15	11.6	436.1	51.0	7.8	541.4	29.7	1.6	−2.7	−0.4
	55	11.0	399.6	41.0	8.3	494.0	26.6	1.8		
	105	12.0	439.2	29.4	9.8	454.5	27.2	4.1	−8.8	−1.3
	155	12.4	472.7	39.9	7.9	557.2	19.6	15.2		
	205	13.3	507.5	55.3	0.8	640.2	11.8	12.0		
	275	11.7	428.3	57.5	3.9	543.4	4.9	35.0		
	325	11.7	417.0	52.3	7.8	539.4	4.3	40.6	−11.0	−1.4
	415	6.0	219.8	19.6	0.4	249.0	2.7	21.0	−5.0	−1.0
	505	12.1	461.4	50.2	4.7	553.3	2.8	57.1	−2.8	−0.3
	575	11.3	392.8	49.6	0.6	478.2	5.6	30.4	−5.4	−1.1
	625	10.9	395.4	53.3	0.8	480.2	3.9	33.9	−4.8	−1.2

Gas hydrate-bearing intervals and bottom water samples containing gas hydrate aggregates are set in bold type.
GHW, gas hydrate water.

(catarometer), and heavy hydrocarbons by means of a flame-ionization detector.

Isotopic analyses were made on 152 selected water samples, 29 samples of carbonate concretions, and nine samples of *Calyptogena* bivalves. Isotope compositions of methane carbon (δ^{13}C (CH$_4$)), homologues and CO$_2$ as well as oxygen (δ^{18}O) and hydrogen (δD) isotopic analyses of the pore water were carried out in the Centre of Isotopic Research of VSEGEI (St Petersburg). Oxygen isotopes were determined using an isotope-ratio-monitoring mass spectrometric (IRM–MS) method by means of a DELTA plus XL mass spectrometer with a Gas Bench II preparative device (ThermoFinnigan production) working on-line and providing isotope exchange between H$_2$O and CO$_2$; a GC/C-III device was used for the gas measurements. The same apparatus was used for

Table 3. *Results of chemical and isotopic measurements of the water samples obtained off NE Sakhalin during cruises 31 and 32 of R/V Akademik M. A. Lavrentyer*

Coring station	Sub-bottom depth (cm)	Concentration, mM							δD, ‰ SMOW	$\delta^{18}O$, ‰ SMOW
		K^+	Na^+	Mg^{2+}	Ca^{2+}	Cl^-	SO_4^{2-}	HCO_3^-		
					Kitami site					
LV32-09GC	5	12.4	433.5	59.6	9.0	527.7	29.4	2.0	−7.8	−0.5
	23	10.8	389.7	49.4	14.5	472.2	25.3	0.6	−9.1	−1.0
	37	10.9	381.2	47.4	9.0	480.1	16.3	2.6	−3.6	0.1
	52	11.0	389.5	51.4	5.5	494.0	4.5	10.9	−3.6	0.2
	67	*11.6*	*432.1*	*50.3*	*9.1*	*541.8*	*6.6*	*6.2*	*−2.7*	*0.1*
	82	*11.6*	*418.8*	*54.9*	*2.0*	*521.9*	*7.3*	*3.9*	*−10.0*	*−0.8*
	103	8.7	294.3	31.9	6.5	392.8	7.5	6.3	3.7	1.2
GHW	*105*	*2.7*	*89.7*	*11.2*		*93.1*	*4.7*	*3.7*	*13.0*	*1.9*
GHW	*107*	*3.8*	*135.8*	*11.8*	*7.8*	*134.6*	*8.6*	*10.8*	*10.0*	*1.7*
GHW	*130*								*9.4*	*1.7*
	135	8.9	301.3	13.9	13.7		8.1	4.4	−3.6	0.6
GHW	*150*								*11.0*	*1.6*
	155	*9.9*	*374.2*	*42.3*	*5.1*	*374.5*	*5.6*	*13.8*	*1.8*	*1.4*
	167	12.3	424.2	51.7	9.8		25.5	5.4	−6.2	0.3
	182	*10.6*	*385.5*	*48.8*	*9.4*	*467.3*	*26.6*	*1.8*	*−2.9*	*0.1*
LV32-12HC	5	10.6	390.3	48.6	9.4	467.3	25.3	2.0	−3.7	−0.3
	35	11.8	404.8	51.1	9.8	471.2	26.6	2.0	−3.8	−0.4
	65	12.4	441.9	51.3	10.6	534.6	27.0	1.8	−4.4	−0.4
	95	10.6	406.4	43.9	9.9	499.0	24.7	2.0	−3.2	−0.3
	115	12.1	429.0	51.0	9.4	526.7	23.5	8.0	−7.7	−0.7
	145	10.4	399.9	46.3	7.1	487.1	14.2	11.6	−3.0	−0.1
	175	11.2	374.2	43.5	3.9	471.2	3.1	20.8	−2.4	0.9
	205	12.1	414.5	48.8	4.9	510.8	1.1	24.0	−3.5	−0.3
	245	12.3	446.6	52.3	2.0	550.4	3.3	22.4	−5.2	−0.4
	275	11.5	412.9	47.1	8.0	521.0	8.2	4.8	−4.7	−0.4
	315	12.6	467.7	54.4	3.2	569.1	7.5	17.4	−9.3	−0.9

Gas hydrate-bearing intervals and bottom water samples containing gas hydrate aggregates are set in bold type.
GHW, gas hydrate water.

measurements of carbon ($\delta^{13}C$ ($CaCO_3$)) and oxygen ($\delta^{18}O$ ($CaCO_3$)) isotopes of the carbonates.

Hydrogen isotope determinations of water samples were carried out using a classic two-channel DELTAplus mass-spectrometer. The water samples were decomposed with chromium at 850 °C in an H-Device preparative chamber (ThermoFinnigan production). The isotope analyses of $\delta^{18}O$ and $\delta^{13}C$ of most carbonate samples were done in the Laboratory for the Isotopic Geology of Fluids (SPbSU) by a fully streamlined bi-collector mass spectrometer MS-20 (AEI production). The carbonate samples were heated in a furnace at a temperature of less than 200 °C and then decomposed by *ortho*phosphoric acid under vacuum. The condensed CO_2 subsample was refined and sealed up under vacuum for subsequent analysis. The error of measurement was in the range of 0.1–0.2‰ for carbon and of 0.1–0.4‰ for oxygen.

Random error during determination of carbon and oxygen isotopic compositions (1σ) in the different substances was in the range 0.1–0.2‰, and for hydrogen from water, 0.4‰. The results of isotopic measurements are represented in per-mil delta notations (‰) relative to SMOW and PDB standards.

Results

The waters sampled have been subdivided in five groups depending on their source:

(1) seawaters from the near-bottom water layer (bottom water);
(2) waters released from hydrate aggregates (hydrate water);
(3) pore waters extracted from sediments containing gas hydrates (water from hydrate-bearing intervals);
(4) waters extracted from core intervals overlying or/and underlying those recovered gas hydrates (water from hydrate-free intervals);

(5) water extracted from sediment cores which did not recover gas hydrates (water from hydrate-free cores).

Hereafter, we discuss water chemistry and isotope data for these five water groups relative to the data from the reference core.

Chloride contents of waters

The composition of pore fluids, particularly Cl^- distributions, is known to be a useful indicator of fluid sources and migration patterns in venting sites, as chloride can be considered geochemically conservative (Martin *et al.* 1996; Aloisi *et al.* 2004; Haese *et al.* 2006). Chloride ion contents in the groups of waters studied are different, and are characterized by considerable fluctuations compared with those in the reference core LV31-41GC (530–570 mM; Tables 1–4, Fig. 5a). In the venting sites Cl^- concentrations vary strongly: 249–640 mM in hydrate-free cores, 148–565 mM in gas-hydrate bearing intervals and 472–561 mM in hydrate-free intervals. The highest Cl^- concentrations measured in waters from hydrate-free and hydrate-bearing intervals are higher compared with those in the bottom water (519–536 mM). Cl^- concentrations measured in waters from hydrate-free cores are also characterized by considerable variations. The waters released from the hydrate aggregates were not fresh (10.8–257 mM), contrary to what one might expect. It is well known that, when water molecules are incorporated into the gas hydrate structure, salts remain in solution. The elevated Cl^- concentrations measured in hydrate waters might be a result of either contamination of the hydrate water samples by pore water from host sediments during sampling or the presence of saline intercrystal solutions in the hydrates.

When plotting Cl^- concentrations v. depth for the CHAOS, Hieroglyph and Kitami sites, the Cl^- distribution between the sites also appears different. As a whole, the waters from the CHAOS site are characterized by a narrow range of Cl^- concentrations, and the chlorinity of the pore waters from the hydrate-bearing intervals has elevated value compared with those from two other sites. The Cl^- profiles for the CHAOS site show increasing and decreasing trends, both in hydrate-bearing and

Table 4. *Results of chemical and isotopic measurements of the water samples obtained off NE Sakhalin during cruises 31 and 32 of R/V Akademik M. A. Lavrentyev*

Coring station	Sub-bottom depth (cm)	Concentration, mM							δD, ‰ SMOW	$δ^{18}O$, ‰ SMOW
		K^+	Na^+	Mg^{2+}	Ca^{2+}	Cl^-	SO_4^{2-}	HCO_3^-		
				Reference core						
LV31-41GC	5	12.6	459.7	51.0	25.1	530.6	33.2	1.2	−0.4	0.2
	50	12.5	472.6	53.1	22.7	546.5	28.8	3.0		
	95	12.6	462.9	52.2	24.3	542.5	24.6	6.6	−0.4	0.4
	155	12.6	466.1	52.2	20.4	548.5	16.1	17.0		
	195	12.8	462.9	54.1	7.8	546.5	13.6	19.4	−0.6	0.2
	245	10.9	462.9	53.1	5.1	540.5	8.0	23.0		
	295	10.8	462.9	53.9	5.9	542.5	4.5	27.0	−0.6	0.2
	345	10.8	462.9	57.1	4.3	542.5	5.5	31.4		
	395	13.0	490.3	62.4	3.9	570.2	3.1	32.8	−0.8	0.1
	445	12.6	461.3	58.0	4.3	546.5	2.1	33.0		
	495	12.6	467.7	58.8	4.7	542.5	1.2	39.0	−0.5	0.2
	525	13.0	475.8	60.8	4.7	554.4	1.4	43.0	−0.3	0.2
				Bottom water						
LV31-24GC	**0**	**10.2**	**438.7**	**51.7**	**11.8**	**510.8**	**28.7**	**4.4**	**−5.0**	**−0.7**
LV31-34GC	0	10.0	448.4	55.3	19.6	536.6	30.6	2.0		
LV32-13GC	0	10.0	418.0	52.8	11.37	519.2	28.3	0.6		
LV32-09GC	0	9.8	432.2	52.5	10.6	520.7	28.3	1.7	−2.6	−0.1
LV31-41GC	0	10.8	448.4	52.9	26.7	530.6	28.4	4.0	−2.1	−0.1
LV32-12HC	0	10.2	440.3	52.7	11.4	525.0	28.6	1.7	−1.5	−0.1
LV31-28GC	0	9.8	424.5	51.6	13.5	523.9	28.3	0.8	−3.2	−0.3
LV31-30GC	0								−1.0	−0.3
LV32-21GC	0								−1.2	−0.3

Gas hydrate-bearing intervals and bottom water samples containing gas hydrate aggregates are set in bold type.
GHW, gas hydrate water.

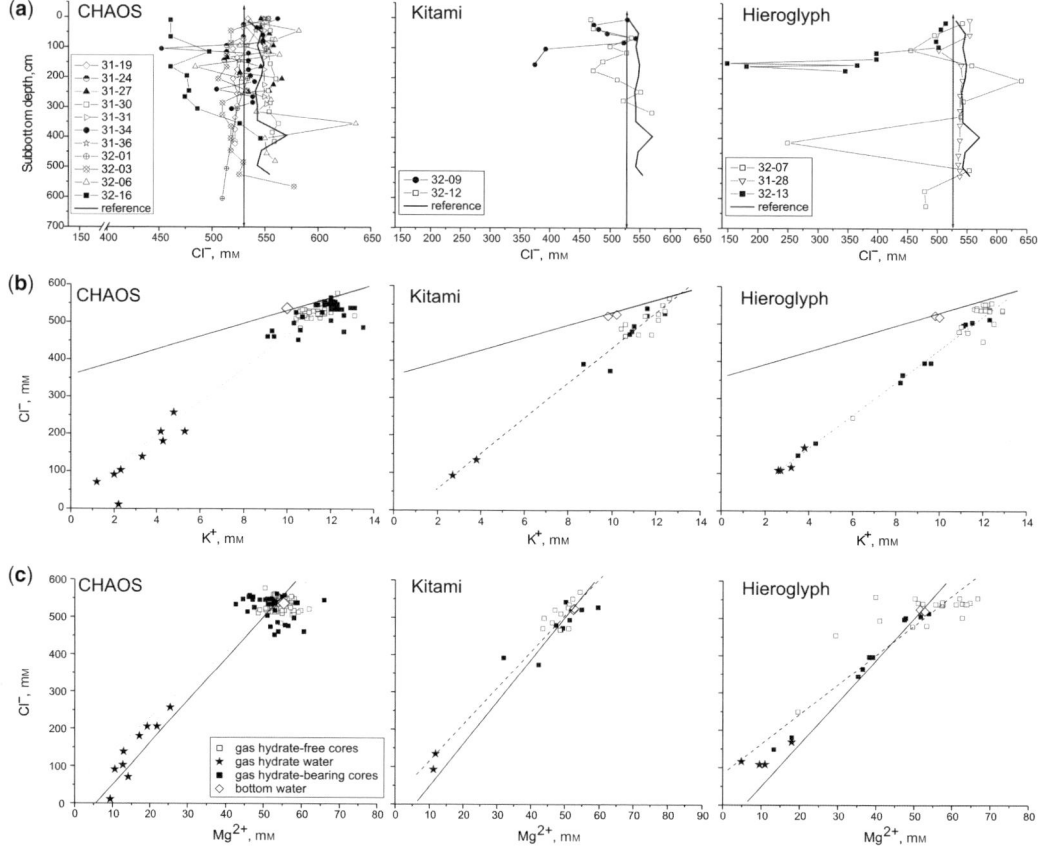

Fig. 5. Characteristic plots showing differences in the composition of studied waters from the studied vents. (**a**) Cl⁻ distribution with depth; open symbols correspond to hydrate-free cores, filled symbols are hydrate-bearing cores. Note that core LV31-24 retrieved gas hydrates in the corer head only – data marked by a circle with its upper half filled; seawater concentration is shown by vertical lines terminated with arrows. (**b, c**) Cl⁻ v. K⁺ and Mg^{2+} concentrations; solid lines represent constant seawater ratios, dashed lines are linear regression lines.

hydrate-free cores. In particular, for core LV31-27HC (Fig. 5a, Table 1) the chlorinity of pore water from the hydrate-bearing interval is higher than that in the hydrate-free interval and higher than in the bottom water. The chloride ion concentrations in water from the hydrate-bearing intervals of the CHAOS site did not show strong negative deviations related to gas hydrate decomposition as was expected based on visual hydrate content estimations (20–60% from the sediment volume).

The Cl⁻ concentration profiles from the Hieroglyph and the Kitami sites are characterized by strong negative and positive deviations relative to the reference core, both in hydrate-containing cores and in those cores where gas hydrates were not observed. In the following sections we discuss that the variations in the Cl⁻ content in the different water groups may be caused not only by the presence of gas hydrates, but also by some other factors.

Correlation between major anions and cations

Na–Cl, Ca–Cl and Mg–Cl ratios generally follow the seawater distribution trend. In the samples from the venting area some deviations from the seawater line are observed. The major deviation occurs for K–Cl ratios (Fig. 5b) that appeared to be much higher than in bottom water for all the vents studied. This observation is in agreement with data for hydrate-containing sediments from the Obzhirov gas vent (Matveeva *et al.* 2003) situated within the studied area about 10 km SW from the sites discussed here. Assuming that Cl⁻ is conservative, the elevated K–Cl ratio indicates that at depth K⁺ is released into the pore fluid. There are a number of processes responsible for variations of potassium concentrations in marine sediments. This may be due to diagenetic alteration via hydration of analcite, clinoptilolite and phillipsite resulting from volcanic ash transformation at great depth and at temperatures of 25 °C (Egeberg & Leg 126 Shipboard Scientific Party 1990). However, these hydration reactions will lead to the removal of K^+ and Mg^{2+} along with the increased Cl^- concentration. Other processes are dissolution of K-bearing minerals such as K-feldspar or release of K^+ from more deeply buried sediments. The last cause seems the most credible. The local increase of Na^+ and Mg^{2+} concentrations (Fig. 5c) above the hydrate-bearing interval apparently resulted from concentration of Na and Cl species in the residual pore water during rapid hydrate formation.

The calcium ion concentration profiles v. depth in the studied cores are typical for the re-distribution of Ca^{2+} and HCO_3^- due to precipitation of authigenic carbonates from excess carbonate generated by the anaerobic microbial oxidation of methane (Ritger et al. 1987). This process is represented by the following diagenetic reactions (Martin et al. 1996):

$$Ca^{2+} + 2HCO_3^- \rightarrow CACO_3 + CO_2 + H_2O \quad (1)$$

$$Ca^{2+} + Mg^{2+} + 2HCO_3^- \longleftrightarrow (Ca, Mg)CO_3 + CO_2 + H_2O \quad (2)$$

Isotope characteristics

The results of isotopic composition determinations of oxygen and hydrogen in water are presented in Tables 1–4 and in Figures 6 & 7. Extreme and average values of isotopic composition of studied waters are represented in Table 5.

δD and $\delta^{18}O$ values form a normal correlation line with end-members $\delta D = -11.0\permil$ and $\delta^{18}O = -1.5\permil$ and $\delta D = +14\permil$ and $\delta^{18}O = +2.4\permil$ (Fig. 6a). The good agreement between δD and δO^{18} values suggests the presence of some common effect influencing both δD and δO^{18}. Isotopic fractionation during capture of water molecules from the liquid phase and creation of a gas hydrate crystalline structure can be considered as one of the reasons for this. It is known that, during this process, gas hydrates do not only exclude Cl^- (and others ions) from their lattice structure, but also incorporate preferentially the heavier isotopes of oxygen and hydrogen in their water cage. The coefficients of fractionation obtained experimentally for methane hydrate in saline water (3 wt% NaCl) at a constant temperature of 273.5 K are 1.014–1.022 for D/H and 1.0023–1.0032 for $^{18}O/^{16}O$ (Maekawa 2004). These values define a difference between isotope compositions of the original 'hydrate-forming' (0% of hydrate) and the 'entirely hydrate' (100% of hydrate) waters averaging about 3‰ for oxygen and 18‰ for hydrogen. The fractionation process is strongly supported by natural observations (Ginsburg et al. 1999; Wallmann et al. 2000; Matveeva et al. 2003). In fact, a share of the 'entirely hydrate' water comprises only some part (see below) of the water released from the studied hydrate samples (the hydrate water). Therefore, isotope distinction of the hydrate water from the original water should be below the values obtained experimentally. However, the measured δD and $\delta^{18}O$ values reported here (25‰

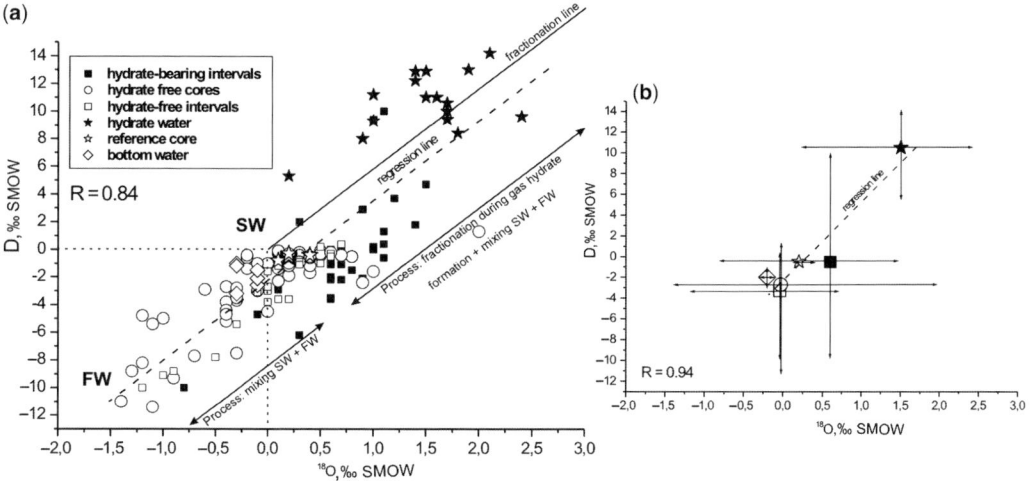

Fig. 6. Correlation of $\delta^{18}O$ with δD for the entire dataset (**a**) and for the average values for all types of waters (**b**). Lines with arrows indicate different processes affecting the isotope characteristics of the studied groups of water. FW (fluid water) and SW (seawater) represent waters of the two sources discussed. See explanations in the text.

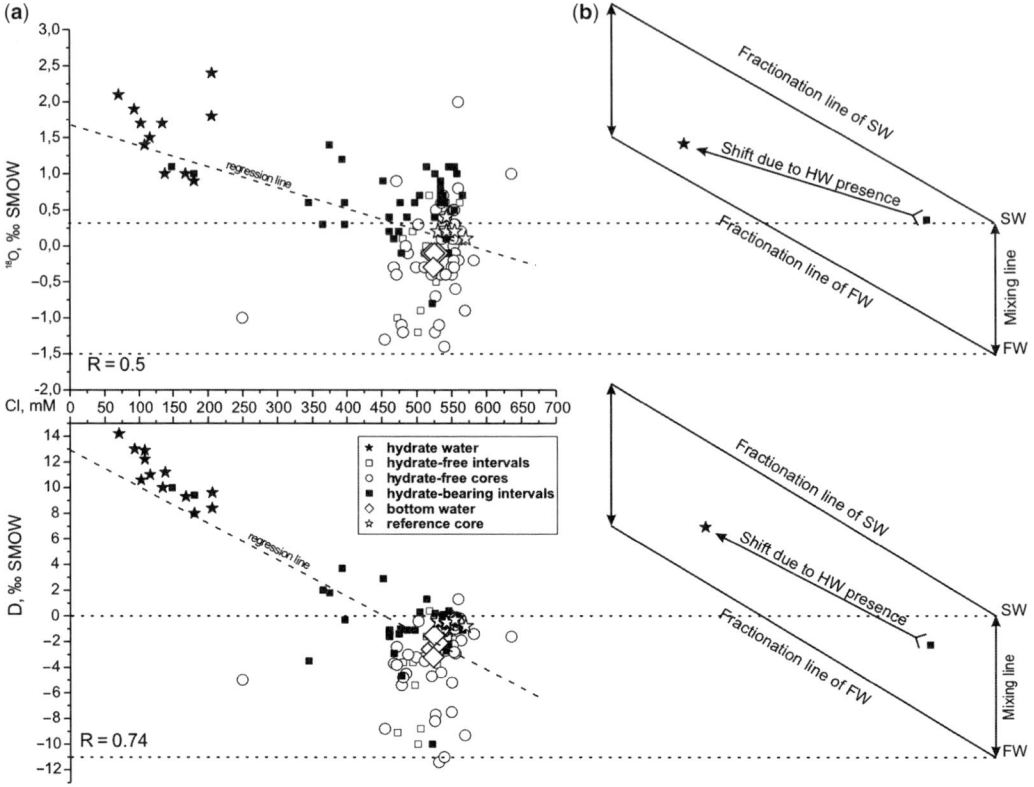

Fig. 7. Water oxygen and hydrogen isotopic composition versus chloride ion concentration for the different types of water (**a**) and the isotopic model of infiltrating fluids (**b**). A mixing line of the fluid water (FW) and the seawater (SW) connects coordinates of two sources. HW is hydrate water. See explanations in the text.

Table 5. *Extreme and average values of oxygen and hydrogen isotopic composition measured in different types of water*

Type of samples	Number of samples	Maximum (‰, SMOW)		Minimum (‰, SMOW)		Average (‰, SMOW)		Standard deviation	
		δD	$\delta^{18}O$	δD	$\delta^{18}O$	δD	$\delta^{18}O$	δD	$\delta^{18}O$
Hydrate water	16	14.2	2.4	5.3	0.2	10.5	1.5	2.3	0.5
Pore water from hydrate-bearing intervals	41	10	1.5	−10	−0.8	−0.5	0.6	3.43	0.4
Pore water from hydrate-free intervals	23	0.4	0.7	−10	−1.2	−3.3	−0.04	3.1	0.5
Bottom water	6	−1	−0.1	−3.2	−0.3	−2.0	−0.2	0.9	0.1
Pore water from gas hydrate-free cores	57	1.3	2	−11.4	−1.4	−2.7	−0.03	2.9	0.6
Pore water from reference core	7	−0.3	0.4	−0.8	0.1	−0.5	0.2	0.2	0.1

and 3.9‰, respectively see Table 5) exceed the maximum possible range of isotopic fractionation during hydrate formation (18 and 3‰ for δD and $\delta^{18}O$, respectively) as referred to above, possibly suggesting the presence of other processes as responsible for the isotopic variations found.

Hydrate gas compositions

Analysis of the hydrate gases shows that methane is the dominant gas (from 99.04 to 99.29% of the total hydrocarbons). Carbon dioxide ranges from 0.5 to 0.95%, ethane (0.0001–0.0025%), propane (0.0013–0.0018%), i-butane (0–0.0004%), n-butane (0–0.0004%) and He + Ne (0.0001%). The predominance of methane (more than 99%), and a $C_1-C_2 + C_3$ ratio higher than 20 000, indicate a biogenic source of the original gas. The isotopic compositions of $\delta^{13}C$-CH_4 from the gas hydrates vary from −63.7 to −65.5‰ (Table 6). It is suggested that the methane taking part in gas hydrate formation within gas venting sites off NE Sakhalin is produced predominantly during the biological reduction of organic matter by microbes in the anaerobic environment below the zone of sulfate reduction.

The $\delta^{13}C$ and $\delta^{18}O$ of authigenic carbonates

The authigenic carbonates in the seepage structures of the Derugin Basin are mainly composed of low- and high-Mg calcite with an impurity of aragonite in some samples (Derkachev et al. 2002; Krylov et al. 2007). The isotopic compositions of carbon 13 ($\delta^{13}C(CaCO_3)$) in the studied samples of authigenic carbonates vary considerably and range from −9.7 to −46.2‰ PDB, whereas $\delta^{18}O(CaCO_3)$ of the same samples ranges from +1.4 to +5.9‰ PDB (Fig. 8). The isotope values of $\delta^{13}C$ and $\delta^{18}O$ in buried *Calyptogena* bivalves are as follows: from −1.7 to −37.7‰ PDB and from −0.2 to +3.5‰ PDB, respectively. The 'light' carbon isotopic composition suggests that the formation of all the authigenic carbonates mentioned above was caused by irreversible microbial methane oxidation.

The wide range of $\delta^{13}C$ values indicates that carbon from other sources also takes part in carbonate formation (Formolo et al. 2004).

There are some scenarios for how carbon is taking part in authigenic carbonate precipitation. The isotope composition of carbon is defined by the dissolved inorganic carbon isotope composition. The isotopic composition of carbon will also depend on the condition of its formation:

(1) $\delta^{13}C$ of carbon evolved during anaerobic methane oxidation ranges from −90 to −30‰, according to Claypool & Kaplan (1974);
(2) $\delta^{13}C$ value of carbon formed during organic matter degradation is about −20‰;
(3) $\delta^{13}C$ of inorganic carbon in the seawater varies from +0.5 to +2‰; data from Irwin et al. (1977).

Gas migration (free or dissolved) may also take part in the formation of the methane-derived carbonates.

The anaerobic destruction of organic matter in the subsurface sediments described by reactions (3) and (4) will supply bicarbonate with isotopically heavy carbon into the pool of dissolved inorganic carbon.

$$2CH_2O + SO_4^{2-} \longrightarrow 2HCO_3^- + H_2S \quad (3)$$
carbon oxidation

$$2CH_2O + H_2O \longrightarrow HCO_3^- + CH_4 + H^+ \quad (4)$$
methane generation

Theoretical $\delta^{18}O$ values for carbonates precipitating in the present Derugin Basin bottom waters have been calculated by Krylov et al. (2007) according to the following experimental equation for Mg-calcite and assuming a 5.2–16.2% $MgCO_3$ content:

$$1000 \ln \alpha = 2.78 \times 10^6 T^{-2} - 2.89 + 0.06 \, Mg \, (mol\%) \quad (5)$$

(Tarutani et al. 1969)

Table 6. *Composition of the hydrated gases*

Venting site	Core number, interval (cm)	$\delta^{13}C(CH_4)/\delta^{13}C(CO_2)$, ‰, PDB	Molecular compositions of hydrated gases (assuming hydrocarbons + CO_2 = 100%)					
			CH_4	C_2H_6	C_3H_8	i-C_4H_{10}/n-C_4H_{10}	He + Ne	CO_2
Kitami	LV32-09GC, 167	−65.0/−15.4	99.44	0.0025	0.0013	0/0	0.0001	0.5
Hieroglyph	LV32-13GC, 160	−64.6/−18.4	99.34	0.0024	0.0018	0.0003/0.0003	0.0001	0.65
Chaos	LV32-16GC, 490	−64.4/−4.8	99.04	0.0018	0.0015	0.0003/0.0002	0.0001	0.95
Chaos	LV31-34GC, 160	−64.0/−8.7	99.29	0.0001	0.0015	0.0004/0.0004	0.0001	0.7

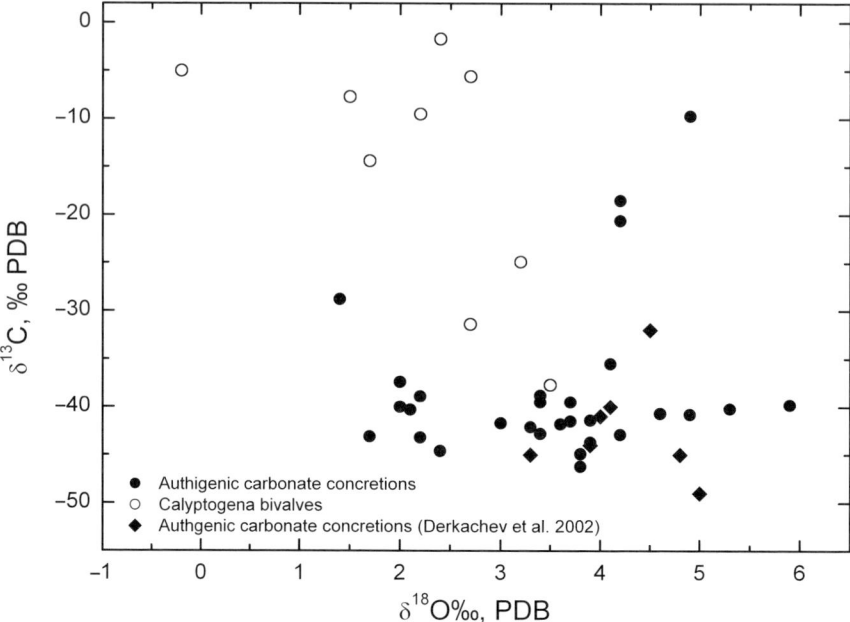

Fig. 8. Isotope composition of $\delta^{13}C$ and $\delta^{18}O$ of the studied authigenic carbonates and published isotope characteristics for authigenic carbonate concretions obtained from the gas vents off NE Sakhalin within the study area.

where α is the isotope fractionation factor and T is the temperature in Kelvin.

If the measured bottom water temperature value is $+2\,°C$ (Matveeva et al. 2005) and $\delta^{18}O$ of the water is $-0.2‰$ standard mean ocean water (SMOW), then the resulting theoretical equilibrium $\delta^{18}O$ values for methane-derived carbonates vary from $+3.6$ up to $+4.2‰$ PDB, depending on the $MgCO_3$ content. This is in good accordance with our $\delta^{18}O$ measurements.

Discussion

Fluid infiltration is one of principal process in the formation of gas hydrate accumulations (Ginsburg 1998; Ginsburg & Soloviev 1998). Gas hydrate accumulations are formed as a result of two main types of fluid infiltration: dispersed and focused. Gas hydrates related to the focused type of infiltration are associated with fluid venting on the seafloor (Soloviev & Mazurenko 2003). With respect to hydrate formation mechanisms, it is convenient to consider different vents via the fluid components: venting of free gas, gas saturated water and gas-saturated mud flows (mud domes and mud volcanoes). Whilst the venting structures off Sakhalin are mainly known to discharge free gas, there is some evidence of ascending water flows in the study area (Matveeva et al. 2003; Kulikova et al. 2007).

Residual salts

The decrease of Cl^- concentrations in hydrate bearing intervals (see Fig. 5a) is in some cases due to hydrate decomposition in the samples. This observation mainly relates to the Kitami and Hieroglyph sites. Despite the fact that samples taken from hydrate-containing sediment at the CHAOS site should be depleted in Cl^-, the water squeezed from sediments overlaying or underlying the hydrate-rich zones appeared to be enhanced in Cl^- (see Table 1). This suggests strong concentration of salts in the residual pore water, which probably equals and even exceeds freshening. This can be explained by very rapid and (what is more important) recent hydrate accumulation (in probably days or hours) under conditions of very high gas fluxes. Otherwise, residual salts in such an open system would come to equilibrium with the surrounding pore water of the host sediment. Suess et al. (2001) reported such chloride enrichments in fluids recovered at a sediment depth of 1.2 m from the southern summit of Hydrate Ridge. The chloride anomalies observed by drilling at ODP Site 1249 confirm the presence of gas hydrate-generated brine at this location (Torres et al. 2002). At the same time, a slight freshening of pore water is observed due to the decomposition of gas hydrates from some gas-hydrate-bearing sediment samples, and is characterized by decreasing concentration of most ionic species measured.

Spatial pattern of water flow

A more detailed analysis shows that a pattern of chlorinity distribution can be recognized and interpreted as a function of fluid regime (the presence of ascending water and probably the character and intensity of gas flux).

The positive Cl^- trends of the water from hydrate-free cores LV32-12HC (Kitami site) and LV32-03GC and from hydrate-bearing core LV32-16GC (CHAOS site) suggest the discharge of fluids with compositions different from seawater. The Cl^- distribution profile for core LV32-16GC is most remarkable because gas hydrates were recovered there from the sediment–water interface to the core catcher. The sediment recovered contained abundant and large gas hydrate aggregates and was considerably fluidized owing to hydrate decomposition during core retrieval. Despite the high hydrate content, the increasing chlorinity and total salinity with depth for this core suggest intensive water up flow. These data suggest that in the cores from the venting structures there is some water with a composition different to that of seawater and from the 'normal' interstitial fluids of the studied basin. One can expect that the locations of most intensive ascending water flow should be characterized by highest gradient of chlorinity (salinity). The chlorinity gradient was calculated for the cores from the CHAOS site (Fig. 9). To avoid inaccuracy from an influence of freshening owing to hydrate decomposition in the samples, we only took into consideration data on intervals underlying those containing hydrate. The one exception is the core LV32-16GC, which represents a considerable chlorinity trend against the background of hydrate-induced freshening of the pore waters. One can expect a much higher chlorinity gradient at this location. Although we acknowledge that Figure 9 is constrained by relatively few data points, and that the exact location of gradient boundaries should be treated with caution, it does appear to represent an upflow distribution mode within the CHAOS site and suggests an upward water infiltration of more saline water in the vicinity of

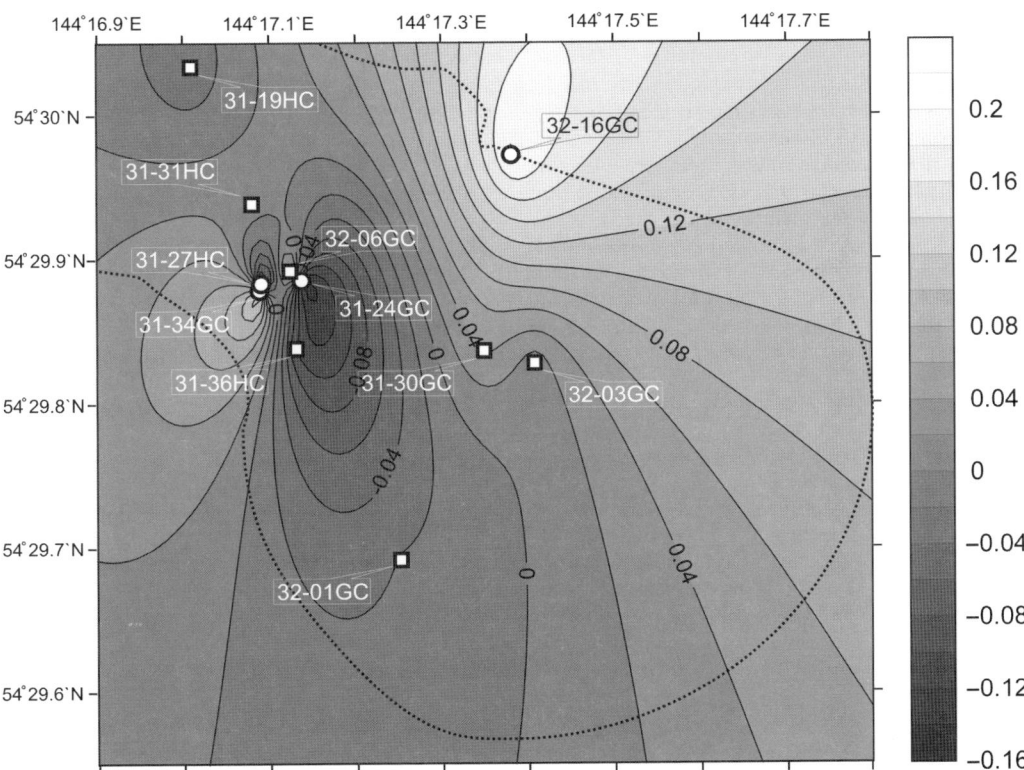

Fig. 9. Map of chlorinity (G_{Cl}) gradient (mM/cm) distribution within the southern part of the CHAOS structure. The structure boundary is shown by a dotted line. The cores that recovered gas hydrates are represented by circles; hydrate-free cores are represented by squares.

stations LV32-16GC, LV31-27HC and LV31-34GC (all recovered gas hydrates) and relatively desalinated water around stations LV31-24HC, LV31-19GC and LV31-36HC. The last cannot be explained by freshening resulting from gas hydrate decomposition in samples, because the freshening is also observed in hydrate-free cores. Reduced pore water salinities may also be attributed to clay membrane ion filtration, clay mineral dehydratation, and anaerobic methane oxidation affecting on the water isotope composition in different ways (see discussion below). The Cl^- profiles from the Hieroglyph and the Kitami sites are characterized by strong negative and positive deviations relative to the reference core, both in hydrate-containing cores and in those cores where gas hydrates were not observed. These deviations in hydrate-bearing cores result from the hydrate dissociations as supported by visual observations and water isotope data (Tables 2 and 3). The local depletions in δD isotopes accompanying reduced chlorinity (salinity) observed in water from hydrate-free sediments (Table 2) may be explained by anaerobic methane oxidation. Since bacterial methane is highly depleted in hydrogen (δD c. $-200‰$; Sassen et al. 1998), anaerobic methane oxidation generates water that is rather light in δD and reduces water salinity. High methane concentrations measured in the sediments from the studied vents (up to 122 nl l^{-1}) and in the water above the structures (up to 260,857 nl l^{-1}; Obzhirov & Vereschagina 2005) support this supposition.

Rapid gas hydrate formation at the CHAOS site appears to occur at locations of most intensive saline water upflow (Fig. 9) and is probably a function of gas solubility in water in equilibrium with hydrate. The influence of salts on methane solubility within hydrate stability zones was considered by Handa (1990), Zatsepina & Buffet (1998), and later by Davie et al. (2004) from a theoretical point of view. The fugacity of methane dissolved in saline, gas-saturated water in equilibrium with hydrate is higher than that in less saline water, even though its solubility is lower (Handa 1990). Therefore, if a gradient of water salinity exists under conditions of hydrate stability, diffusion of methane induces hydrate formation on one side of the boundary between fresher water and more saline water (Ginsburg & Soloviev 1998). We suppose that this is responsible for hydrate formation just at the locations of the saline water upflow (other conditions being equal).

Isotopic model of infiltrating fluids

In order to estimate the share of gas hydrate water, we propose a model on the basis of the observed correlation between chlorinity and isotopic composition of studied waters (see Fig. 7). The model implies that the pure hydrate water has zero chlorinity, the fluid water is characterized by a chlorinity of 550 mM and that residual water is not present in the sample. The original isotopic composition of water from hydrate samples was calculated on the basis of isotope differences of 18 and 3‰ for hydrogen and oxygen (see Table 7). The calculated share of the hydrate water in the studied samples is in the range of 63–87%, and average values of the isotope characteristics for the hypothetical original hydrate-forming water are: $\delta D = (-2.2 \pm 0.4)‰$ and $\delta^{18}O = (-0.7 \pm 0.14)‰$. The model does not take into account the presence of residual water and represents only a common case of hydrate formation observed at the Kitami and Hieroglyph sites. It is unlikely that the residual water causes considerable errors in our estimations as it is supported by coincidence of the average δD and $\delta^{18}O$ values for the waters from hydrate-free intervals and hydrate-free cores (Fig. 6b, Table 5). This correlation for 80 water samples (open circles and squares in Fig. 6a) cannot be explained by the hydrate formation process. Most likely, the observed variations in the isotopic composition of water in porous media are caused by different contributions of waters of different origin (two sources). This idea is readily illustrated by Figure 7. The hydrate-free samples lie along vertical lines that are the mixing lines of waters with similar chlorinity values from two sources (the fluid water (FW) and the seawater (SW)). The presence of the hydrate water (HW) in the samples results in an upward-left shift of the points. The coordinates of these two sources and the fractionation ranges (3 and 18‰) define a parallelogram on the graph (Fig. 7b). From Figures 6 & 7, it follows that the coordinates $\delta D_1 \approx -11‰$, $\delta O_1 \approx -1.5‰$ and $\delta D_2 \approx 0‰$, $\delta O_2 \approx +0.3‰$ are apparently the best approximation to the isotope characteristics of these two types of waters. The coordinates of one of the sources closely approximate the isotope characteristics of bottom waters taken above the studied structures and pore waters from the reference core. Obviously, this source is seawater. The second source is evidently a deep fluid discharged on the seafloor at the study area and having a 'lighter' oxygen and hydrogen composition.

An attempt has been made to estimate the contribution of each source to the hydrate water (Table 7). Calculations for oxygen have shown that the contribution of the 'fluid' water in these samples ranges from 0 to 90% and on average is about 60%; for hydrogen the contribution ranges from 0 to 40% and is 20% on average. The differences are probably induced by inaccuracy during determination of the share of hydrate water and other uncertainties of

Table 7. *Measured δD and δ¹⁸O values in hydrate water samples and calculated composition of original hydrate-forming water*

Sub-bottom depth, cm	Cl^-, mM	Share of hydrate water (SHW)	Measured values		Calculated values		Share of fluid water (SFW)	
			δD, ‰ SMOW	$δ^{18}O$, ‰ SMOW	δD, ‰ SMOW	$δ^{18}O$, ‰ SMOW	δD	$δ^{18}O$
207	138.6	0.75	11.2	1	−2.3	−1.3	0.21	0.87
100	180.2	0.67	8.0	0.9	−4.1	−1.1	0.37	0.73
190	—	—	12.9	1.5	—	—		
247	205.9	0.63	9.6	2.4	−1.7	0.5	0.15	−0.33
130/1	70.6	0.87	14.2	2.1	−1.5	−0.5	0.14	0.33
130/2	257.0	0.53	—	—	—	—		
370/1	102.8	0.81	10.6	1.7	1.1	−0.7	0	0.47
370/2	90.9	0.83	—	—	—	—		
370/3	10.8	0.98	—	—	—	—		
380	205.5	0.63	8.4	1.8	−2.9	−0.1	0.26	0.07
107	108.2	0.80	12.9	1.4	−1.5	−1.0	0.14	0.67
127	108.2	0.80	12.2	1.4	−2.2	−1.0	0.20	0.67
158	168.3	0.69	9.3	1	−3.1	−1.1	0.28	0.73
170	116.2	0.79	11.0	1.5	−3.2	−0.9	0.29	0.60
105	93.1	0.83	13.0	1.9	−1.9	−0.6	0.17	0.40
107	134.6	0.75	10.0	1.7	−3.5	−0.6	0.32	0.40
150	—	—	11.0	1.6	—	—		
285	—	—	5.3	0.2	—	—		
130	—	—	9.4	1.7	—	—		
Average					−2.2 ± 0.4	−0.7 ± 0.14		

Note: SHW = 1 − Cl^- (mM)/550 (mM), assuming that the pure hydrate water has zero chlorinity, the fluid water is characterized by a chlorinity of 550 mM and residual water is not present in the sample.
Calculation of the original isotope composition of water in the hydrate water samples was carried out based on $Δ(δD) = 18$‰ and $Δ(δ^{18}O) = 3$‰:

$$δD_{calculated} = δD_{measured} − 18 × SHW$$

$$δO_{calculated} = δ^{18}O_{measured} − 3 × SHW$$

SFW(D) = $δD_{calculated}/(−11)$; SFW(O) = $δO_{calculated}/(−1.5)$.

the model. Despite this, it suggests that there is a considerable contribution of this 'fluid' water, of at least 30% (the average of oxygen and hydrogen value above would be 40%; we have kept to a slightly lower value to avoid over-estimation). In some hydrate-free samples the contribution is much higher (Fig. 6). The absence of hydrates in these samples probably points to variable degrees of saturation of waters by methane, i.e. by a pulsating gas venting activity or due to different gas fugacity inside and outside ascending water flow.

Conclusions

The hydrate-induced sediment structures at venting sites off NE Sakhalin suggest two mechanisms of hydrate formation:

(1) precipitation from a solution oversaturated with gas due to a decrease of gas solubility associated with changing conditions in filtration pathways;

(2) precipitation from an infiltrating fluid in a fractured reservoir. This implies that not only free gas but also gas-saturated water infiltrates at the vents studied.

Analysis of the water chemistry and oxygen–hydrogen isotope compositions of the different types of water involved suggests the presence of an ascending water with a different chemical and isotope composition to that of seawater and 'normal' pore water of the basin. This locally ascending fluid has a lighter isotope composition and an enrichment of dissolved potassium and chloride. Gas hydrates are formed from the mixture of these upward migrating fluids and *in-situ* pore water, which in the open marine system is equal to the composition of seawater.

Based on the water isotope data and the correlation between chlorinity and isotopic composition, the share of the ascending fluids in the hydrate water was estimated to be at least 30%. Modelling of the isotope characteristics of the original hydrate-forming fluids resulted in values of $\delta D = -2.2‰$ and $\delta^{18}O = -0.7‰$. The isotope compositions of the ascending fluids and seawater taking part in hydrate formation were calculated as -11, $-1.5‰$ and 0, $+0.3‰$ for hydrogen and oxygen respectively.

The presence of residual salts near large gas hydrate inclusions suggests very rapid hydrate accumulation for the CHAOS structure and strong present-day venting and hydrate formation activity. The chlorinity gradient calculations indicate that hydrate formation is focused at locations of intensive ascending flow of water enriched by salts. Gas solubility in water in equilibrium with hydrate probably determines the pattern of hydrate formation.

Based on the composition of hydrate gas and of carbon–oxygen isotopes from authigenic carbonates, we assume that gas involved in the formation of the gas hydrates has different sources and the relative quantities of these gases are responsible for the variations in the isotope composition.

This paper is one of the last ones written by Dr. Leonid Mazurenko, who was one of the active members of the CHAOS project. Dr Leonid Mazurenko tragically died in July 2007 at the age of 30. Despite his young age, Dr Mazurenko was one of the most prominent figures in the Russian gas hydrate and fluid venting geochemical community. All his friends and colleagues involved in the CHAOS project will never forget him and his contribution to the gas hydrate research and will miss him greatly.

This work was supported by the OSL Fellowship Program (grants OSL-03-14 and OSL-05-17); Russian Ministry of Natural Resources; Russian Academy of Science; RFBR Grant 05-05-66860-MF; the Japan Society for the Promotion of Science, KAKENHI 14254003 and 15550060; the Ministry of Education, Cultural, Sport, Science and Technology, KAKENDI 15760640; Ministry of Maritime Affairs and Fisheries research and development, Korea, grant PM21700; a post-doctoral research assistantship of the Flemish Fund for Scientific Research (FWO-Vlaanderen) for J. Poort; the Bilateral Scientific Coorporation Russia-Flanders; and the Federal Ministry of Education and Research, Germany, KOMEX grant 03G0568. We would like to acknowledge scientists from the Laboratory For Gas Hydrate Geology, VNIIOkeangeologia (St Petersburg, Russia), New Energy Resource Center (Kitami, Japan), Korea Polar Research Institute (Ansan, Korea), scientists from Pacific Oceanological Institute (Vladivostok, Russia), Institute of Oceanology (Moscow, Russia), Reynard Center of Marine Geology (Gent, Belgium), IFM GEOMAR (Kiel, Germany), Limnological Institute (Irkutsk, Russia) and the crew of the 31st and 32nd cruises of the R/V *Akademik M. A. Lavrentyev* (Vladivostok, Russia). We wish to thank Dr A. Krylov (KIT) and Dr A. Stadnitskaia (NIOZ) and anonymous reviewers for the fruitful suggestions greatly improved this manuscript. We would like to express our sincere thanks to Dr C. A. Rochelle for great help in completing this manuscript.

References

ALOISI, G., WALLMANN, K., BOLLWERK, S. M., DERKACHEV, A., BORMANN, G. & SUESS, E. 2004. The effect of dissolved barium on biogeochemical processes at cold seeps. *Geochimica et Cosmochimica Acta*, **68**, 1735–1748.

ASTAKHOV, A., BOTSUL, A. ET AL. 2000. Paleoceanography and sedimentology. *In*: BIEBOW, N., LUDMANN, T., KARP, B. YA. & KULINICH, R. (eds) *KOMEX: Kurile-Sea of Okhotsk Marine Experiment. Cruise Reports: KOMEX V and KOMEX VI, R/V 'Professor Gagarinsky' Cruise 26 and M/V 'Marshal Gelovany' Cruise 1.* GEOMAR Report **88**, Kiel, 189–209.

BARANOV, B., SALOMATIN, A. & SALYUK, A. 2005. Sakhalin slope morphology and structural position of gas venting areas. *In*: MATVEEVA, T., SOLOVIEV, V., SHOJI, H. & OBZHIROV, A. (eds) *Hydro-carbon Hydrate Accumulations in the Okhotsk Sea (CHAOS Project Leg I and Leg II).* Report of RV *Akademik M. A. Lavrentyev* Cruise 31 and 32. VNIIOkeangeologia, St Petersburg, 18–27.

BEZRUKOV, P. L. 1960. Bottom sediments of the Sea of Okhotsk. *Transactions of the Institute of Oceanology*, **32**, 15–97 (in Russian).

BIEBOW, N. & HUTTEN, E. 1999. *Cruise Report KOMEX I and II: RV Professor Gagarinsky Cruise 22, RV Akademik M. A. Lavrentyev Cruise 28.* Report no. **82**, GEOMAR.

BIEBOW, N., KULINICH, R. & BARANOV, B. 2003. *KOMEX Cruise Report: RV Akademik M. A. Lavrentyev Cruise 29, leg I and Leg II. Vladivostok–Pusan–Okhotsk Sea–Pusan–Okhotsk Sea–Pusan–Vladivostok.* Report no. 190, GEOMAR.

BOGOROV, V. G. 1974. *Plankton of the World Ocean.* Moscow, Nauka (in Russian).

BOTSUL, A., BIEBOW, N. ET AL. 1999. Paleoceanography and sedimentology in the Sea of Okhotsk. *In*: BIEBOW, N. & HUTTEN, E. (eds) *KOMEX: Kurile-Sea of Okhotsk Marine Experiment. Cruise Reports: KOMEX I and KOMEX II, R/V Professor Gagarinsky Cruise 22 and R/V Akademik M. A. Lavrentyev Cruise 28.* GEOMAR Report **82**, Kiel, 148–177.

CHERSKY, N. V. & MIKHAILOV, N. E. 1990. Razmer ravnovesnykh kriticheskikh zarodyshei gazovykh gidratov (The size of equilibrium critical nuclei of gas hydrates). *Doklady AN SSSR (Reports of USSR Academy of Sciences)*, **312**, 968–971 (in Russian).

CLAYPOOL, G. E. & KAPLAN, I. R. 1974. The origin and distribution of methane in marine sediments. *In*: KAPLAN, I. R. (ed.) *Natural Gases in Marine Sediments.* Plenum Press, New York, 99–139.

DAVIE, M. K., ZATSEPINA, O. Y. & BUFFETT, B. A. 2004. Methane solubility in marine hydrate environments. *Marine Geology*, **203**, 177–184.

DERKACHEV, A. N., OBZHIROV, A. I., BOHRMANN, G., GREINERT, J. & ZUSS, E. 2002. Authigenic mineral formation in the cold gas-seep structures within the Sea of Okhotsk. *In*: *Conditions of the Generation of Bottom Sediments and related Mineral Deposits within Marginal Seas.* Vladivostok, Dalnauka, 47–60 (in Russian).

DICKENS, G. R. 1999. The blast in the past. *Nature*, **401**, 752–755.

DMITRIEVSKY, A. N., BALANYUK, I. E. & KARAKIN, A. V. 2004. Hydrothermal mechanism of gas hydrate formation in mid-ocean ridges. *Gas Industry Journal*, **5**, 50–54.

DULLO, W.-CHR., BIEBOW, N. & GEORGELEIT, K. (eds). 2004. Cruise Report *SO178-KOMEX Cruise Report*, Kiel, 125.

EGEBERG, P. K. & LEG 126 SHIPBOARD SCIENTIFIC PARTY. 1990. Unusual composition of pore waters found in the Izu–Bonin fore-arc sedimentary basins. *Nature*, **244**, 215–218.

FORMOLO, M. J., LYONS, T. W., ZHANG, C., KELLEY, C., SASSEN, R., HORITA, J. & COLE, D. R. 2004. Quantifying carbon sources in the formation of authigenic carbonates at gas hydrate sites in the Gulf of Mexico. *Chemical Geology*, **205**, 253–264.

FOURNIER, M., JOLIVET, L., HUCHON, P., SERGEYEV, K. F. & OSCORBIN, L. S. 1994. Neogene strike–slip faulting in Sakhalin and the Japan Sea Opening. *Journal of Geophysical Research*, **99**, 2701–2725.

GINSBURG, G. D. 1998. Gas hydrate accumulation in deep-water marine sediments. *In*: HENRIET, J.-P. & MIENERT, J. (eds) *Gas Hydrates: Relevance to World Margin Stability and Climate Change*. Geological Society, London, Special Publications, **137**, 51–63.

GINSBURG, G. D. & SOLOVIEV, V. A. 1998. *Submarine Gas Hydrates*. VNIIOkeangeologia, St Petersburg, 216.

GINSBURG, G. D., SOLOVIEV, V. A., CRANSTON, R. E., LORENSON, T. D. & KVENVOLDEN, K. A. 1993. Gas hydrates from the continental slope, offshore Sakhalin Island, Okhotsk Sea. *Geo-Marine Letters*, **13**, 41–48.

GINSBURG, G. D., MILKOV, A. V. ET AL. 1999. Gas hydrate accumulation at the Haakon Mosby Mud Volcano. *Geo-Marine Letters*, **19**, 57–67.

GNIBIDENKO, H. S. & KHVEDCHUK, I. I. 1982. The tectonics of the Okhotsk Sea. *Marine Geology*, **50**, 155–198.

GORBARENKO, S. A. 1991. The stratigraphy of the upper Quaternary sediments in the central part of the Sea of Okhotsk and its paleoceanology according to data obtained by the $\delta^{18}O$ and other methods. *Oceanology*, **31**, 761–766.

GORBARENKO, S. A., KOVALYUKH, N. N., ODINOKOVA, L. Yu., RYBAKOV, V. F., TOKARCHUK, T. N. & SHAPOVALOV, V. V. 1990. Upper Quaternary sediments of the Sea of Okhotsk and the reconstruction of paleoceanologic conditions. *Geology Pacific Ocean*, **6**, 309–330.

HAESE, R., HENSEN, C. & DE LANGE, G. J. 2006. Pore water geochemistry of eastern Mediterranean mud volcanoes: Implications for fluid transport and fluid origin. *Marine Geology*, **225**, 191–208.

HANDA, Y. P. 1990. Effect of hydrostatic pressure and salinity on the stability of gas hydrates. *Journal Physics Chemistry*, **94**, 2652–2657.

HENSEN, C., NUZZO, M. ET AL. 2007. Sources of mud volcano fluids in the Gulf of Cadiz – indications for hydrothermal imprint. *Geochimica et Cosmochimica Acta*, **71**, 1232–1248.

IRWIN, H., CURTIS, C. & COLEMAN, M. 1977. Isotopic evidence for source of diagenetic carbonates formed during burial of organic-rich sediments. *Nature*, **269**, 209–213.

KOBLENTS-MISHKE, O. I. 1967. *Primary Production of the Pacific Ocean, Oceanology: Biology of the Pacific Ocean*. Nauka, Moscow, part 1, 62–65.

KRYLOV, A., MAZURENKO, L. ET AL. 2007. Sediments and authigenic carbonates related to gas-hydrates in the Sea of Okhotsk: First results from the CHAOS 2005 expedition. *In*: TSUNEMOTO, H., SHOJI, H. & YAMASHITA, S. (eds) *Gas Hydrates for the Future Energy and Environment. Proceedings of the 2nd International Workshop on Gas Hydrate Studies and Other Related Topics, October 22–23, 2005, Kitami*. Kitami Institute of Technology, 49–54.

KULIKOVA, M., POORT, J., MATVEEVA, T., MAZURENKO, L., JIN, Y. K. & SHOJI, H. 2007. The features of temperature and acoustic fields related to the gas hydrate formation at the fluid discharge structures. *International Conference on Gas Hydrate Studies*, Irkutsk, 35.

LISITSYN, A. P. 1978. *Protsessy okeanskoi sedimentatsii. Litologiya i geokhimiya (Oceanic Sedimentation Processes: Lithology and Geochemistry)*. Nauka, Moscow (in Russian).

LUFF, R. & WALLMANN, K. 2003. Fluid flow, methane fluxes, carbonate precipitation and biogeochemical turnover in gas hydrate-bearing sediments at Hydrate Ridge, Cascadia Margin Numerical modeling and mass balances. *Geochimica et Cosmochimica Acta*, **67**, 3403–3421.

MAEKAWA, T. 2004. Experimental study on isotopic fractionation in water during gas hydrate formation. *Geochemical Journal*, **38**, 129–138.

MARTIN, J. B., KASTNER, M., HENRY, P., LE PICHON, X. & LALLEMENT, S. 1996. Chemical and isotopic evidence for sources of fluids in a mud volcano field seaward of the Barbados accretionary wedge. *Journal of Geophysical Research*, **101**, 20325–20345.

MATVEEVA, T., SOLOVIEV, V. ET AL. 2003. Geochemistry of gas hydrate accumulation offshore NE Sakhalin Island (the Sea of Okhotsk): Results from the KOMEX-2002 cruise. *Geo-Marine Letters*, **23**, 278–288.

MATVEEVA, T., SOLOVIEV, V., SHOJI, H. & OBZHIROV, A. (eds). 2005. *Hydro-carbon Hydrate Accumulations in the Okhotsk Sea (CHAOS Project Leg I and Leg II)*. Report of R/V Akademik M. A. Lavrentyev Cruise 31 and 32. VNIIOkeangeologia, St Petersburg, 164.

MAZURENKO, L., KAULIO, V., GRINEVA, E. & SIGACHEVA, A. 2005. Gas hydrates. 6.1. Results obtained during Leg 1. *In*: MATVEEVA, T., SOLOVIEV, V., SHOJI, H. & OBZHIROV, A. (eds) *Hydro-carbon Hydrate Accumulations in the Okhotsk Sea (CHAOS Project Leg I and Leg II)*. Report of R/V Akademik M. A. Lavrentyev Cruise 31 and 32. VNIIOkeangeologia, St Petersburg, 47–60.

MAZURENKO, L., JIN, Y. K., SHOJI, H., OBZHIROV, A. & SHIPBOARD SCIENTIFIC PARTY OF THE 36 CRUISE OF R/V 'AKADEMIK M. A. LAVRENTYEV'. 2006. New gas hydrate observations in the Sea of Okhotsk: Results from the CHAOS-2005 Cruise. *International Conference, Minerals of the Ocean-3: Future*

Developments. St Petersburg, VNIIOkeangeologia, 19–22 June 2006. VNIIOkeangeologia, St Petersburg, 89–91.

MELNIKOV, V. & NESTEROV, A. 1996. Modeling of gas hydrate formation in porous medium. *Second International Conference, Natural Gas Hydrates, Proceedings*, Toulouse, 541–548.

OBZHIROV, A. I. 1992. Gas-geochemical manifestations of gas-hydrates in the Sea of Okhotsk. *Alaska Geology*, **21**, 1–7.

OBZHIROV, A. & VERESCHAGINA, O. 2005. Geochemical studies of gases. *In*: MATVEEVA, T., SOLOVIEV, V., SHOJI, H. & OBZHIROV, A. (eds) *Hydro-carbon Hydrate Accumulations in the Okhotsk Sea (CHAOS Project Leg I and Leg II). Report of R/V Akademik M. A. Lavrentyev Cruise 31 and 32*. VNIIOkeangeologia, St Petersburg, 67–84.

REZNIKOV, A. A. & MULIKOVSKAYA, E. P. 1956. Analysis of natural waters and brines. *In*: KNIPOVICH, YU. & MORACHEVSKY, YU. (eds) *Analysis of Minerals*. State Chemical Publishing, Leningrad, 872–1047 (in Russian).

RITGER, S., CARSON, B. & SUESS, E. 1987. Methane derived authigenic carbonates formed by subduction-induced pore water expulsion along the Oregon, Washington margin. *Geological Society American Bulletin*, **98**, 147–156.

ROZHDESTVENSKY, V. S. 1986. Evolution of the Sakhalin fold system. *Tectonophysics*, **127**, 331–339 (in Russian).

SANCETTA, C. 1981. Oceanographic and ecologic significance of diatoms in surface sediments of the Bering and Okhotsk seas. *Deep Sea Research*, **28**, 789–817.

SASSEN, R., MACDONALD, I. R. ET AL. 1998. Bacterial methane oxidation in sea-floor gas hydrate. Significance to life in extreme environments. *Geology*, **26**, 289–293.

SAVOSTIN, L. A., ZONENSHAIN, L. & BARANOV, B. 1983. Geology and plate tectonics of the Sea of Okhotsk. *In*: HILDE, T. W. C. & UYEDA, S. (eds) *Geodynamics of the Western Pacific–Indonesian Region*. Geodynamics Series 11, Washington, DC, 189–222.

SHIGA, K. & KOIZUMI, I. 2000. Latest Quaternary oceanographic changes in the Sea of Okhotsk based on diatoms. *Marine Micropaleontology*, **38**, 91–117.

SHOJI, H., SOLOVIEV, V. ET AL. 2005. Hydrate-bearing structures in the Sea of Okhotsk. *EOS*, **86**, 13–24.

SOLOVIEV, V. & MAZURENKO, L. 2003. Sea floor venting and gas hydrate accumulation *In*: MAX, M. (eds) *Natural Gas Hydrate in Oceanic and Permafrost Environments*. Kluwer Academic, Dordrecht, A1–A8.

SOLOVIEV, V. A., GINSBURG, G. D. ET AL. 1994. Gas hydrates of the Okhotsk Sea. *Otechestvennaya Geologia*, **2**, 10–16 (in Russian).

SUESS, E., TORRES, M., BOHRMANN, G., COLLIER, R., RICKERT, D. & GOLDFINGER, C. 2001. Sea floor methane hydrates at Hydrate Ridge, Cascadia margin. *In*: PAULL, C. & DILLON, W. (eds) *Natural Gas Hydrates: Occurrence, Distribution, and Detection*, **124**. American Geophysics Union, 87–98.

TARUTANI, T., CLAYTON, R. N. & MAYEDA, T. K. 1969. The effect of polymorphism and magnesium substitution on oxygen isotope fractionation between calcium carbonate and water. *Geochimica et Cosmochimica Acta*, **33**, 987–996.

TORRES, M. E., MCMANUS, J. ET AL. 2002. Fluid and chemical flux in and out of sediments hosting methane hydrate deposits on Hydrate Ridge, OR, I: Hydrological provinces. *Earth and Planetary Science Letters*, **201**, 525–540.

WALLMANN, K., BOLLWERK, S., KOLEVICA, A. & SHULGA, Y. 2000. Pore water geochemistry. *In*: BIEBOW, N., LUDMANN, T., KARP, B. & KULINICH, R. (eds) *Cruise Report KOMEX V and VI: RV Professor Gagarinsky Cruise 26, MV Marshal Gelovany Cruise 1*. Report no. 88. GEOMAR, 180–181 and 153–172.

WONG, J. Y., STABNIKOVA, O., IVANOV, V., TAY, S. T. L. & TAY, J. H. 2003. Intensive aerobic bioconversion of sewage sludge and fecal waste into fertilizer. *Waste Management Research*, **21**, 405–415.

YOUSIF, M. H. & SLOAN, E. D. 1991. Experimental investigation of hydrate formation and dissociation in consolidated porous media. *SPE Reservoir Engineering*, 452–458.

ZATSEPINA, O. Y. & BUFFETT, B. A. 1998. Conditions for the stability of gas hydrate in the seafloor. *Journal of Geophysics Research*, **103**, 24127–24139.

ZONENSHAYN, L. P., MURDMAA, I. O. ET AL. 1987. An underwater gas source in the Sea of Okhotsk. *Oceanology*, **27**, 598–602.

Hydrate occurrences in the Namibe Basin, offshore Namibia

ROGER SWART*

Blackgold Geosciences CC, P.O. Box 24287, Windhoek, Namibia
Corresponding author (e-mail: rogerswart@mweb.com.na)

Abstract: Seismic data has long suggested the presence of methane hydrates offshore Namibia. The seismic data shows the presence of well-developed bottom-simulating reflectors which can be mapped over a large area. A recent seabed coring programme has confirmed the presence of hydrates in the Namibe Basin. The hydrates appear to be associated with slump features and are a drilling hazard as well as a potential resource. The methane hydrate appears to originate from both biogenic and thermogenic sources.

Methane hydrates occur widely on all deep-water continental and insular margins (Kvenvolden 1998; Booth *et al.* 1998) but their occurrence on the Namibian continental margin is not yet widely known and has largely been inferred either from the presence of bottom-simulating reflectors (BSRs) or indirect evidence from slump features. Summerhayes *et al.* (1979) described slumps from near the Walvis Ridge which he interpreted as possibly the result of slumping associated with hydrates. Similarly, from further south in Namibia slumps described by Dingle (1980) could be the result of hydrate dissolution. Ben Avraham *et al.* (2002) also inferred the presence of hydrates in the Orange Basin of South Africa from the presence of BSRs and mud volcanoes.

The presence of BSRs alone is however not a reliable indicator of the presence or not of methane hydrates. Not only can BSRs be formed by other methods, e.g. silica diagenesis (Kvenvolden 1998), but more importantly, as methane hydrate BSRs form at the interface between hydrate and free gas, then in the absence of free gas a BSR will not develop, even if hydrates are present (Haq 1998; Paull *et al.* 1996). In Leg 175 of the ODP project specific tests for the presence of methane hydrates were done, specifically at Hole 1077 in the Congo Canyon region, as well-developed BSRs had been recognized in that region. However, their conclusion was that there was no physical evidence for the presence of methane hydrates (Meyers & Shipboard Scientific Party 1998) and the BSR was attributed by them to the presence of dolostones. This conclusion was applied to the entire area north of the Walvis Ridge. Brooks *et al.* (1994), however, reported the physical recovery of hydrate samples at several locations in the Niger Delta. Cunningham and Lindholm (2000) interpreted the presence of BSRs on the continental slope off the Niger and Congo River deltas as firm evidence for the presence of hydrates, an occurrence confirmed by other authors (Sultan *et al.* 2004).

The presence of well-developed BSRs observed on recently acquired two- and three-dimensional seismic data in northern Namibia which are underlain by numerous gas chimneys and bright amplitude reflectors suggested that aerially extensive methane hydrates may occur north of the Walvis Ridge (Fig. 1). This has since been confirmed by seafloor sampling undertaken in January–February 2004 by TDI-Brooks on behalf of Namcor, and will be discussed in more detail later in this paper.

The mapping and understanding of the Namibian hydrate occurrences are of critical importance as they are an unquantified potential hazard in the drilling of future exploration wells in the area in which they occur. The main occurrence of hydrates described here occurs directly above a large four-way dip closure that is currently the focus of intense exploration efforts. Instability of the margin is evident in the presence of large slump features identified during data acquisition for Namibia's extended continental shelf claim and is a major concern for drilling. The potential of the hydrates as a possible future resource is also not insignificant.

Regional setting

The Namibian continental margin formed in the Late Jurassic–Early Cretaceous during the breakup of West Gondwana. The Namibe Basin developed north of the Walvis Ridge and is conjugate to the Santos Basin of Brazil, the site of several recent major hydrocarbon discoveries. The Namibe Basin is one of the last frontier basins left and is as yet completely undrilled. The age of the post-rift fill is believed to range from the Late Jurassic/Early Cretaceous to Recent. The oceanic circulation of the area is today characterized by the organically highly productive, cold Benguella Upwelling System. Biogenic methane may be a significant contributor to the hydrate component and needs to be assessed in future work.

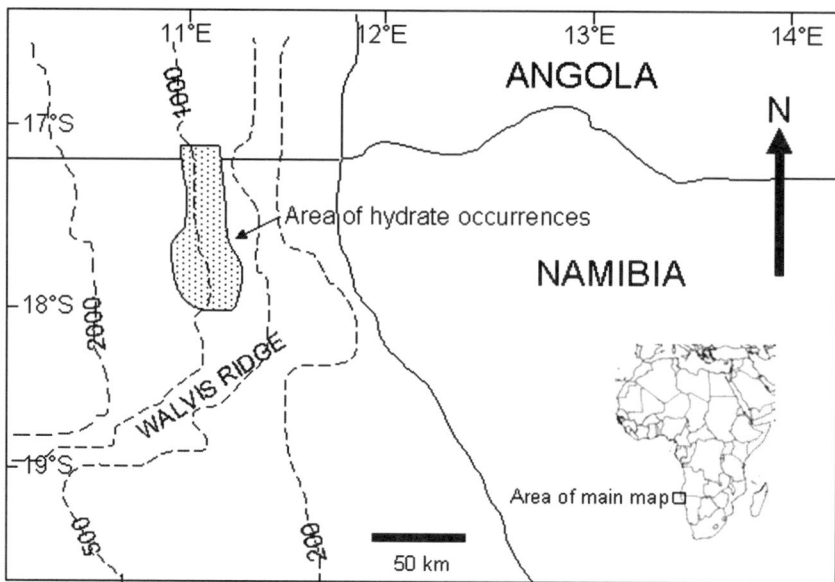

Fig. 1. Locality map showing the main area of occurrence of the methane hydrate in the Namibe Basin, offshore Namibia. The southern limit of the basin is defined by the Walvis Ridge and it continues northwards into Angola.

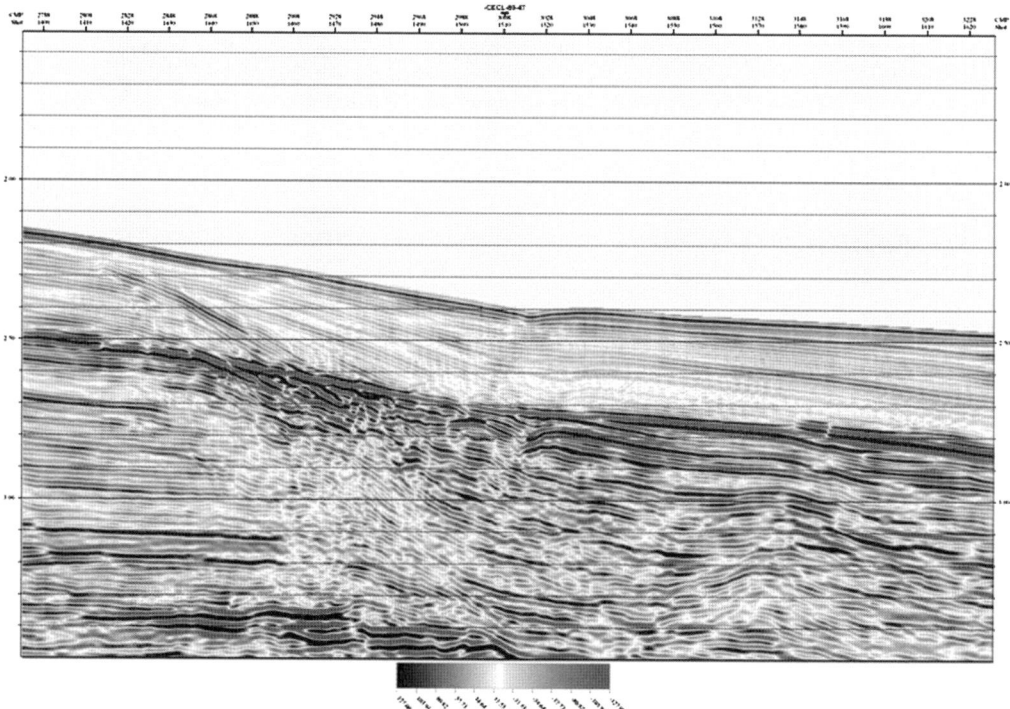

Fig. 2. Seismic line showing BSR features. The BSR has the reverse polarity to that of the seafloor and cross cuts the seismic layering.

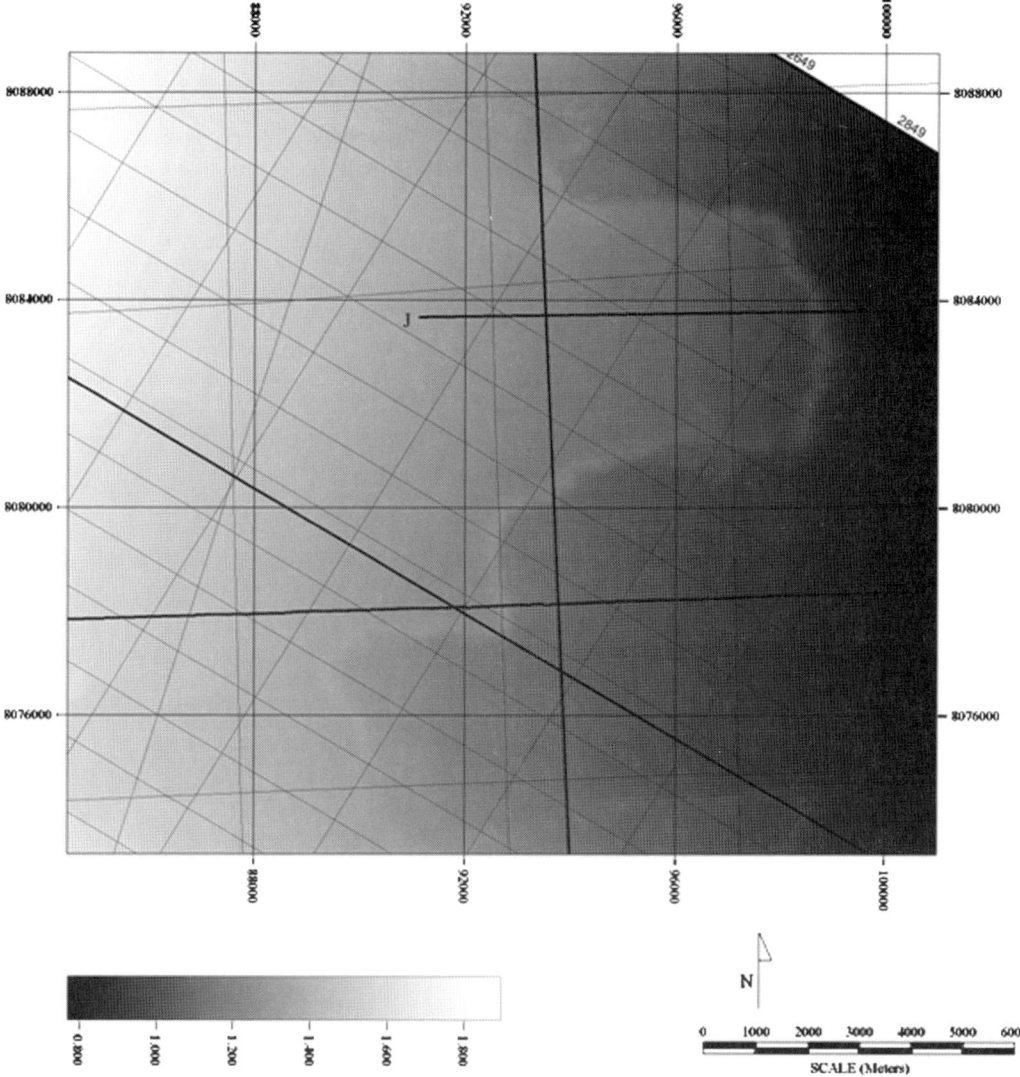

Fig. 3. Time map of seabed showing the orthogonal nature of the moat feature depicted in Figures 4 and 5. This feature indicates an incipient slide (see text for further discussion).

Evidence for hydrates

ODP/DSDP drilling

The objective of Leg 175 of the ODP programme was to evaluate the history of the Benguela Current. Specific tests were done at Hole 1077b after holes at two previous sites (six holes in total) had shown little or no direct evidence for the presence of hydrate layers. The additional tests included sampling of headspace gases at 3 m intervals between 100 and 130 metres below seafloor (mbsf). Methane levels in the headspace gases were found to vary between 7000 and 41 000 ppmv (Meyers & Shipboard Scientific Party 1998). None of the other features known to be associated with the presence of hydrates from previous DSDP cruises (Kvenvolden & Barnard 1983) was found – these include frothing of the sediments, disruption of sedimentary structures, frosting of the outside of the core liner and dilution of interstitial chloride accompanying the decomposition of hydrates (Meyers & Shipboard Scientific Party 1998). Their conclusion that the BSR was possibly caused by the presence of dolostones and that the gases present were of microbial origin was based on the absence of recovered hydrate and drilling refusal at shallow depths (Meyers & Shipboard Scientific Party 1998).

Seismic surveys

Early seismic surveys showed that BSRs were present offshore western Africa in the Benguela Current area. Modern exploration multi-channel two-dimensional seismic data has been acquired since 1989 in the Namibe Basin for hydrocarbon exploration and, although focused on targets at much greater depths than the BSRs, superb images of BSRs were obtained (Fig. 2). The thickness of the transparent blanking layer above the BSR is typically 250–350 mstwtt. The BSR can be mapped over an area of at least 2000 sq km, but it is uncertain for the reasons given above if hydrates are present all over this area. The BSRs all show a reverse polarity to that of the seafloor.

Interpretation of the two-dimensional surveys led to the recognition of a large four-way dip closure that was identified at depth and this structure became the focus of heightened petroleum exploration. A 650 sq km three-dimensional survey was acquired over the structure (named Kunene) in 2000 and several features were immediately apparent from the data. The BSR was evident and numerous gas chimneys were recognized in the data. These were sourced from at least as deep as the syn-rift interval and terminated below the BSR as strong amplitude reflectors. The BSR covers more than 1500 sq km and extends beyonds the limits of the seismic data available in Namibia.

The morphology of the seafloor as mapped from the three-dimensional seismic was also unusual (Fig. 3) in that an orthogonal moat was developed over the area of the hydrates (Figs 4 & 5). The moat is up to 80 mstwtt (c. 30 m) deep and up to 110 mstwtt (c. 42 m) higher on the inboard side than the outside. There is no evidence of faulting immediately underlying the moat (Figs 4 & 5). The orthogonal character of the moat precludes it from being a drainage channel similar to those described by Fichler et al. (2004). Detailed examination of the block downdip of the moat suggests that it is an incipient slide. Crinkling of the reflectors

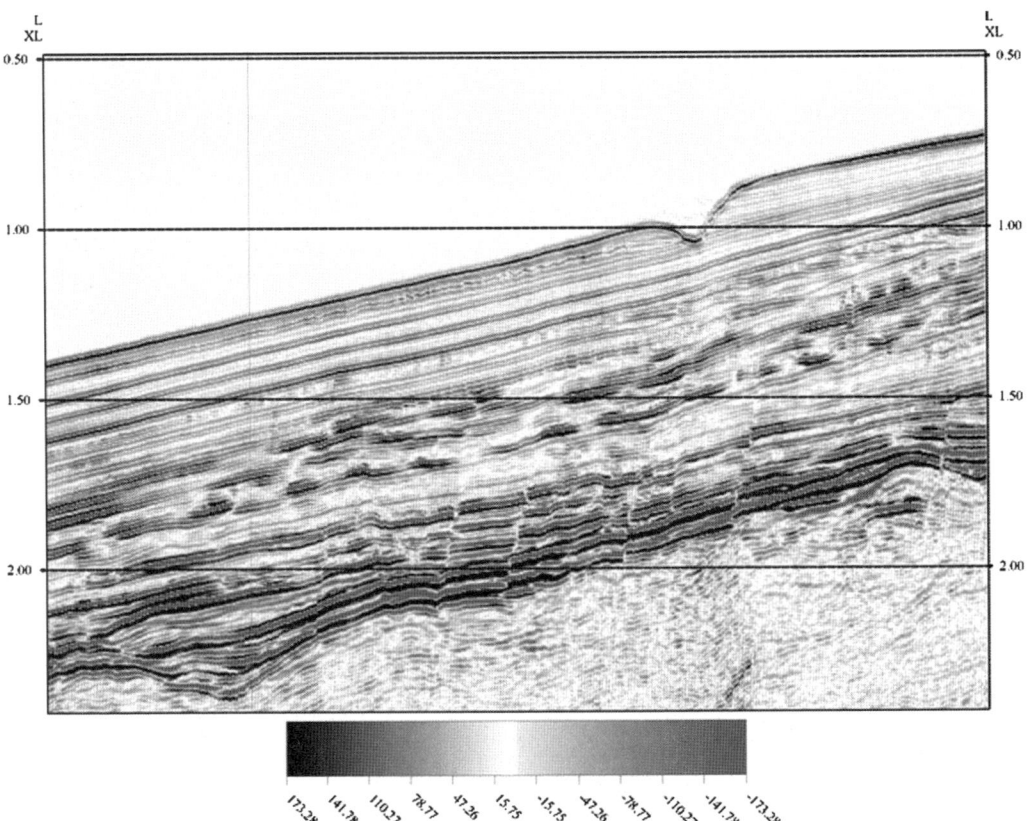

Fig. 4. Seismic line showing moat feature developed on the seafloor. This feature is always higher on the eastern (inboard) margin.

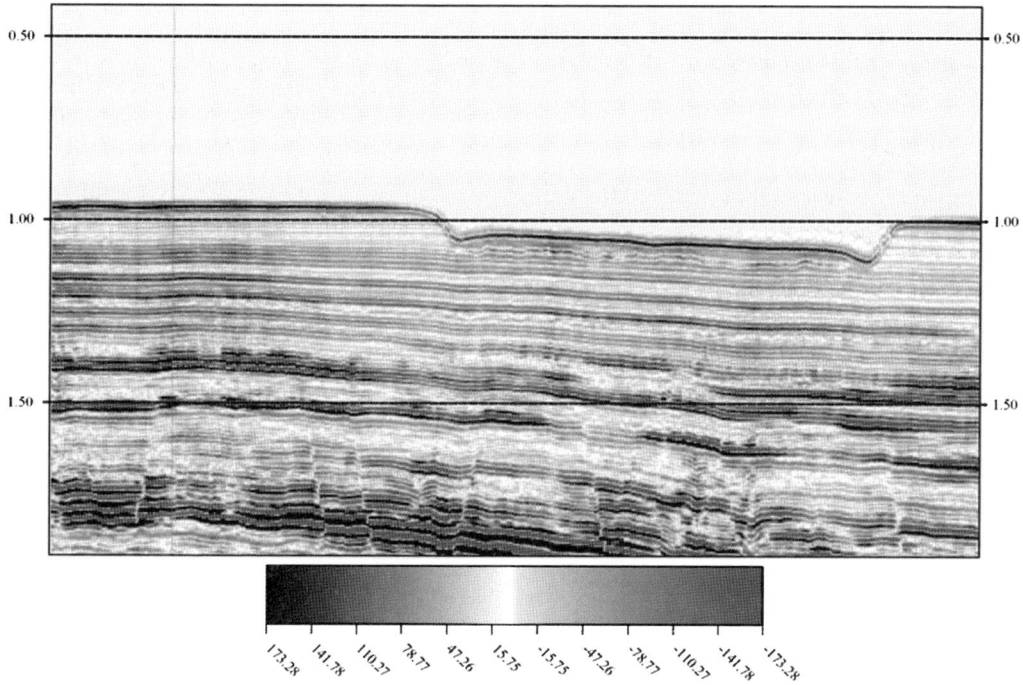

Fig. 5. Strike section through the orthogonal feature in the NE part of the map shown in Figure 3.

immediately below the seafloor downdip of the moat suggests compression of the reflectors (Fig. 6). The crinkling tends to die out downdip. Low-angle thrusts may also be present (Fig. 7), but are not common. Sections parallel to the coast show that the seafloor and all the underlying sequence is domed above the Kunene structure (Fig. 8). The slides appear to dip away from the crest of the dome.

Seabed sampling

In January–February 2004 Namcor undertook a seabed sampling survey over the whole of offshore

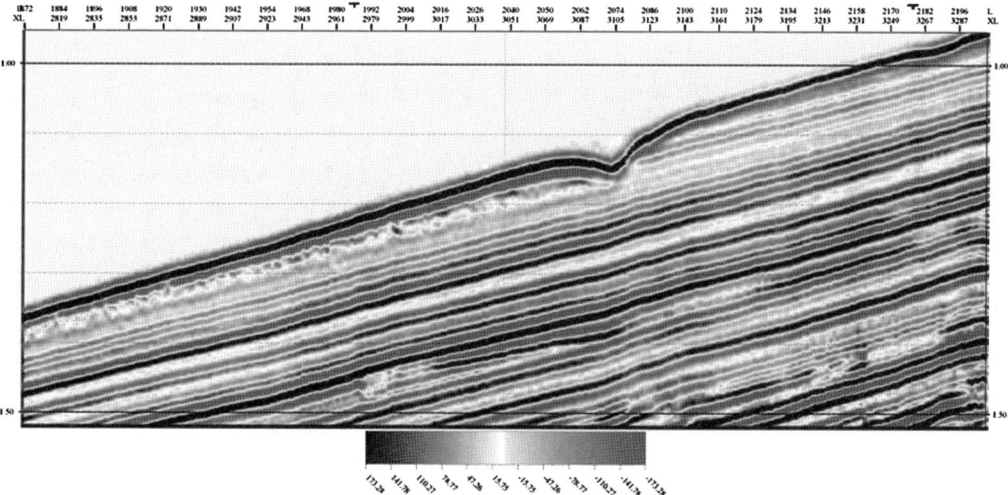

Fig. 6. Dip section showing well developed crinkling down dip from the moat, indicating compression and incipient sliding of the sediment mass. Further downdip there are large slumps and debris flow deposits.

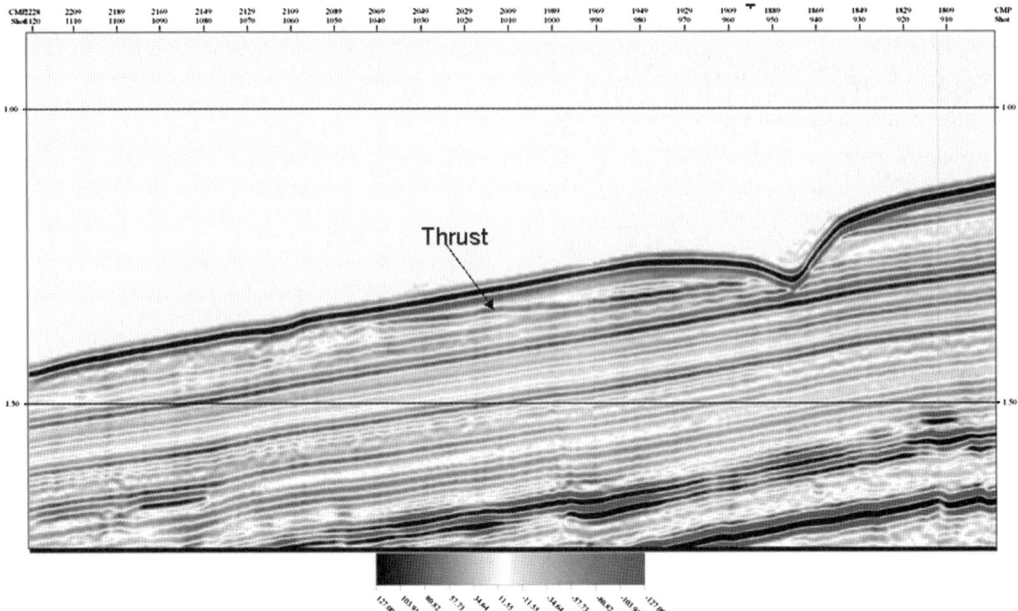

Fig. 7. Dip section showing a possible thrust feature downdip of the moat, again indicating compression and incipient sliding.

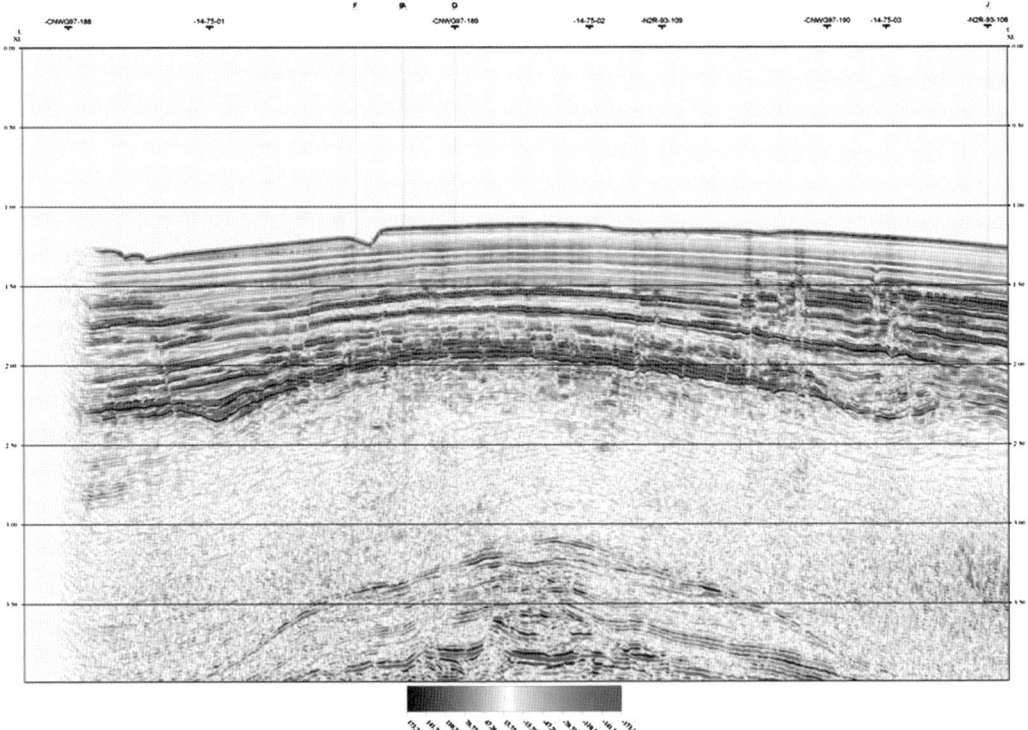

Fig. 8. Regional seismic line parallel to the coast showing how the seabed and all underlying units are domed above the major structure found at a depth of 3.2 stwt. The sliding appears to dip away from this bulge.

Fig. 9. Photograph of hydrate sample obtained during a sea bed geochemical sampling campaign in 2004. The hydrate is the paler coloured material and occurs as clots of hydrate in a muddy matrix.

Namibia, including several sites in the Namibe Basin. Heat flow measurements were also made at sites away from the geochemical sites. TDI-Brooks undertook the study on Namcor's behalf. No sampling of hydrates was planned but a hydrate was recovered from the bottom of a 5 m drop core (Fig. 9). The hydrate was formed as solid clots in a muddy matrix. No age data for the sediments are available, nor is the depositional environment known. The location of the sample was at 17°51.2613′S, 11°20.3308′E on a two-dimensional seismic line, NWG97-201 (Fig. 10) in approximately 1500 m water. The locality is inboard of the moat shown in Figure 6. The BSR and transparent zone are clearly evident in this data. Unfortunately no facilities were present on the vessel for long-term preservation of the hydrate. Consequently no isotope data was obtained. Heat flow measured in the basin was in the range 43.0–45.9 mW m^{-2} and sea-bottom temperature varied from 4.4 to 4.6 °C.

Discussion and conclusions

The clear association of the hydrate sampled with the BSR and transparent zone proves that, at least in part, the BSR occurrences offshore Namibia are associated with hydrates, as has been increasingly evident elsewhere in West Africa. A clear indication of the volume of hydrate that is present is difficult because of the uncertainties about whether all BSRs are associated with hydrates and also the thickness of hydrate above the BSR. The lack of evidence for hydrates in the ODP wells drilled in the 1990s

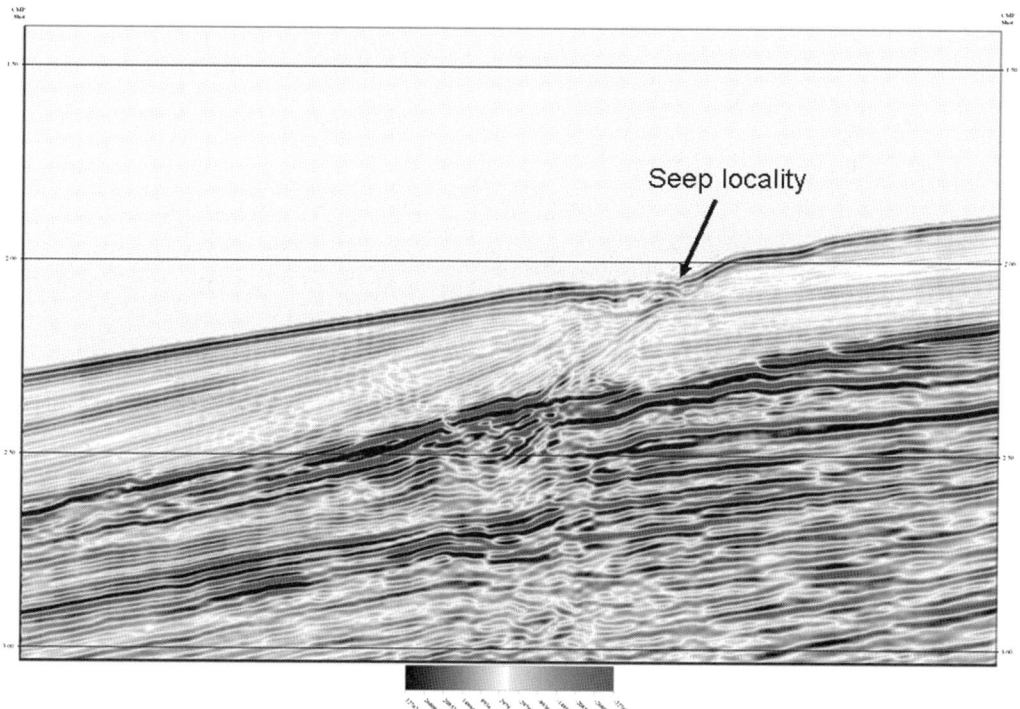

Fig. 10. Seismic line showing the location of the hydrate sample shown in Figure 9. The BSR is well developed and there is also an irregularity on the sea bed.

confirms the difficulties in estimating how much hydrate is present. More focused sampling needs to be done to confirm the distribution of hydrates and the volume in place. It is clear from the seismic data that there is a considerable supply of gas from the syn-rift interval and it is possible that a significant potential hydrate resource exists in the Namibe Basin in addition to conventional hydrocarbon resources.

Seabed morphology features evident from three-dimensional surveys are comparable with features seen on margins elsewhere where hydrates are an integral part of the development of the slope character. The slightly crinkled layer (Fig. 6) downdip of the moat is probably indicative of compression, reflecting an incipient slide. The low angled thrust feature visible in Figure 7 also supports this. Large slump features developed in the deep offshore that were imaged during Namibia's UNCLOS survey in 2004 confirm that the margin is unstable, although the continuity of the data is at present not good enough to prove a relationship of these slumps with BSRs and the known areas of hydrate occurrence. Upward doming of the seafloor may be the result of recent movement and has contributed further to the regional instability. The faults that are associated with the sea-bottom flexure may have provided further conduits for gas migration.

The gas is probably of mixed thermogenic and biogenic origin. The high organic productivity of the Benguela Ecosystem provides significant potential for the production of biogenic methane by anaerobic decay. During ODP Leg 175 only biogenic methane was identified from the high C1 : C2 ratio. There is, however, also abundant evidence for the presence of thermogenic methane in the stratigraphic column. Over 170 gas chimneys have been mapped on the three-dimensional seismic and these can be seen to be feeding the high-amplitude, free gas zones below the BSR. Although isotope analysis was not done on the hydrate sampled in 2004 as the sample was unexpected, further understanding of the origin of the hydrates and the quantity of hydrate present should be the major focus of any future research.

Namcor is thanked for allowing the author to attend the Geological Society Conference on hydrates while he was in their employ. WesternGeco and CGGVeritas are thanked for the data that used here. Adam Heffernan is thanked for the photograph of the hydrate shown in Figure 9.

References

BEN AVRAHAM, Z., SMITH, G., RESHEF, M. & JUNGSLAGER, E. 2002. Gas hydrate and mud volcanoes on the southwest African continental margin. *Geology*, **30**, 927–930.

BOOTH, J. S., WINTERS, W. J., DILLON, W. P., CLENNELL, M. B. & ROWE, M. M. 1998. Major occurrences and reservoir concepts of marine clathrate hydrates: Implications of field evidence. *In*: HENRIET, J.-P. & MIENERT, J. (eds) *Gas Hydrates: Relevance to World Margin Stability and Climate Change*. Geological Society, London, Special Publications, **137**, 113–127.

BROOKS, J. M., ANDERSON, A. L., SASSEN, R., MACDONALD, I. R., KENNICUTT II, I. C. & GUINASSO, L. 1994. Hydrate occurrences in shallow subsurface cores by continental slope sediments. *In*: SLOAN, E. D., HAPPEL, J. & HNATOW, M. A. (eds) *International Conference on Natural Gas Hydrates*. New York Academy of Sciences, New York, **715**, 381–391.

CUNNINGHAM, R. & LINDHOLM, R. M. 2000. Seismic evidence for widespread gas hydrate formation, offshore West Africa. *In*: MELLO, M. R. & KATZ, B. J. (eds) *Petroleum Systems of South Atlantic Margins*. American Association of Petroleum Geologists Memoirs, **73**, 93–105.

DINGLE, R. V. 1980. Large allochthonous sediment masses and their role in the construction of the continental slope and rise off southwestern Africa. *Marine Geology*, **37**, 333–354.

FICHLER, C., HENRIKSEN, S., RUESLATTEN, H. & HOVLAND, M. 2004. North Sea Quaternary morphology from seismic and magnetic data: Indications for gas hydrates during glaciation? *Petroleum Geoscience*, **11**, 331–337.

HAQ, B. U. 1998. Natural gas hydrates: Searching for the long-term climatic and slope-stability records. *In*: HENRIET, J.-P. & MIENERT, J. (eds) *Gas Hydrates: Relevance to World Margin Stability and Climate Change*. Geological Society, London, Special Publications, **137**, 303–318.

KVENVOLDEN, K. A. 1998. A primer on the geological occurrence of gas hydrates. *In*: HENRIET, J.-P. & MIENERT, J. (eds) *Gas Hydrates: Relevance to World Margin Stability and Climate Change*. Geological Society, London, Special Publications, **137**, 9–30.

KVENVOLDEN, K. A. & BARNARD, L. 1983. Hydrates of natural gas in continental margins. *In*: WATKINS, J. & DRAKE, C. (eds) *Studies in Continental Margin Geology*. American Association Petroleum Geologists, Memoirs, **34**, 631–640.

MEYERS, P. A. & SHIPBOARD SCIENTIFIC PARTY. 1998. Microbial gases in sediments from the southwest Africa margin. *In*: WEFER, G., BERGER, W. H. ET AL. (eds) *Proceedings of the Ocean Drilling Program Initial Reports*, **175**, 555–560.

PAULL, C. K., MATSUMOTO, R., WALLACE, P. J. & SHIPBOARD SCIENTIFIC PARTY. 1996. *Proceedings ODP Initial Reports*, **164**. Ocean Drilling Program, College Station, TX.

SULTAN, N., FOUCHER, J. P. ET AL. 2004. Dynamics of gas hydrate: Case of the Congo Continental slope. *Marine Geology*, **206**, 1–18.

SUMMERHAYES, C. P., BORNHOLD, B. D. & EMBLEY, R. W. 1979. Surficial slides and slumps on the continental slope and rise of South West Africa: A reconnaissance study. *Marine Geology*, **31**, 265–277.

Mapping hydrate stability zones offshore Scotland

A. P. CAMPS[1,2]*, D. LONG[3], C. A. ROCHELLE[1] & M. A. LOVELL[2]

[1]*The British Geological Survey, Keyworth, Nottinghamshire NG12 5GG, UK*
[2]*The University of Leicester, Department of Geology, Leicester LE1 7RH, UK*
[3]*The British Geological Survey, Edinburgh EH9 3LA, UK*
*Corresponding author (e-mail: apcamps@bgs.ac.uk)

Abstract: One practical method to reduce environmentally damaging greenhouse gas emissions is through the geological storage of carbon dioxide. Deep, warm storage of carbon dioxide is currently taking place at Sleipner, North Sea and Weyburn, Canada. It is, however, also possible to store carbon dioxide as a liquid and hydrate in cool, sub-seabed sediments. Offshore north and west of Scotland seafloor pressures and temperatures are suitable for hydrate formation. In addition to the possibility of natural methane hydrate being present in this region, conditions may also be favourable for carbon dioxide storage as a liquid and hydrate. A computer program has been developed to calculate the depth to the base of the carbon dioxide and methane hydrate stability zones in two offshore regions: the Faeroe–Shetland Channel and the northern Rockall Trough. Results predict that methane hydrate remains stable to a maximum depth of 650 m below the seabed in the Faeroe–Shetland Channel, and 600 m below the seabed in the northern Rockall Trough; the carbon dioxide hydrate stability zone extends below the seabed to a depth of 345 and 280 m, respectively. No physical evidence for the existence of natural hydrate in these regions has been confirmed. Suitable conditions for carbon dioxide storage as a liquid and hydrate exist, and should this storage method be developed further, a more refined program and greater offshore investigations to improve data sets would be necessary to scope the full potential.

Prediction of hydrate stability has been used as a useful tool in borehole drilling to determine the depths at which hydrate formation may become a problem. In more recent years, hydrate stability mapping has been used to identify regions for exploration and hydrate sampling. Miles (1995) provided the first map of the depth to the base of the methane hydrate stability zone off the European continental shelf, using a quadratic equation fitted to $P-T$ stability curves and a thinned crust model to estimate geothermal gradients. However, no such model exists to estimate carbon dioxide hydrate stability zones. The majority of natural hydrates are dominated by methane, though natural carbon dioxide hydrates have been discovered in the Okinawa Trough, offshore Japan (Sakai *et al.* 1990), and they may exist elsewhere in the solar system, such as on Mars or Jupiter's moon Europa (Miller & Smythe 1970; Duxbury *et al.* 2001, 2004; Prieto-Ballesteros *et al.* 2005). In addition, carbon dioxide hydrates are being researched for important greenhouse gas mitigation strategies (Kiode *et al.* 1997; House *et al.* 2006; Rochelle *et al.* 2009).

Greenhouse gas emissions continue to rise at an alarming rate, and at the current rate, global temperatures are predicted to rise by a total of 5.8 °C this century (Department of Trade and Industry 2003). The Kyoto Protocol in 1997 set targets to stabilize global temperatures, aiming to cap temperatures to 2 °C above pre-industrial levels. The UK agreed to aim for a reduction in greenhouse gas emissions by 12.5% below 1990 levels by 2008–2012; however, recent government announcements suggest these targets will not be met. Other than a global change in attitude and lifestyle, methods must be put into place to reduce atmospheric emissions. Although renewable energy sources are available, worldwide dependence on energy from fossil fuels does not appear to be in decline. A practical method of reducing environmentally damaging carbon dioxide emissions into the atmosphere is to store carbon dioxide in geological reservoirs (Holloway *et al.* 1996; IPCC 2005; Rochelle *et al.* 2009). This is a safe and effective method of preventing this greenhouse gas entering the atmosphere, storing it for thousands of years, and possibly over even longer time scales. The technology has been proven and deep, warm storage of supercritical carbon dioxide is already taking place at the Sleipner gas field in the North Sea (Baklid *et al.* 1996; IEA GHG 1998) and at the Weyburn field in Canada (Malik & Islam 2000; Wilson & Monea 2004). There is also another potential storage method that could offer additional storage sites: storage as a liquid and hydrate in cold sub-seabed sediments, beneath deep oceanic waters (Kiode et al. 1997; Rochelle et al. 2009).

In deep-water offshore regions, cold bottom-water temperatures could be favourable for storage of liquid carbon dioxide and solid carbon dioxide hydrate. Injection of slightly buoyant liquid carbon dioxide below its hydrate stability zone could lead to the formation of a hydrate 'cap', which would create an impermeable trapping mechanism.

Pressures and temperatures are suitable for hydrate formation offshore north and west of Scotland, therefore these offshore regions may also have suitable conditions for carbon storage as a liquid and hydrate. In addition to favourable conditions for carbon dioxide hydrate formation, natural methane hydrate may be present. These have received diverse attention, including their role in slope stability mechanisms, as a drilling hazard and as a possible future energy resource (Collett et al. 2000; Chatti et al. 2004). To determine the full potential of carbon stores, both natural and artificial, data have been collected for the northern Rockall Trough and the Faeroe–Shetland Channel as input into a computer program. This has been developed to calculate the depth to the base of the methane and carbon dioxide hydrate stability zones. The predicted hydrate thicknesses are discussed with a view to utilizing this approach elsewhere for the storage of carbon dioxide.

Oceanographic setting

The Rockall Trough lies to the west of the UK (British Isles) and is an intra-continental basin of deep water with depths increasing southwards to over 3500 m. The bathymetry is interrupted by igneous seamounts, such as Rosemary Bank and the Anton Dohrn seamount (Howe et al. 2006). To the NE of the Rockall Trough lies the Faeroe–Shetland Channel, separating the Faeroese plateau from the UK Continental Shelf. This channel reaches depths of 1500–2000 m in its northern entrance and is separated from the Rockall Trough by the Wyville–Thomson Ridge.

The oceanography of the study region is complex. The Faeroe–Shetland Channel surface waters consist of the North Atlantic Water (NAW). These warmer, more saline waters possibly originate from the Eastern North Atlantic Waters (ENAW) of the Rockall Trough, and therefore have similar temperatures (Fig. 1). Intermediate waters, lying at approximately 400–800 m, consist of Arctic Intermediate Water (AIW) originating

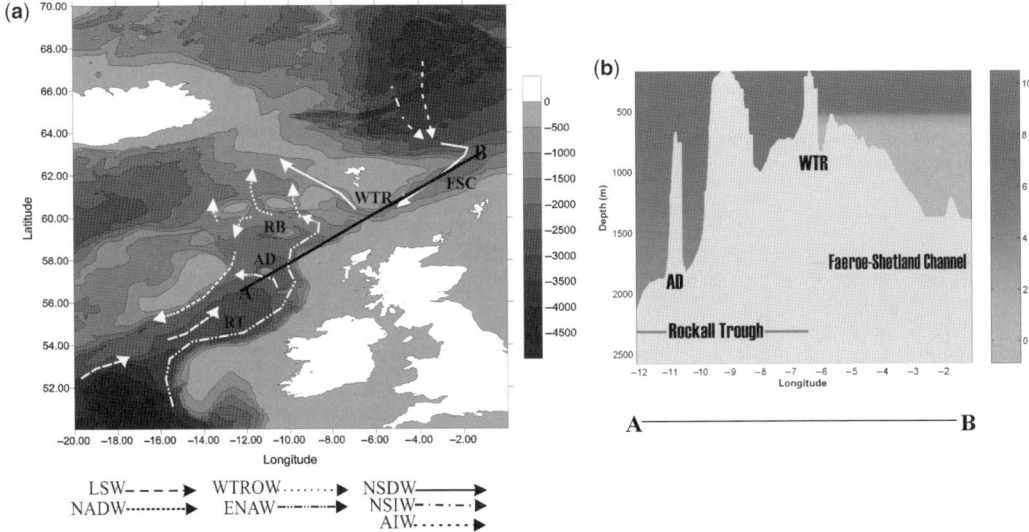

Fig. 1. (a) Bathymetry of the British and Icelandic continental shelves, showing the line of cross section, and regional water masses with current pathways (adapted from Howe et al. 2006). RB, Rosemary Bank seamount; AD, Anton Dohrn seamount; WTR, Wyville–Thomson Ridge, FSC, Faeroe–Shetland Channel; RT, Rockall Trough; LSW, Labrador Sea Water; NADW, North Atlantic Deep Water; WTROW, Wyville–Thomson Ridge Overflow Water; ENAW, East North Atlantic Water; NSDW, Norwegian Sea Deep Water; NSIW, Norwegian Sea Intermediate Water; AIW, Arctic Intermediate Water. (b) Cross section of bottom-water temperatures across the offshore study region, created from CTD data used in program predictions.

from north of the Iceland/Faeroe Ridge (Meincke 1978; Turrell et al. 1999), and Norwegian Sea Intermediate Water (NSIW), with a possible origin in the Arctic zone of the Iceland and Greenland Seas. The channel's bottom-waters consist of Faeroe Shetland Channel Bottom Water (FSCBW), which itself is formed from Norwegian Sea Deep Water (NSDW), originating from waters lying below 800 m in the Norwegian Sea, and overlying intermediate waters (Turrell et al. 1999). These waters are less saline and much colder, causing bottom-water temperatures to fall below 0 °C (Fig. 1). Minimal overflow of these colder waters may occur over the Wyville–Thompson Ridge, entraining warmer Atlantic waters, but the overwhelming volume leaves via the Faeroe Bank Channel.

The Rockall Trough surface waters consist of ENAW, extending to a depth of approximately 1200 m. Below this lies North Atlantic Deep Water (NADW), which itself consists of Labrador Sea Water (LSW), Antarctic Bottom Water (AABW) and NSDW. NADW forms by overflowing NSDW across the Wyville–Thompson Ridge, which entrains westward-flowing ENAW and northward-flowing LSW and AABW (Howe et al. 2006). These Atlantic water temperatures and salinities are considerably higher than those in the Faeroe–Shetland Channel, with bottom-waters reaching a minimum of approximately 2.6 °C.

Hydrate stability program construction

A simple empirical relationship between pure methane and pure water hydrate was provided in the JOIDES Pollution Prevention and Safety Panel Report (1992), where equilibrium is governed by the equation:

$$\ln P = A - B/T$$

where P is the pressure in kilopascals, T the temperature in Kelvin, and A and B are constants derived from experimental hydrate stability data. The JOIDES report provided A and B values of 38.53 and 8386.8, respectively, for a pressure range of 2505–11,556 kPa, and values of 46.74 and 10,748.1, respectively, for a pressure range of 11,556–80,000 kPa. Using this relationship a Visual Basic computer program, hydcalc.exe, was developed to enable calculation of the depth to the base of the hydrate stability zone (Long 1996, 2000). This program uses water depth (m), bottom-water temperature (°C) and geothermal gradient (°C/km), with an assumed hydrostatic pressure of 10.17 kg m^{-2}. A bottom water temperature offset of 1.1 °C was also included in the model to account for the reduction in hydrate stability when formed in water of seawater salinities (Dickens & Quinby-Hunt 1994). This program developed for methane hydrate has been modified to calculate carbon dioxide hydrate stability.

To modify hydcalc.exe, carbon dioxide hydrate stability $P-T$ diagrams were produced using the stability program CSMHYD.EXE of Sloan (1998) in order to determine the constants required for calculations. The natural log of pressure and the reciprocal of temperature were plotted against one another from data produced by Sloan's hydrate stability relationships, and the data divided into pressure ranges for linear constants to be determined. Constants derived for carbon dioxide hydrate stability are as follows:

1. Between 1317.17 and 3627.23 kPa, $A = 40.495$, $B = 9114.3$.
2. Between 3627.23 and 41,203.53 kPa, $A = 129.31$, $B = 34161$.
3. Between 41,203.53 and 114,431.6 kPa, $A = 60.566$, $B = 14372$.

Using these constants, the previous assumed hydrostatic pressure of 10.17 kg m^{-2} and assumed 1.1 °C temperature offset, a program to calculate the predicted depth to the base of the carbon dioxide hydrate stability zone has been developed: hycalcCO$_2$.exe.

Data acquisition

Data for stability calculations for the Faeroe–Shetland Channel and the Rockall Trough were provided by Dr David Long of the British Geological Survey. These deep water offshore regions were chosen as they provide the greatest potential for hydrate existence below waters around the UK, and data for these areas could be easily accessed. This data included GEBCO (Global Bathymetric Chart of the Oceans) bathymetry (IOC, IHO & BODC 2003), which is a 1 minute grid based largely on contours contained within the GEBCO digital atlas. The derivation of bathymetry from contours expressed problems in shallower areas, particularly at depths less than 100 m. However, both methane and carbon dioxide hydrates are not stable in such shallow waters, and therefore limit any errors present in this data. BGS regional surveys, including DigBath 250, were used to supplement GEBCO bathymetry, improving the data set locally. British Oceanographic Data Centre (BODC) with various other literature identified conductivity–temperature–depth (CTD) casts within the study area, which have been used to derive bottom-water temperatures (Long 2000). This provided a dataset with fairly consistent temperatures, expressing a standard deviation of less than 0.5 °C (Long 2000).

Geothermal gradient has been derived from downhole temperature measurements from released well data on the West Shetland Shelf and Slope. The calculated gradients in this region show some variation and appear to have a fairly random distribution, however not enough data were available to construct a grid of geothermal gradients for the study region. Using the available data a 30 °C/km geothermal gradient seemed to approximate observations, and was used in early estimates of hydrate for the Faeroe–Shetland Channel (Long 1996). Although it is acknowledged that this provides limitations in the model, this estimated geothermal gradient was used throughout the calculated stability predictions.

Predicted hydrate stability zones

The region of the Faeroe–Shetland Channel studied covers an offshore area between 59.5–62.5°N and 0–7°W (Fig. 2). The Rockall Trough study area is larger between 56–61°N and 6–14°W (Fig. 3).

Data used to calculate hydrate stability zones in the Faeroe–Shetland Channel have a minimum bottom-water temperature of −0.85 °C, and a maximum water depth of 1763 m. This compares with a minimum bottom-water temperature of 2.62 °C in the Rockall Trough, and a maximum water depth of 3397 m. Bottom-water temperatures and water depths have been plotted as filled contour plots to enable comparison with predicted hydrate thickness (Figs 2 & 3).

Methane hydrate stability

The predicted methane hydrate stability zones for these two offshore regions have been plotted using the contouring package SURFER (Fig. 4). In the Faeroe–Shetland Channel the methane hydrate stability model estimated a maximum depth of

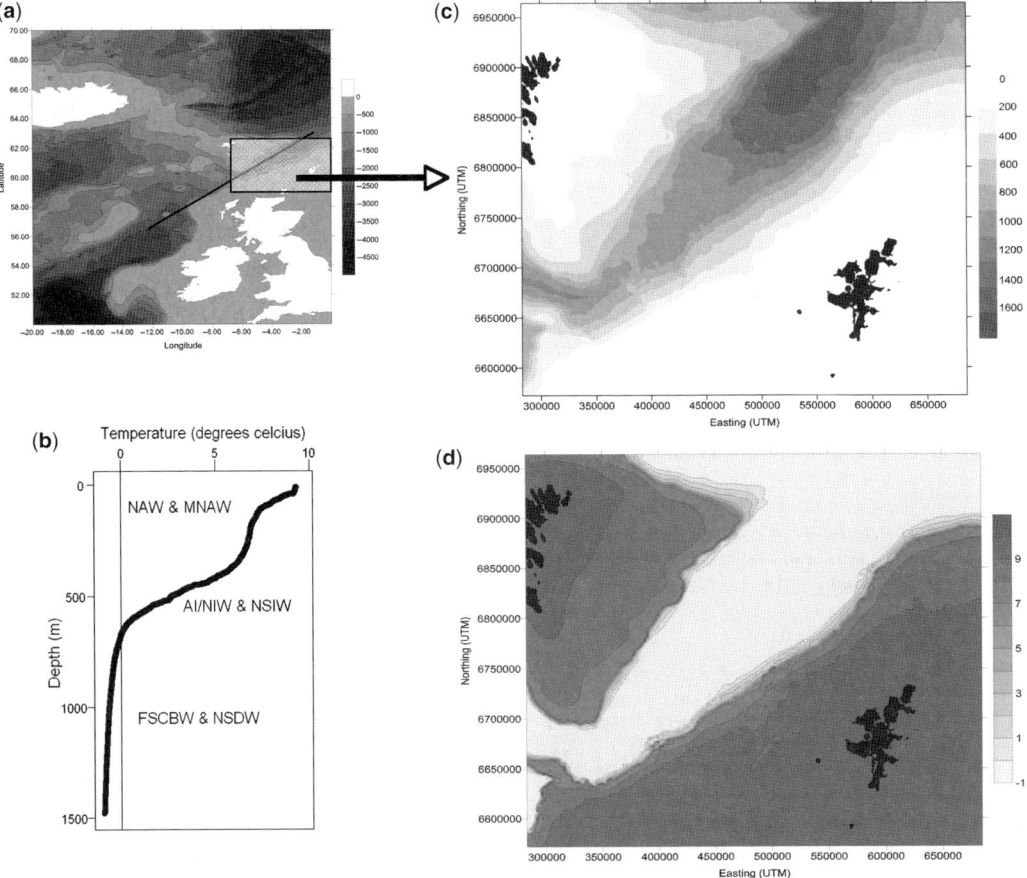

Fig. 2. (**a**) Bathymetric map offshore UK showing the position of the study area north of Scotland. (**b**) Faeroe–Shetland Channel CTD temperature profile with annotated regional water masses. (**c**) Faeroe–Shetland Channel contoured bathymetry. (**d**) Contoured bottom-water temperatures taken from used dataset.

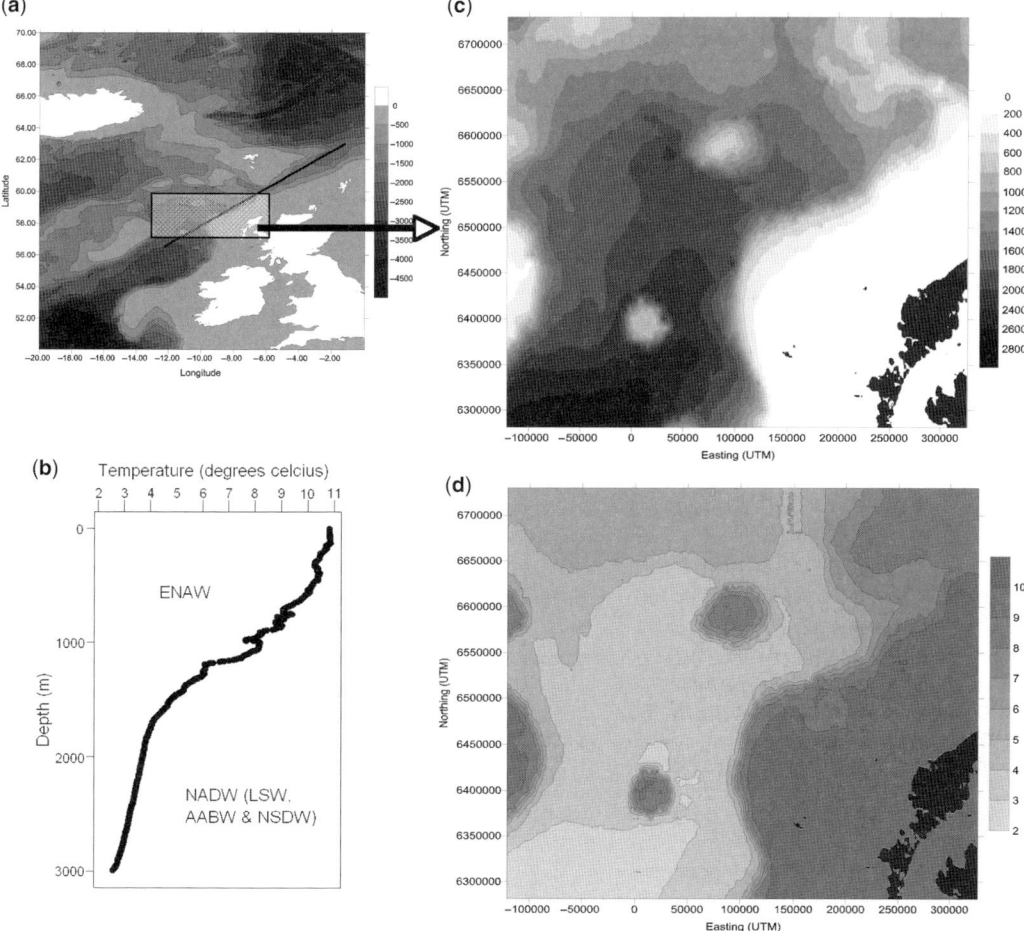

Fig. 3. (**a**) Bathymetric map offshore UK showing the position of the study area west of Scotland. (**b**) Northern Rockall Trough CTD temperature profile with annotated regional water masses. (**c**) Northern Rockall Trough contoured bathymetry. (**d**) Contoured bottom-water temperatures taken from Rockall Trough dataset.

682 m below the seabed to the base of the hydrate stability zone. This compares with a maximum depth of 668 m below the seabed in the Rockall Trough study region.

Carbon dioxide hydrate stability

The predicted carbon dioxide hydrate stability zones for the two offshore regions show a maximum depth of 346 m below the seabed to the base of the hydrate stability zone, in the Faeroe–Shetland Channel. This compares with a maximum depth of 281 m below the seabed in the Rockall Trough (Fig. 5).

Carbon dioxide and methane hydrate stability zones extend to greatest depth at the northern end of the Faeroe–Shetland Channel and the southern end of the Rockall Trough. Here water depths increase and bottom-water temperatures decrease. Although water depths are greater in the Rockall Trough, the stability zones extend to greater depths in the Faeroe–Shetland Channel due to much colder bottom-waters north of the Wyville–Thomson Ridge.

Although carbon dioxide hydrate is often considered to be more stable than methane hydrate in the laboratory environment, and is easier to produce than methane hydrate, it has a narrower predicted hydrate stability zone. Carbon dioxide hydrate is more stable at low pressures and temperatures, but with increase in pressure, and especially temperature, with depth below the seabed, methane hydrate becomes relatively more stable (Fig. 9), explaining the difference in predicted hydrate thicknesses displayed (Figs 4 & 5).

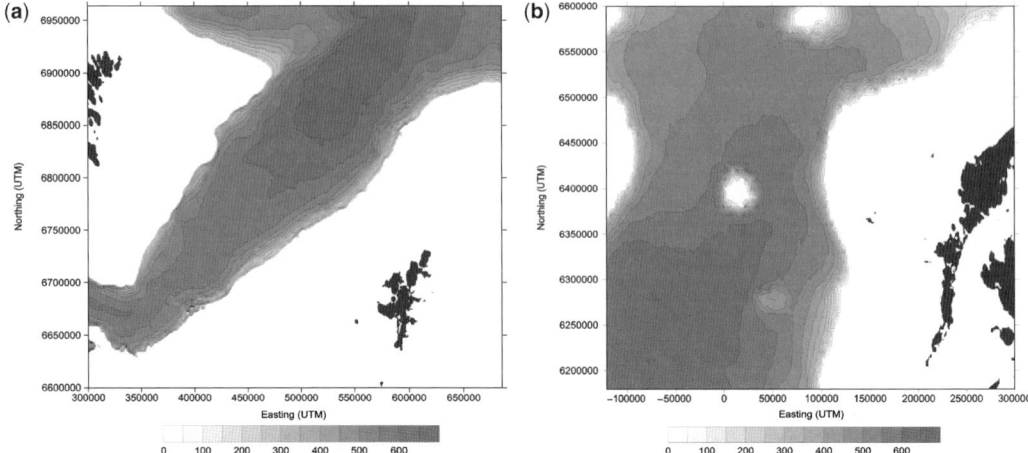

Fig. 4. Depth to the base of the methane hydrate stability zone below the seabed (m) in the Faeroe–Shetland Channel (**a**) and the Rockall Trough (**b**). Contours are spaced at 50 m intervals.

Discussion

Program predictions for the depth to the base of the methane hydrate stability zone show direct comparison with previous calculations. Miles (1995) predicted the base of the methane hydrate to lie between 200 and 250 m, using a thinned crust model to estimate continental geothermal gradient (approximately 80 °C/km offshore north and west of Scotland). In this study a geothermal gradient of 30 °C was estimated. If the developed model hydcalc.exe uses a geothermal gradient of 80 °C/km, the maximum methane hydrate thickness is predicted to be 236 m, within the same range as that predicted by Miles (1995). However, although the methane hydrate algorithm compares well with other estimates above, it is important to consider any inaccuracies that may affect stability zone calculations. There is no direct comparison for carbon dioxide hydrate thickness predictions, although rough approximations can be calculated using dissociation data and predictions appear relatively accurate.

Use of CSMHYD for pressure predictions

The basis for Sloan's hydrate prediction model (Sloan 1998) originates from the basic model

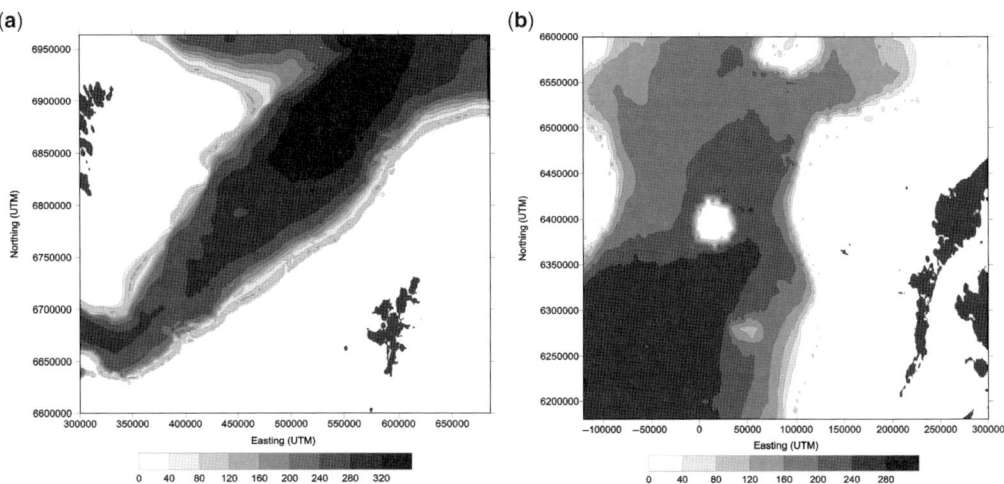

Fig. 5. Depth to the base of the carbon dioxide stability zone below the seabed (m) in the Faeroe–Shetland Channel (**a**) and the Rockall Trough (**b**). Contours are spaced at 40 m intervals.

produced by van der Waals & Platteeuw (1959), in which the phase equilibrium of gas hydrate was derived from statistical thermodynamics (later used for prediction of dissociation pressures of gas hydrates; Parrish & Prausnitz 1972). The Parrish and Prausnitz model was simplified by introducing universal reference properties for each type of hydrate structure (Holder *et al.* 1980). All models produced from these origins give relatively consistent results, and are comparable with experimental data (Sloan 1998). Model assumptions may introduce some errors. For example, the model assumes spherical molecules, and although the model incorporates algorithms that provide better predictions for non-spherical molecules, it tends to overestimate cage occupancy (Sun & Duan 2005). Sloan predicts an average pressure prediction deviation of not less than $\pm 7\%$ for hydrate results using CSMHYD, and any such inaccuracies are within an acceptable range for many calculations (Sloan 1998). Figure 6 compares carbon dioxide hydrate $P-T$ predictions using CSMHYD with experimental data from various sources. A degree of variation occurs, but true inaccuracy of model predictions cannot be fully determined as experimental errors must also be taken into account.

Constants derived from CSMHYD

In order to use the JOIDES linear equilibrium equation data for the calculation of hydrate thickness, linear regression of the carbon dioxide hydrate logarithmic pressure predictions for a range of temperature reciprocals was necessary. The original methane constants used for hydrate thickness predictions were derived from two linear equations, as the logarithm of methane stability data provides values which lie on a linear line of best fit. However, the logarithm of carbon dioxide hydrate pressure predictions is more complicated, and therefore linear regression is more difficult (Fig. 7).

The hydrate stability data was divided into three pressure ranges, providing three linear regression equations. As these linear equations were derived from lines of best fit, it is possible that errors were introduced at this point. These were plotted as error bars on pressure predictions derived from CSMHYD.exe (Fig. 7). The maximum logarithmic pressure derivation is $+0.37$, lying within the middle pressure range, or second linear equation. This compares with a maximum pressure derivation of $+3947$ kPa lying within data for the third pressure range. However, although these pressure derivations appear to be significant, both positive and negative differences can be seen between the model $P-T$ equilibrium and that of CSMHYD.exe, therefore both overestimate and underestimate hydrate stability zone thickness at different $P-T$ conditions. As the model overestimates and underestimates hydrate stability zone thickness, the differences may average out, reducing the effect of equilibrium errors over large regions.

Sensitivity to pressure, temperature and geothermal gradient

The carbon dioxide hydrate prediction program is sensitive to temperature changes, calculating a 36 m decrease in hydrate thickness per 1 °C increase in temperature, when well within its stability field. The methane hydrate prediction model shows a sensitivity of -38 m per 1 °C increase in temperature.

Hydrate stability sensitivity increases with decreasing pressure. If the pressure decreases from 3800 to 3700 kPa the program predicts a 4 m decrease in carbon dioxide hydrate thickness. However, decreasing the pressure from 2200 to 2100 kPa decreases the carbon dioxide hydrate thickness by 7 m. This compares with 7 and 10 m, respectively, for methane hydrate.

The developed models show that hydrate stability is highly sensitive to the geothermal gradient and therefore any increase would significantly reduce the resulting thickness of the hydrate layer. However, the model's sensitivity is highly variable. For the original methane hydrate stability equation it was proposed that there was an uncertainty of $\pm 1\%$ in the total (water and sediment) depth estimate for ± 0.1 °C (JOIDES Pollution Prevention and Safety Panel Report 1992). Using hydcalc.exe, and altering the data file for the Faeroe–Shetland Channel to account for a 1° rise in geothermal gradient per kilometre, the average reduction in the methane hydrate stability zone would be 3%, with a standard deviation of 0.3. In this case, this reduction is the

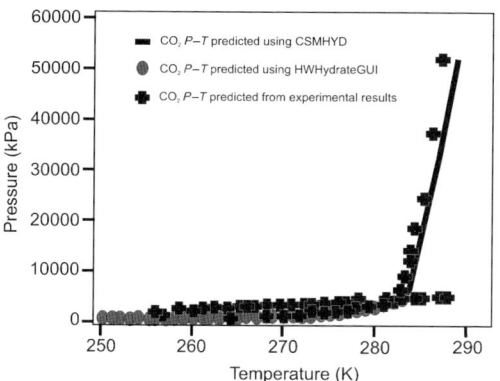

Fig. 6. Graphical comparison of carbon dioxide $P-T$ predicted using CSMHYD, experimental results from various sources provided in Sloan (1998) and HWHydrateGUI.

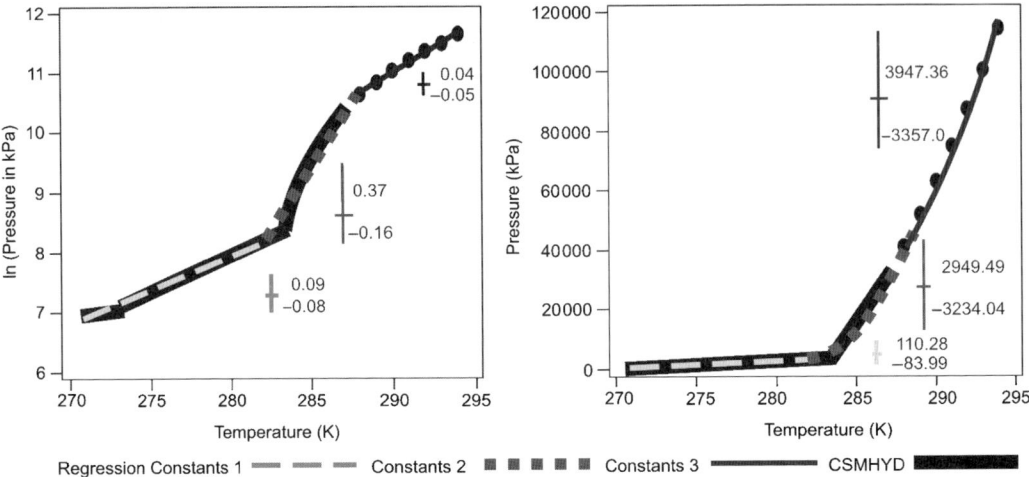

Fig. 7. Pressure predictions for carbon dioxide hydrate results from Sloan's model CSMHYD and their comparison with constants derived from regression of such data used to construct carbon dioxide hydrate stability model. Error bars have been added to express the maximum and minimum pressure derivation of regressed equation data from original model predictions.

equivalent of an average 5.6 m, with a standard deviation of 7.7.

Geothermal gradient

Geothermal gradients collected from downhole temperature measurements (measured at less than 1 km from the seafloor) show wide variation with a fairly random distribution, ranging from 28 to 67 °C/km (Fig. 8). Re-examination of the available geothermal gradients after calculations used to produce maps in this study, with the additional use of further downhole measurements, determined an average gradient of 45 °C/km, with a standard deviation of 11 °C/km. Although the 30 °C/km gradient used in these calculations lies within 2 standard deviations of the mean, it does lie at the lower end of the main range of measured geothermal gradients

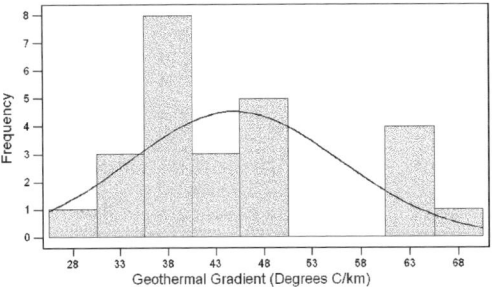

Fig. 8. Histogram of calculated geothermal gradients offshore north and west of Scotland with an added normal curve.

(Fig. 8) and the consequence of using this single value would be that calculations would tend to overestimate the thickness of the hydrate stability zones. More recent research (Long et al. 2005) suggests that a geothermal gradient of 45 °C/km may be more appropriate. Re-calculation of hydrate thickness using the average regional geothermal gradient would decrease the predicted hydrate stability zones. The depth to the base of the methane hydrate stability zone would reduce from 682 to 436 m in the Faeroe–Shetland channel, and from 668 to 435 m in the Rockall Trough at its most southerly point. This would compare with depths to the base of the carbon dioxide hydrate stability zone reducing from 346 to 225 m and from 281 to 184 m, respectively.

As the program is extremely sensitive to geothermal gradient variations the accuracy of these values continues to represent the greatest level of uncertainty in the model, and may be particularly significant at a localized scale. A map of accurate geothermal gradient in the study area, at the appropriate depth below seafloor, would be necessary for improved hydrate thickness predictions, although this is beyond the scope of this project. Such a map would be extremely difficult to derive due to the lack of detailed geothermal gradient data, and therefore the hydrate model used within this study is still a useful predictive tool.

1.1 °C offset

The model incorporates a previously assumed temperature offset defined for methane hydrate stability in seawater as compared with pure water. This 1.1 °C offset was determined from the experimental dissociation pressure of methane hydrate in the pressure range of 2.75–10.0 MPa (Dickens & Quinby-Hunt 1994). However, by using CSMHYD.exe (Sloan 1998) to determine carbon dioxide hydrate stability in seawater (3.5% by weight), it was found that a variable temperature offset would be necessary to simulate the depression in dissociation temperature with the increase in salt content for carbon dioxide (Fig. 9).

Initially, in Figure 9, the difference in the dissociation temperature is minimal, but this increases as carbon dioxide hydrate reaches its liquid-water–liquid hydrocarbon phase. At this point (about 450 m depth), the temperature offset becomes fairly constant at approximately 1.9 °C.

Using previous estimates of program temperature sensitivity, this increase in temperature offset would account for an approximate decrease of 29 m in hydrate stability, at depths greater than 450 m (with presented results therefore overestimating hydrate thickness). Future model refinement should incorporate regression of data derived from carbon dioxide hydrate formation in seawater.

Evidence for natural methane hydrate

There appears to be no direct evidence for natural methane hydrate in the study area, but this may be a consequence of a lack of pressurized core sampling (Long 2000). Mud diapirs have been discovered in the northern end of the Faeroe–Shetland Channel (Long 2000; Holmes et al. 2003), and similar structures have been associated with gas hydrates in other explored regions. No classic bottom-simulating reflectors have been discovered in the Faeroe–Shetland Channel or the Rockall Trough; however, an anomalous reflector has been mapped in the northern section of the Faeroe–Shetland Channel. This has been found to cross-cut other reflectors and appears at approximately 300–400 m below the seabed (Long & Holmes 1998). Doubts as to whether this reflector could be attributed to gas trapped below methane hydrate have proven to be correct with Davies & Cartwright (2002) demonstrating that the cross-cutting seismic reflector is an example of a diagenetic fossilized Opal A to Opal C/T (Cristobalite/Tridymite) transition. There have been instances of hydrate formation on the seabed resulting from artificial releases of deep methane gas (Long et al. 2005). This confirms that conditions are suitable for hydrate formation, as they are in most deep-water areas, although natural samples remain undiscovered.

A complicated network of currents present in this offshore region creates variation in bottom-water temperatures (and salinities), which could influence hydrate stability if natural methane hydrate existed in near-surface seafloor sediments. Mixing water masses, particularly between Faeroese intermediate and bottom-waters, can lead to the development of internal eddies and waves, causing large temperature changes. Long-term oceanographic monitoring also indicates compositional changes in the Faeroe–Shetland Channel bottom water over time scales of as little as

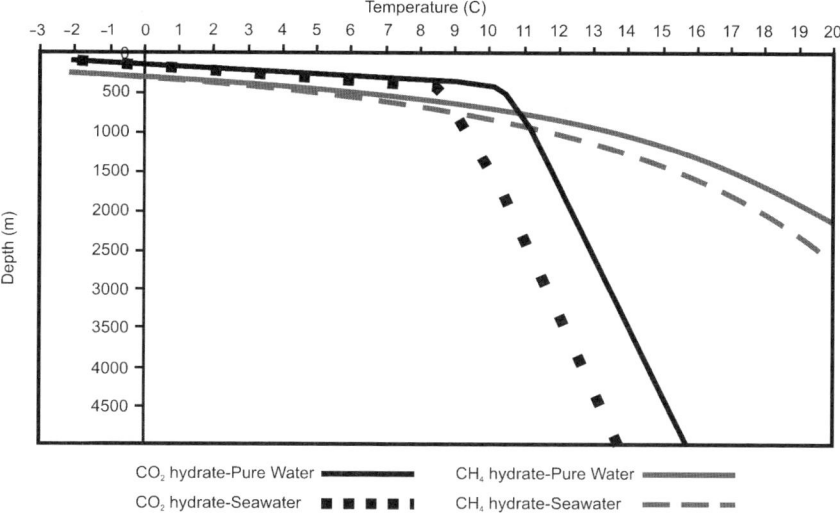

Fig. 9. Depth–temperature stability diagram for CO_2 hydrate and CH_4 hydrate in pure water and seawater (pressure represented as depth, m), comparing the difference in $P–T$ equilibrium with pure water and seawater, and the changing offset with pressure (Sloan 1998).

10 years, with a NSDW reduction of approximately 20% in bottom waters (Turrell et al. 1999). The Wyville–Thompson Ridge overflow also shows variation with the volume of water overflowing it from the Faeroe–Shetland channel to the Rockall Trough; the water alters considerably, and mixing in descending overflow plumes can cause a bottom-water temperature increase of c. 3 °C (Sherwin & Turrell 2005). Such fluctuations would influence the stability of hydrate at the seabed, and therefore it is unlikely that natural hydrate would be found in areas with a shallow hydrate stability zone.

Although pressures and temperatures are suitable for hydrate formation and calculations predict a thick methane gas hydrate stability zone, greater exploration in the Faeroe–Shetland Channel and the Rockall Trough would be necessary to fully determine whether natural gas hydrate is present. Possible methane sources in the two regions have not been investigated within this study, and may help to pin-point suitable areas for further exploration.

Carbon dioxide storage potential

Maximum predicted carbon dioxide hydrate thickness in the Rockall Trough and the Faeroe–Shetland Channel are 280 and 345 m, respectively for a 30 °C/km geothermal gradient. Therefore, pressures and temperatures do appear to be in the range required for carbon dioxide storage as a liquid and hydrate, and a significant layer of carbon dioxide hydrate could be formed if all conditions were favourable. Borehole records show that sediments within and near the studied areas may have suitable porosity for hydrate formation (BGS pers. comm. 2004). Research shows that a reduced porosity limits hydrate formation (Booth et al. 1998; Clennell et al. 2000), and a sandstone reservoir would be advantageous for storage. Sandy/sandstone horizons do appear on borehole logs within the predicted hydrate stability zone, though more thorough investigation would be necessary to fully comprehend the storage potential.

The studied regions are a fair distance offshore, and pipeline cost from the coast would be expensive. However, should offshore platforms be constructed, these potential storage sites would increase their appeal. Alternative regions with storage potential may also occur offshore Western Europe, with thick hydrate stability zones relatively close to shore (see Rochelle et al. 2009), and therefore more easily accessible.

Conclusions

Models have been developed to calculate the predicted depth to the base of the carbon dioxide hydrate and methane hydrate stability zone, and to input into the developed program, have been collected within two regions offshore the north and west of Scotland: the Faeroe–Shetland Channel and the Rockall Trough. The methane hydrate stability zone is predicted to reach a maximum of 682 m below the seabed in the Faeroe–Shetland Channel, and 668 m in the northern Rockall Trough. No physical evidence has been determined to support the existence of natural methane hydrate stores in these regions, although the presence of mud diapirs in the Faeroe–Shetland channel may indicate otherwise.

The carbon dioxide hydrate stability zone is predicted to extend to a depth of 346 m beneath the seafloor in the Faeroe–Shetland Channel, and to a depth of 281 m in the northern Rockall Trough. Therefore, conditions may be suitable for carbon dioxide storage as a liquid and hydrate. This storage method offers many advantages over existing methods, including a reduced chance of migration to surface sediments. Should this storage method be further developed, a more refined stability model and greater exploration of study areas would be necessary to scope the full potential of offshore zones.

A. P. Camps acknowledges the Natural Environment Research Council for funding under grant NER/S/A/2003/11923, the Society of Petrophysicists and Well-Log Analysts (SPWLA) and the BGS–University Collaboration Advisory Committee. This paper is published with the permission of the executive director of the British Geological Survey.

References

BAKLID, A., KORBØ, L. R. & OWREN, G. 1996. Sleipner Vest CO_2 disposal, CO_2 injection into a shallow underground aquifer. *Society of Petroleum Engineers*, **36600**, 269–277.

BOOTH, J. S., WINTERS, W. J., DILLON, W. P., CLENNELL, M. B. & ROWE, M. M. 1998. Major occurrences and reservoir concepts of marine clathrate hydrates: Implications of field evidence. *In*: HENRIET, J.-P. & MIENERT, J. (eds) *Gas Hydrates, Relevance to World Margin Stability and Climatic Change*. Geological Society, London, Special Publications, **137**, 113–127.

CHATTI, I., DELAHAYE, A., FOURNAISON, L. & PETITET, J-P. 2004. Benefits and drawbacks of clathrate hydrates: A review of their areas of interest. *Energy Conservation and Management*, **46**, 1333–1343.

CLENNELL, M. B., HENRY, P., HOVLAND, M., BOOTH, J. S., WINTERS, W. J. & THOMAS, M. 2000. Formation of natural gas hydrates in marine sediments: Gas hydrate growth and stability conditioned by host sediment properties. *In*: HOLDER, G. D. & BISHNOI, P. R. (eds) *Gas Hydrates: Challenges for the Future*. Annals of the New York Academy of Sciences, **912**, 887–896.

COLLETT, T. S., LEWIS, R. & UCHIDA, T. 2000. Growing interest in gas hydrates. *Oilfield Review*, Summer, 42–57.

DAVIES, R. J. & CARTWRIGHT, J. 2002. A fossilized Opal A to Opal C/T transformation on the northeast Atlantic margin: Support for a significantly elevated

Palaeogeothermal gradient during the Neogene? *Basin Research*, **14**, 467–486.

DICKENS, G. R. & QUINBY-HUNT, M. S. 1994. Methane hydrate stability in seawater. *Geophysical Research Letters*, **21**, 2115–2118.

DEPARTMENT OF TRADE AND INDUSTRY. 2003. *Energy White Paper: Our Energy Future – Creating a Low Carbon Economy*. The Stationary Office, London.

DUXBURY, N. S., NEALSON, K. H. & ROMANOVSKY, V. E. 2001. On the possibility of clathrate hydrates on the Moon. *Journal of Geophysical Research*, **106**, 27, 811–827.

DUXBURY, N. S., ABYZOV, S. S., ROMANOVSKY, V. E. & YOSHIKAWA, K. 2004. A combination of radar and thermal approaches to search for methane clathrate in the Martian subsurface. *Planetary and Space Science*, **52**, 109–115.

HOLDER, G. D., CORBIN, G. & PAPADOPOULOS, K. D. 1980. Thermodynamic and molecular properties of gas hydrates from mixtures containing methane, argon and krypton. *Industrial and Engineering Chemical Fundamentals*, **19**, 282–286.

HOLLOWAY, S., HEEDERICK, J. P. ET AL. 1996. *The Underground Disposal of Carbon Dioxide: Summary Report*. British Geological Survey, Keyworth.

HOLMES, R., HOBBS, P. R. N. ET AL. 2003. *DTI Strategic Environmental Assessment Area 4(SEA4): Geological Evolution of Pilot Whale Diapirs and Stability of the Seabed Habitat*. British Geological Survey Commercial Report **CR/03/082**.

HOUSE, K. Z., SCHRAG, D. P., HARVEY, C. F. & LACKNER, K. S. 2006. Permanent carbon dioxide storage in deep-sea sediments. *Proceedings of the National Academy of Science, USA*, **103**, 12291–12295.

HOWE, J. A., STOKER, M. S. ET AL. 2006. Seabed morphology and the bottom-current pathways around Rosemary Bank seamount, northern Rockall Trough, North Atlantic. *Marine and Petroleum Geology*, **23**, 165–181.

IEA GHG. 1998. Sleipner aquifer storage of CO_2. *Greenhouse Issues*, **34**, 4.

IOC, IHO & BODC. 2003. *Centenary Edition of the GEBCO Digital Atlas*. Published on CD-ROM on behalf of the Intergovernmental Oceanographic Commission and the International Hydrographic Organization as part of the General Bathymetric Chart of the Oceans. British Oceanographic Data Centre, Liverpool.

IPCC. 2005. *Carbon Dioxide Capture and Storage*. Cambridge University Press, New York.

JOIDES POLLUTION PREVENTION AND SAFETY PANEL REPORT. 1992. Ocean Drilling Program Guidelines for Pollution Prevention and Safety. *Joides Journal* **18** (Special Issue No. 7).

KIODE, H., TAKAHASHI, M. ET AL. 1997. Hydrate formation in sediments in the sub-seabed disposal of CO_2. *Energy*, **22**, 279–283.

LONG, D. 1996. *Hydrates and their Potential Existence in the Faeroe–Shetland Channel*. British Geological Survey Technical Report **WB/96/34**.

LONG, D. 2000. *The Methane Hydrate Stability Zone North and West of Scotland*. British Geological Survey Technical Report **CR/00/104C**.

LONG, D. & HOLMES, R. 1998. Methane hydrates in the Faeroe–Shetland Channel? Abstract volume. *Geoscience*, **98**, 28.

LONG, D., JACKSON, P. D., LOVELL, M. A., ROCHELLE, C. A., FRANCIS, T. J. G. & SCHULTHEISS, P. J. 2005. Methane hydrates: Problems in unlocking their potential. *In: Petroleum Geology: North–West Europe and Global Perspectives – Proceedings of the 6th Petroleum Geology Conference*, **1**, 723–730

MALIK, Q. M. & ISLAM, M. R. 2000. CO_2 injection in the Weyburn field of Canada: Optimization of enhanced oil recovery and greenhouse gas storage with horizontal wells. *Society of Petroleum Engineers*, **59327**.

MEINCKE, J. 1978. On the distribution of low salinity intermediate water around the Faroes. *Deutsche Hydrographische Zeitschrift*, **31**, 50–64.

MILES, P. R. 1995. Potential distribution of methane hydrate beneath the European continental margins. *Geophysical Research Letters*, **22**, 3179–3182.

MILLER, S. L. & SMYTHE, W. D. 1970. Carbon dioxide clathrate in the Martian Ice Cap. *Science*, **170**, 531–533.

PARRISH, W. R. & PRAUSNITZ, J. M. 1972. Dissociation pressures of gas hydrates formed by gas mixtures. *Industrial and Engineering Chemistry Process Design and Development*, **11**, 26–35.

PRIETO-BALLESTEROS, O., KARGEL, J. S., FERNANDEZ-SAMPEDRO, M., SELSIS, F., MARTINEZ, E. S. & HOGENBOOM, D. L. 2005. Evaluation of the possible presence of clathrate hydrates in Europa's icy shell or seafloor. *Icarus*, **177**, 491–505.

ROCHELLE, C. A., CAMPS, A. P. ET AL. 2009. Can CO_2 hydrate assist in the underground storage of carbon dioxide? *In*: LONG, D., LOVELL, M. A., REES, J. G. & ROCHELLE, C. A. (eds) *Sediment-Hosted Gas Hydrates: New Insights on Natural and Synthetic Systems*. Geological Society, London, Special Publications, **319**, 171–183.

SAKAI, H., GAMO, T. ET AL. 1990. Venting of carbon dioxide-rich fluid and hydrate formation in Mid-Okinawa Trough backarc basin. *Science*, **248**, 1093–1095.

SHERWIN, T. J. & TURRELL, W. R. 2005. Mixing and advection of a cold water cascade over the Wyville Thomson Ridge. *Deep-Sea Research I*, **52**, 1392–1413.

SLOAN, E. D., JR. 1998. *Clathrate Hydrates of Natural Gases*. Marcel Dekker, New York.

SUN, R. & DUAN, Z. 2005. Prediction of CH_4 and CO_2 hydrate phase equilibrium and cage occupancy from ab initio intermolecular potentials. *Geochimica et Cosmochimica Acta*, **69**, 4411–4424.

TURRELL, W. R., SLESSER, G., ADAMS, R. D., PAYNE, R. & GILLIBRAND, P. A. 1999. Decadal variability in the composition of Faroe Shetland Channel bottom water. *Deep-Sea Research I*, **46**, 1–25.

VAN DER WAALS, J. H. & PLATTEEUW, J. C. 1959. Clathrate solutions. *Advances in Chemical Physics*, 1–57.

WILSON, M. & MONEA, M. 2004. IEA GHG Weyburn CO_2 Monitoring and Storage project Summary Report 2000–2004. *Proceedings of the 7th International Conference on Greenhouse Gas Control Technologies*, 5–9 September 2004, Vancouver, **3**.

The pore-scale distribution of sediment-hosted hydrates: evidence from effective medium modelling of laboratory and borehole seismic data

T. A. MINSHULL[1]* & S. CHAND[2]

[1]*National Oceanography Centre, European Way, Southampton SO14 3ZH, UK*

[2]*Geological Survey of Norway (NGU), Tromsøkontoret, Polarmiljøsenteret, 9296, Tromsø, Norway*

*Corresponding author (e-mail: tmin@noc.soton.ac.uk)

Abstract: Much of our knowledge on hydrate distribution in the subsurface comes from interpretations of remote seismic measurements. A key step in such interpretations is an effective medium theory that relates the seismic properties of a given sediment to its hydrate content. A variety of such theories have been developed; these theories generally give similar results if the same assumptions are made about the extent to which hydrate contributes to the load-bearing sediment frame. We have further developed and modified one such theory, the self-consistent approximation/differential effective medium approach, to incorporate additional empirical parameters describing the extent to which both the sediment matrix material (clay or quartz) and the hydrate are load-bearing. We find that a single choice of these parameters allows us to match well both *P* and *S* wave velocity measurements from both laboratory and *in situ* datasets, and that the inferred proportion of hydrate that is load-bearing varies approximately linearly with hydrate saturation. This proportion appears to decrease with increasing hydrate saturation for gas-rich laboratory environments, but increases with hydrate saturation when hydrate is formed from solution and for an *in situ* example.

In order to assess the significance of gas hydrate as a resource, as a hazard and as an agent in climate change, it is necessary to determine how much hydrate is present in subsurface sediments. The most reliable measure of hydrate saturation comes where there is direct sampling and core material is recovered under pressure; hydrate saturations may then be estimated accurately from the volume of methane evolved during decompression (e.g. Dickens *et al.* 1997). Such direct sampling provides essential 'ground truth' for geophysical estimates of hydrate saturation, but is expensive and can only be carried out in a very few locations that may not be representative. Therefore, for most of our information on hydrate volumes we must rely on remote geophysical methods. The replacement of conductive (normally saline) pore water with resistive hydrate in the pore space can lead to significant anomalies in electrical resistivity, and some progress has been made in the measurement of such anomalies (e.g. Schwalenberg *et al.* 2005). In principle, hydrate saturations may also be estimated using sensitive measurements of the response of the seafloor to tidal variations in seafloor pressure (Latychev & Edwards 2003). However, seismic techniques remain the primary remote methods used to determine hydrate content of the subsurface.

With an appropriately designed experiment, detailed knowledge may be obtained of *P* and *S* wave velocities within the hydrate stability zone (e.g. Singh *et al.* 1993; Hobro *et al.* 2005; Westbrook *et al.* 2005).

A key step in the process of remotely determining hydrate content is a quantitative relationship between that content and the physical properties measured, namely seismic velocities. The elastic properties of the individual components of hydrate-bearing sedimentary rocks (water, hydrate, quartz, clay minerals, etc.) are generally well known, although those of clay minerals are difficult to measure and normally must be determined indirectly (e.g. Hornby *et al.* 1994). Methods that combine these component properties into the properties of the composite material are called 'effective medium theories', and a key factor in the predictions of such theories is the extent to which hydrate is assumed to contribute to the strength of the sediment frame. In this paper we discuss briefly the advantages and limitations of such theories and then focus on the application of one particular theoretical approach to a range of datasets where the hydrate content is known from independent observations. The approach is described in more detail by Chand *et al.* (2006). Here we summarize the results of Chand *et al.* (2006) and further develop some of their ideas.

Effective medium theories

The most widely used effective medium theories were reviewed and compared by Chand *et al.* (2004). The simplest of such theories are essentially empirical correlations, such as the 'weighted equation' approach applied to hydrate-bearing sediments by Lee *et al.* (1996). An advantage of this approach is that it can be readily adjusted to match any dataset, but because it has no physical basis, its application leads to little understanding of the physics involved and it is difficult to apply to areas away from or sediment types distinct from those in which the correlations have been developed. Therefore many authors prefer more sophisticated approaches that have some physical basis. Ideally such rock physics-based approaches would have no adjustable parameters other than the unknown hydrate content. However, in the case of real sedimentary rocks, even if parameters such as composition and porosity are perfectly known, assumptions must be made about the way the different components are organized and generally these assumptions must be modified in some way to fit real data – hence these approaches retain empirical elements.

The rock physics-based approaches reviewed by Chand *et al.* (2004) include the self-consistent approximation/differential effective medium (SCA/DEM) approach of Jakobsen *et al.* (2000), the three-phase effective medium model (TPEM) of Ecker *et al.* (1998) and Helgerud *et al.* (1999), and the three-phase Biot theory (TPB) developed by Gei & Carcione (2003). Each of these approaches involves different simplifying assumptions regarding the shapes of individual sediment components and the way in which they interact with each other. All assume that, on the scale of a seismic wavelength, there is a degree of uniformity in the hydrate distribution, and that hydrate is disseminated in some way through the pore space. Hence none of these approaches copes well if hydrate occurs dominantly in nodules or veins, a form that may be important in some fine-grained sediments (e.g. Holland *et al.* 2006).

Jakobsen *et al.* (2000) developed two versions of the SCA/DEM approach: one in which the SCA is applied initially to a clay–water mixture, to simulate a microstructure in which hydrate is formed inside pores and is not load-bearing, and a second in which the SCA is applied initially to a clay–hydrate mixture, to simulate a microstructure in which the hydrate forms at pore throats and is load-bearing. Here we use the term 'clay' to indicate clay minerals, rather than as a grain-size descriptor.

Similarly, Helgerud *et al.* (1999) describe two versions of the TPEM approach: one in which hydrate forms part of the pore fluid and one in which hydrate forms part of the sediment frame.

Because of the difficulty of recovering and performing experiments on hydrate-bearing sediments without causing hydrate dissociation, direct observations of the pore-scale distribution of hydrates under *in situ* conditions are sparse. Sophisticated techniques such as thermal imaging have been developed to determine the hydrate distribution within pressurized cores (Expedition 311 Scientists 2005; Weinberger *et al.* 2005), but such techniques do not resolve down to the scale of individual pores. Tohidi *et al.* (2001) showed that hydrate grown from solution on a two-dimensional glass substrate formed preferentially in the centres of pores, but the extent to which this idealized environment simulates nature is unclear.

As discussed in more detail by Chand *et al.* (2004), the models that assume hydrate forms part of the pore fluid result in similar predictions of *P* and *S* wave velocity as a function of hydrate saturation (Fig. 1). Similarly, models that assume that the hydrate is load-bearing may result in similar predictions (Fig. 2), but for most hydrate saturations the predicted velocities are significantly higher than

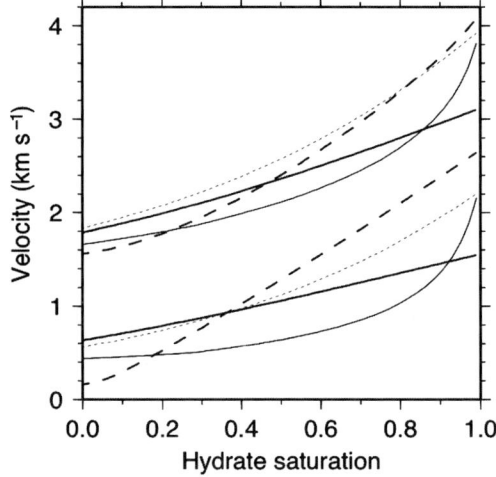

Fig. 1. Variation of *P* wave (larger values) and *S* wave velocities with hydrate saturation at fixed porosity (50%) and clay content (50%), for the four models discussed by Chand *et al.* (2004); other parameters required by the models are given by Chand *et al.* (2004). Thick solid line corresponds to the SCA/DEM model of Jakobsen *et al.* (2000) with a clay–water starting model. Thin solid line corresponds to the TPEM model of Helgerud *et al.* (1999) with hydrate forming part of the pore fluid. Dashed line corresponds to the TPB model of Gei & Carcione (2003). Dotted line corresponds to the weighted equation of Lee *et al.* (1996).

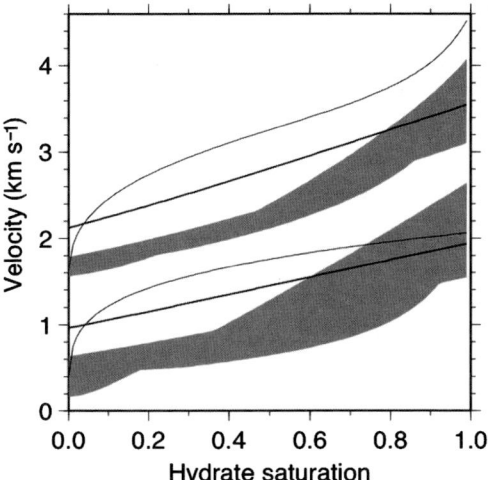
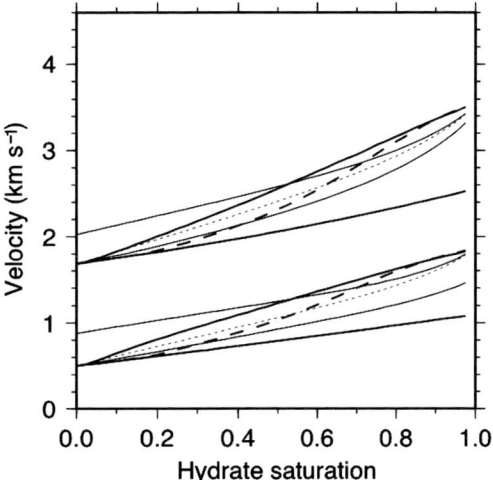

Fig. 2. Variation of P and S wave velocity with hydrate content for same materials as in Figure 1, but for models in which hydrate forms part of the sediment frame. Thick line corresponds to the SCA/DEM model with a clay–hydrate starting model. Thin line corresponds to the load-bearing hydrate model of Helgerud *et al.* (1999). Grey area marks the range of predictions from the rock physics based models (i.e. excluding the weighted equation) of Figure 1.

Fig. 3. Results from applying the SCA/DEM approach of Jakobsen *et al.* (2000) in several different ways to a real continental margin sediment. Sediment properties are those described by Hobro *et al.* (2005). Upper set of curves correspond to P wave velocities and lower set of curves correspond to S wave velocities. Thin solid lines mark results from using clay–water (lower curves) and clay–hydrate (upper curves) starting model. Dotted curves mark results from the model used by Hobro *et al.* (2005) in which the clay–water and clay–hydrate models are linearly mixed. Other lines mark results from the 'variable cementation' approach of Chand *et al.* (2006). Thick solid lines mark results from assuming that the hydrate is 0% (lower curves) and 100% (upper curves) load-bearing. Dashed line marks result from assuming that the proportion of hydrate that is load-bearing is equal to the hydrate saturation.

those of any of the models in which hydrate forms part of the pore fluid. The two models of Helgerud *et al.* (1999) converge with each other at low hydrate saturations and diverge at high saturations. Conversely, the two models of Jakobsen *et al.* (2000) converge at high saturations and diverge at low saturations, so that they predict different results when no hydrate is present – a clearly unphysical result. This result led Chand *et al.* (2006) to develop a modification to Jakobsen *et al.*'s basic approach in which hydrate is split into two component parts. For clay-rich sediments, a clay–water starting model is used throughout and the part of the hydrate that is not load-bearing is treated as inclusions (rather than replacing water in the clay–water starting model), while the part that is load-bearing replaces clay so that it properly forms part of the sediment frame. The velocities predicted by this modified approach, for the end member cases of all and none of the hydrate being load-bearing, converge at low hydrate saturations and diverge at high saturations (Fig. 3). This is a physically more realistic result that is also compatible with results from the model of Helgerud *et al.* (1999).

Effective medium inversion

The models described above can be turned around to infer hydrate saturations from observed seismic velocities, provided that other key parameters (porosity, composition and the elastic properties of the components) are known. A formal inverse approach to this problem was developed by Chand *et al.* (2006). The inversion is able to use measurements of P and S wave velocity, and also, if available, P and S wave attenuation. Formal uncertainties are also estimated, though these take into account only the uncertainties in the input parameters; it is difficult to account for the uncertainties in the effective medium models themselves.

Figure 4 illustrates the results of applying such an approach to a dataset from offshore Vancouver Island (Hobro *et al.* 2005). In this case there is information on porosity and lithology from Ocean Drilling Program Site 889, which lies within the survey area; the inversion used an exponential decay of porosity with depth and a mean sediment composition based on data from this site. Velocity information comes from a low-resolution P-wave tomographic study. The resolution of such studies

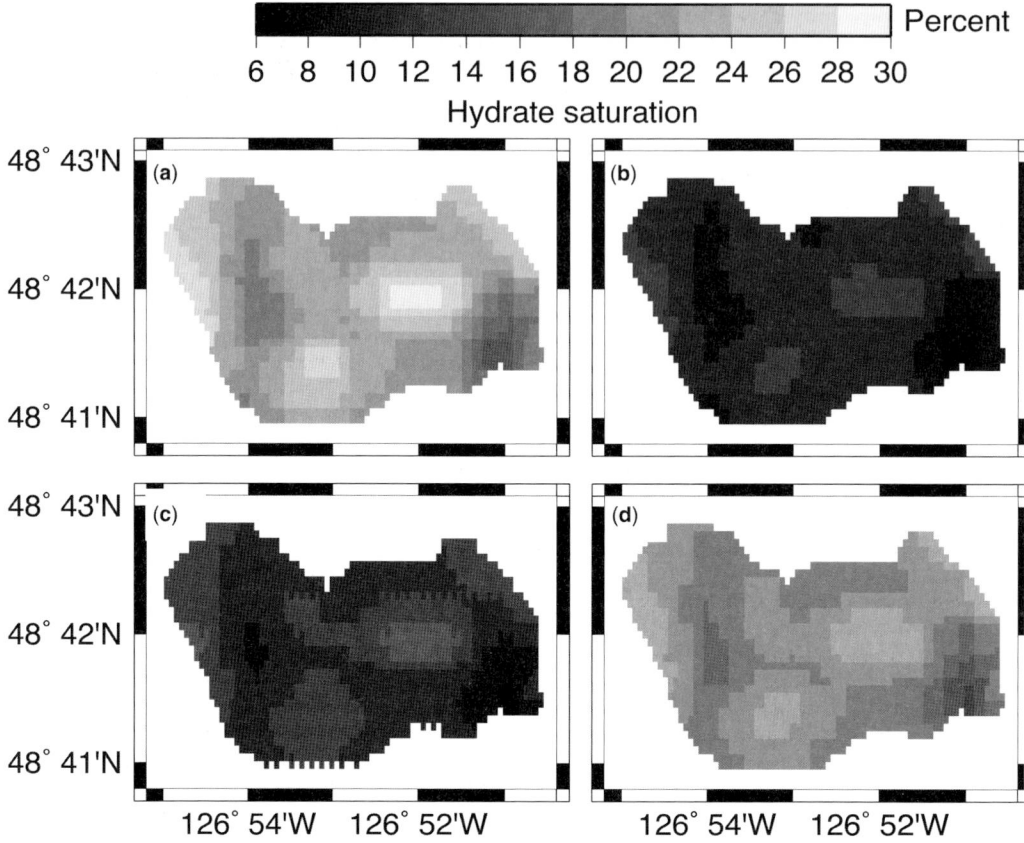

Fig. 4. Estimates of hydrate saturation at the base of the hydrate stability field from the tomographic velocity model of Hobro *et al.* (2005): (**a**) using the approach of Chand *et al.* (2006) and assuming that none of the hydrate is load-bearing; (**b**) using the approach of Chand *et al.* (2006) and assuming that all of the hydrate is load-bearing; (**c**) using the approach of Hobro *et al.* (2005); (**d**) using the approach of Chand *et al.* (2006) and assuming that the proportion of hydrate that is load-bearing is equal to the hydrate saturation.

is clearly an issue in this area and probably in most hydrate provinces: recent work in the same area during Integrated Ocean Drilling Program Expedition 311 (Riedel *et al.* 2006) has revealed a highly heterogeneous hydrate distribution. The tomographic model smooths the velocities over length scales of several hundred metres horizontally and several tens of metres vertically (Hobro *et al.* 2005). Unfortunately, the relationship between hydrate saturation and velocity is not linear (Figs 1–3), so even if the effective medium model is perfectly accurate, applying the effective medium inversion to a mean velocity over a given volume does not necessarily yield an accurate value for the hydrate content of that volume. However, a much larger uncertainty comes from the effective medium models themselves. While the model used by Hobro *et al.* (2005) yields a maximum hydrate saturation of *c.* 15%, other models predict saturations up to double this value (Fig. 4). Using the model of Chand *et al.* (2006) with the proportion of hydrate that is load-bearing set equal to the hydrate saturation, inferred saturations are *c.* 50% higher than those of Hobro *et al.* (2005), because at low hydrate saturations, the velocities predicted by this model are significantly lower (Fig. 3). This calculation is purely illustrative: such a relationship between hydrate saturation and hydrate cementation is based loosely on some calibration data that is described below, and may not be valid for the Vancouver Island margin.

The proportion of hydrate that is assumed to be load-bearing is effectively a free parameter in the SCA/DEM model, since it is impossible *a priori* to calculate what its value should be in a given geological environment. Therefore to choose sensibly between the results shown in Figure 4, some calibration data are required for which the hydrate content is known independently. In the next two sections we examine some datasets from both laboratory and field studies that might be used for calibration. An unavoidable limitation of such calibration datasets is that they involve higher seismic frequencies and correspondingly smaller sampling volumes than the field datasets that we wish to interpret. This limitation does not compromise the calibration if heterogeneities are smaller than the seismic wavelength of the calibration data.

Application to laboratory data

Our first calibration dataset is the dataset of Priest *et al.* (2005). These authors prepared methane hydrate-bearing sand samples by melting fine-grained ice particles in the presence of methane gas, following the method of Stern *et al.* (1996). Velocities were measured at seismic frequencies using a resonant column with both torsional and flexural vibrations, to determine P and S wave velocities respectively. Hydrate saturations were controlled by controlling the amount of ice–water in the system and were therefore accurately known. The resulting hydrate-bearing sediment is essentially dry; absence of residual water was confirmed by freezing the samples and checking that there was no significant change in velocities (Priest *et al.* 2005).

Unfortunately, application of the SCA/DEM approach to a quartz–air mixture leads to computed velocities that are much higher than those observed. In order to match these data, Chand *et al.* (2006) further modified the SCA/DEM approach. Instead of incorporating all of the quartz in the bi-connected SCA/DEM model, only a small proportion of the quartz is included at this stage, and the remainder is incorporated as isolated inclusions. Much lower velocities can then be obtained (Fig. 5). Both P and S wave observations may be matched if between 1 and 5% of the quartz is load-bearing (1–5% cementation). Such low percentages are reasonable given the way the samples are made. The percentage increases with differential pressure, which is physically reasonable as areas of grain contact will increase with increasing pressure. This approach gives us a second empirical factor (the degree of cementation of the host matrix) that is required to match real observations.

Having achieved a fit to the data in the absence of hydrate, the observations with hydrate present may than be matched by varying the degree of hydrate cementation as described in the previous section. At very low hydrate saturations, the predicted velocity is insensitive to the degree of cementation. However at hydrate saturations of a few per cent, all of the hydrate must be cementing (as concluded also by Priest *et al.* 2005), while at higher saturations, the degree of hydrate cementation required drops to about 40% (Fig. 6). The fact that, for both empirical parameters, the same value fits both P and S wave velocities indicates that the approach taken is not a bad approximation to the physics involved.

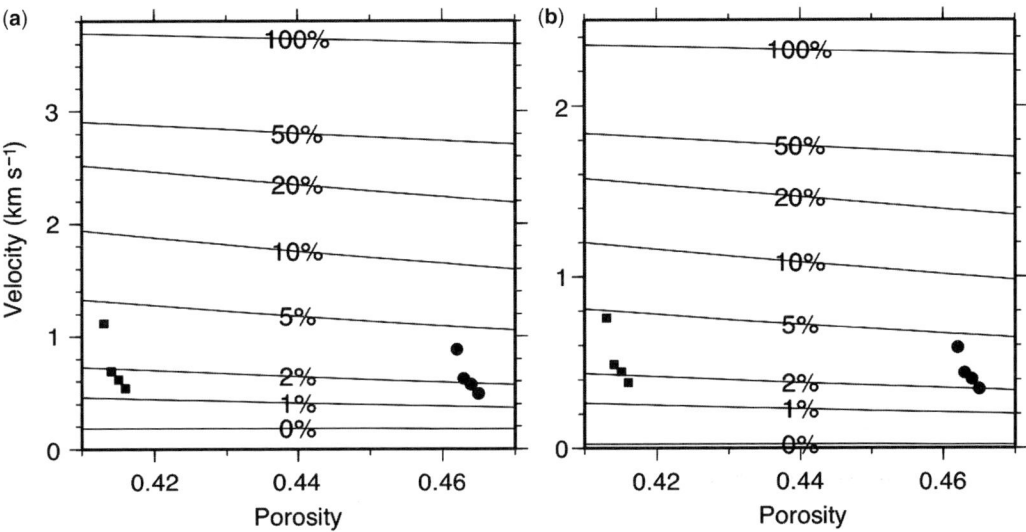

Fig. 5. (**a**) Squares mark P wave velocities derived from resonant column measurements by Priest *et al.* (2005) for loose sand, and circles for tight sand. The higher velocities correspond to higher effective pressures. Curves mark predictions of modified SCA/DEM model for a quartz–air mixture, labelled with the proportion of quartz that is considered load-bearing (see text). (**b**) Same as (**a**) but for S wave velocities.

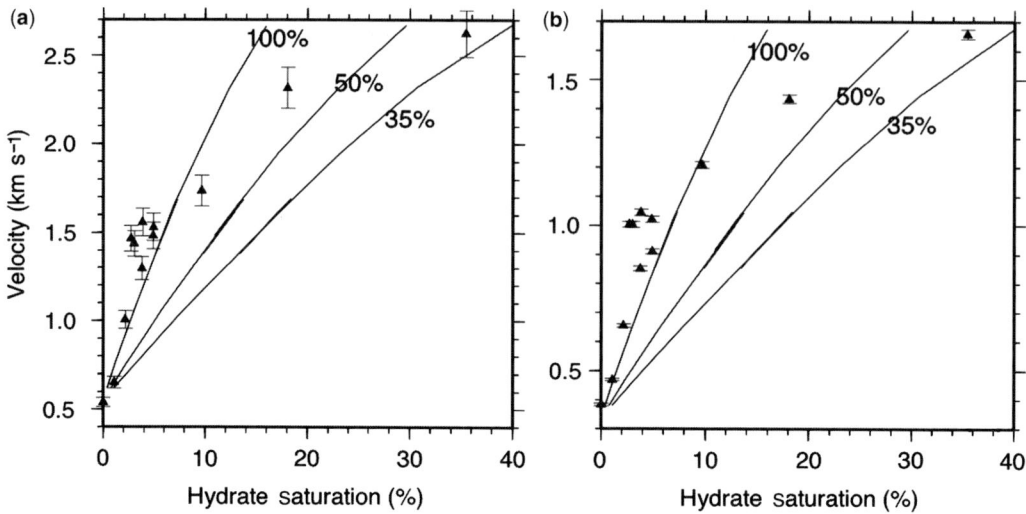

Fig. 6. Triangles with error bars mark velocities derived from resonant column measurements by Priest *et al.* (2005). Curves mark predictions of the modified SCA/DEM model and are labelled with the degree of hydrate cementation, for sand with a porosity of 42.1%, which is the mean porosity of the laboratory samples: (**a**) *P* wave velocities; (**b**) *S* wave velocities (modified from Chand *et al.* 2006).

Application to borehole data

The above laboratory work uses a sediment type (pure sand) that is not normally present in hydrate provinces in nature, and the hydrate is made in a way that does not approximate natural processes of hydrate formation. Therefore the insights obtained regarding the variation of cementation with saturation may have limited applicability. Calibration of the effective medium approach with a real field dataset, for which the hydrate content is known, is therefore desirable. Unfortunately, knowledge of *in situ* hydrate contents requires direct sampling, at pressure, by drilling, and few such datasets are available. However, in several hydrate provinces, estimates of hydrate saturation are available based on borehole resistivity data. Hydrate saturations are commonly derived from such data using Archie's law, an empirical effective medium approach which appears to give reliable estimates of the pore fluid component in a wide range of geological settings. These estimates must be treated with caution because of their empirical origin, but they are at least independent of seismic measurements.

One such dataset that has been widely used is from the Mallik 2L-38 borehole in the Canadian Arctic (Collett *et al.* 1999). A limitation of this dataset for calibration purposes is that porosities are significantly lower than those at equivalent depths in deep marine environments and the clay content is relatively low. Based on resistivity, hydrate saturations reach 80% at some depths in the hole, and both *P* and *S* wave velocities are available from borehole logs. In the absence of hydrate, velocities are around 2.2 km s^{-1} at a porosity of around 38%. These velocities are much higher than those of the hydrate-free laboratory sand samples described above (Fig. 5), which have slightly higher porosity, and also much higher than those of Yun *et al.*'s (2005) sand samples at a similar porosity (37%). Velocities are higher despite the fact that the Mallik material contains about 50% clay minerals, which have lower elastic moduli than quartz. This difference illustrates the importance of cementation: at Mallik the matrix material (clay) is much more strongly connected than the laboratory quartz samples, and in our approach is modelled as 100% load-bearing, in contrast to the 1–5% connected laboratory sand sample described above. The effect of hydrate cementation on velocity is much less in such circumstances than in the case of the laboratory samples.

When plotted as a function of hydrate saturation, *P* and *S* wave velocities are quite scattered (Fig. 7) because they depend also on porosity and composition, which vary through the interval sampled. In general, this scatter is larger than the variation of predicted velocities with degree of cementation for fixed porosity and clay content, though at high hydrate saturations there is some indication that a better fit is achieved to both *P* and *S* wave velocities if the degree of hydrate cementation is high, in contrast to the low degrees of cementation required

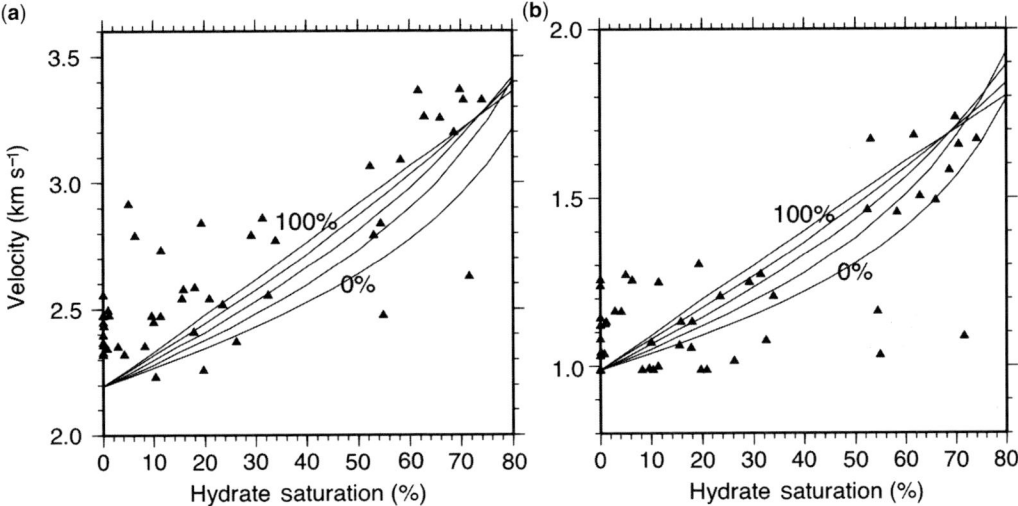

Fig. 7. Filled circles mark velocities from borehole logs as a function of resistivity-derived hydrate saturation. Both parameters are averaged over 10 m intervals down the borehole. Lines mark predicted velocities from the SCA/DEM model for a porosity of 38% and a clay content of 57%, which are mean values for the interval studied. Lines are computed for degrees of hydrate cementation of 0, 25, 50, 75 and 100%. (**a**) *P* wave velocity; (**b**) *S* wave velocity.

for the laboratory data described above. This effect may be seen much more clearly if the degree of hydrate cementation required to match the observations is displayed as a function of hydrate saturation (Fig. 8), since then variations in porosity and composition may be accounted for directly. Degrees of hydrate cementation inferred from *S* wave velocities differ little from those inferred from *P* wave velocities, again suggesting that we are achieving a reasonable approximation to the physics involved.

Other datasets may be represented in the same way. We have also modelled the results of Waite *et al.* (2004), who measured ultrasonic *P* wave velocities on hydrate-bearing samples of partially water-saturated sands. Hydrate was made in these samples by passing gas through the samples at high pressure. As with the resonant column results of Priest *et al.* (2005), these data are matched by a systematic decrease of hydrate cementation with saturation (Fig. 8), though the decrease is not as steep as for the dry samples of Priest *et al.* Finally,

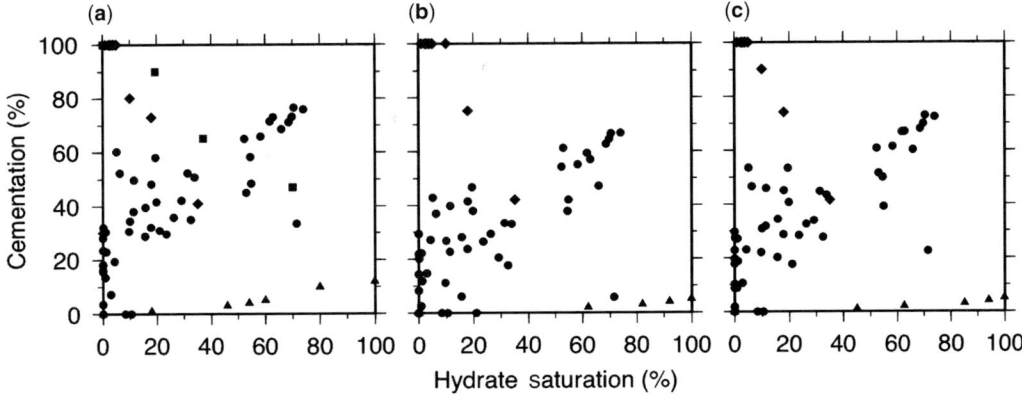

Fig. 8. Degree of cementation required to match observed velocities using the modified SCA/DEM model for a variety of datasets. Circles mark the Mallik 2L-38 dataset (Collett *et al.* 1999); diamonds mark the data of Priest *et al.* (2005); squares mark the data of Waite *et al.* (2004); and triangles mark the data of Yun *et al.* (2005): (**a**) only *P* wave velocities used; (**b**) only *S* wave velocities used; (**c**) degree of hydrate cementation optimized to match both *P* and *S* wave velocities, using the misfit function defined by Chand *et al.* (2006).

we applied our method to measurements of P and S wave velocities made by Yun *et al.* (2005) on hydrate-bearing sand samples made using tetrahydrofuran in solution. In contrast to the other laboratory datasets, these data are matched by very low degrees of hydrate cementation, and the inferred degree of cementation increases systematically with hydrate saturation (Fig. 8).

Discussion

As with all effective medium methods, the SCA/DEM method requires some empirical adjustments to fit real data. The advantage of the approach we describe above is that the adjustments can be related to something physical – the extent to which different components are load-bearing or 'cementing'. Using a variety of published datasets, and assuming that the effect that we model as cementation is indeed cementation, we can determine how cementation varies with hydrate saturation based on independent determinations of the latter. Results are consistent whether we use P or S wave velocities, and the inferred variations of cementation with hydrate content are systematic and approximately linear. Unfortunately, the slopes of the trends in Figure 8 vary in both magnitude and sign depending on how the hydrate has been formed. For samples made in a gas-rich environment, where hydrate may tend to form initially at grain contacts (Priest *et al.* 2005), the inferred hydrate cementation decreases with saturation. For samples formed from solution, the inferred hydrate cementation increases with saturation. For the *in situ* data from Mallik 2L-38, the inferred cementation increases with saturation, but more steeply than for the laboratory samples formed from solution.

In many field situations, an analysis of the type described above will not be possible, because remote seismic observations form the only constraint available on hydrate saturations. In such situations, the following approach may be taken:

1. Define a no-hydrate reference velocity curve based on velocities in regions where no hydrate is thought to be present (e.g. close to the seabed and/or beneath the base of the hydrate stability field), and use this curve and estimates of porosity to determine how the proportion of sediment grains that are load-bearing varies with porosity. This step will be more robust if both P and S wave velocity measurements are available.
2. Generate an empirical, linear or at least monotonic, fit between inferred hydrate cementation and hydrate saturation either using borehole data from the same area (if there is a borehole sampling hydrate) or from similar hydrate-bearing sediments elsewhere.
3. Assume that this empirical fit applies throughout the volume sampled by seismic data and hence infer hydrate saturations.

The above approach is limited by the several assumptions that are required, including the assumption that the hydrate is uniformly disseminated in the pore space and that within the pore space it takes one of the forms described by the DEM theory. Given these assumptions, the main potential for error comes from the second step. A way forward is to develop a larger database of *in situ* seismic velocity measurements where hydrate saturations are known independently. Such a database is gradually emerging through scientific ocean drilling (Tréhu *et al.* 2004; Expedition 311 Scientists 2005 by Riedel *et al.* 2006).

An alternative, more direct approach may be to further develop geophysical techniques that remotely determine other sediment physical properties such as electrical resistivity. The resistivity of hydrate-bearing sediments will depend also on the way the different components are connected and in particular the connectivity of the fluid component. Simultaneous remote measurement of both seismic and electrical properties would yield an additional constraint that may remove some of the ambiguities that come from the use of seismic data alone. Interpretation of such datasets will require a joint effective medium approach that can model both seismic velocity and resistivity (Ellis *et al.* 2005).

Conclusions

From our effective medium calculations and their calibration through laboratory and borehole measurements, we conclude the following:

1. The prediced physical properties of hydrate-bearing sediments depend more strongly on the assumed microstructure than on the particular effective medium model used to approximate them.
2. Both laboratory and borehole measurements of the seismic velocity of hydrate-bearing sediments may be modelled successfully using a modified version of the SCA/DEM approach.
3. For laboratory samples made in the presence of excess gas, the proportion of hydrate that is load-bearing appears to decrease with increasing hydrate saturation.
4. For borehole data from the Canadian Arctic, and for laboratory samples made from solution, this proportion appears to increase with increasing hydrate saturation.
5. Such systematic variations ultimately might be used to infer more accurately hydrate saturations from remote seismic data.

The early part of this work was supported by the European Commission under the contract EVK3-CT-2000-00043 (HYDRATECH). We thank J. Priest, A. Best and W. Waite for access to data and for useful discussions, and P. Jackson and an anonymous reviewer for constructive comments.

References

CHAND, S., MINSHULL, T. A., GEI, D. & CARCIONE, J. M. 2004. Elastic velocity models for gas-hydrate bearing sediments – a comparison. *Geophysical Journal International*, **159**, 573–590.

CHAND, S., MINSHULL, T. A., PRIEST, J. A., BEST, A. I., CLAYTON, C. R. I. & WAITE, W. F. 2006. An effective medium inversion algorithm for gas hydrate quantification and its application to laboratory and borehole measurements of gas hydrate bearing sediments. *Geophysical Journal International*, **166**, 543–552.

COLLETT, T. A., LEWIS, R. E., DALLIMORE, S. R., LEE, M. W., MROZ, T. H. & UCHIDA, T. 1999. Detailed evaluation of gas hydrate reservoir properties using JAPEX/JNOC/GSC Mallik 2L-38 gas hydrate research well down hole well-log displays. *In*: DALLIMORE, S. R., UCHIDA, T. & COLLET, T. S. (eds) *Scientific Results from JAPEX/JNOC/GSC Mallik 2L-38 Gas Hydrate Research Well, Mackenzie Delta, Northwest Territories, Canada*. Geological Survey of Canada, Ottawa, Bulletin, **544**, 295–312.

DICKENS, G. R., PAULL, C. K., WALLACE, P. & THE ODP LEG 164 SCIENTIFIC PARTY. 1997. Direct measurement of *in situ* methane quantities in a large gas-hydrate reservoir. *Nature*, **385**, 426–428.

ECKER, C., DVORKIN, J. & NUR, A. 1998. Sediments with gas hydrates: Internal structure from seismic AVO. *Geophysics*, **63**, 1659–1669.

ELLIS, M. H., MINSHULL, T. A., BEST, A. I., SINHA, M. C. & SOTHCOTT, J. 2005. Joint seismic and electrical measurements of gas hydrates in continental margin sediments, *Proceedings of 5th International Conference on Natural Gas Hydrates*, **2**, paper 2022, 545–554.

GEI, D. & CARCIONE, J. M. 2003. Acoustic properties of sediments saturated with gas hydrate, free gas and water. *Geophysical Prospecting*, **51**, 141–157.

HELGERUD, M. B., DVORKIN, J., NUR, A., SAKAI, A. & COLLETT, T. 1999. Elastic-wave velocity in marine sediments with gas hydrates – effective medium modeling. *Geophysical Research Letters*, **26**, 2021–2024.

HOBRO, J. W. D., MINSHULL, T. A., SINGH, S. C. & CHAND, S. 2005. A three-dimensional seismic tomographic study of the gas hydrate stability field, offshore Vancouver Island. *Journal of Geophysical Research*, **110**, B09102; doi: 10.1029/2004JB003477.

HOLLAND, M., SCHULTHEISS, P., ROBERTS, J. & DRUCE, M., IODP EXPEDITION 311 SHIPBOARD SCIENTIFIC PARTY & NGHP EXPEDITION 1 SHIPBOARD SCIENTIFIC PARTY. 2006. Hydrate-sediment morphologies revealed by pressure core analysis. *EOS, Transactions of the American Geophysical Union, Fall Meeting Supplement*, **OS33B-1689**.

HORNBY, B. E., SCHWARTZ, L. M. & HUDSON, J. A. 1994. Anisotropic effective-medium modelling of the elastic properties of shales. *Geophysics*, **59**, 1570–1581.

JAKOBSEN, M., HUDSON, J. A., MINSHULL, T. A. & SINGH, S. C. 2000. Elastic properties of hydrate-bearing sediments using effective medium theory. *Journal of Geophysical Research*, **105**, 561–577.

LATYCHEV, K. & EDWARDS, R. N. 2003. On the compliance method and the assessment of three-dimensional seafloor gas hydrate deposits. *Geophysical Journal International*, **155**, 923–952.

LEE, M. W., HUTCHINSON, D. R., COLLETT, T. S. & DILLON, W. P. 1996. Seismic velocities for hydrate-bearing sediments using weighted equation. *Journal of Geophysical Research*, **101**, 20347–20358.

PRIEST, J. A., BEST, A. I. & CLAYTON, C. R. I. 2005. A laboratory investigation into the seismic velocities of methane gas hydrate-bearing sand. *Journal of Geophysical Research – Solid Earth*, **110**, B04102; doi: 10.1029/2004JB003259.

RIEDEL, M., COLLETT, T. S., MARONE, M. J. & EXPEDITION 311 SCIENTISTS. 2006. *Proceedings of the Ocean Drilling Program Expedition 311*, Integrated Ocean Drilling Program, Washington, DC.

SCHWALENBERG, K., WILLOUGHBY, E., MIR, R. & EDWARDS, R. N. 2005. Marine gas hydrate electromagnetic signatures in Cascadia and their correlation with seismic blank zones. *First Break*, **23**, 57–63.

SINGH, S. C., MINSHULL, T. A. & SPENCE, G. D. 1993. Velocity structure of a gas hydrate reflector. *Science*, **260**, 204–207.

STERN, L. A., KIRBY, S. H. & DURHAM, W. B. 1996. Peculiarities of methane clathrate hydrate formation and solid state deformation, including possible super heating of water ice. *Science*, **273**, 10299–10311.

TOHIDI, B., ANDERSON, R., CLENNELL, M. B., BURGASS, R. W. & BIDERKAB, A. B. 2001. Visual observation of gas-hydrate formation and dissociation in synthetic porous material by means of glass macromodels. *Geology*, **29**, 867–870.

TRÉHU, A. M., LONG, P. E. *ET AL.* 2004. Three-dimensional distribution of gas hydrate beneath southern Hydrate Ridge: Constraints from ODP Leg 204. *Earth and Planetary Science Letters*, **222**, 845–862.

WAITE, W. F., WINTERS, W. J. & MASON, D. H. 2004. Methane hydrate formation in partially water-saturated Ottawa sand. *American Mineralogist*, **89**, 1202–1207.

WEINBERGER, J. L., BROWN, K. M. & LONG, P. E. 2005. Painting a picture of gas hydrate distribution with thermal images. *Geophysical Research Letters*, **32**, L04609; doi: 10.1029/2004GL021437.

WESTBROOK, G. K., BUENZ, S. *ET AL.* 2005. Measurement of P- and S-wave velocities, and the estimation of hydrate concentration at sites in the continental margin of Svalbard and the Storegga region of Norway. *Proceedings of the 5th International Conference on Natural Gas Hydrates*, **3**, paper 3004, 726–735.

YUN, T. S., FRANCISCA, F. M., SANTAMARINA, J. C. & RUPPEL, C. 2005. Compressional and shear wave velocities in uncemented sediment containing gas hydrate. *Geophysical Research Letters*, **32**, L10609; doi: 10.1029/2005GL022607.

Regional versus detailed velocity analysis to quantify hydrate and free gas in marine sediments: the South Shetland Margin case study

UMBERTA TINIVELLA*, MARIA FILOMENA LORETO & FLAVIO ACCAINO

Istituto Nazionale di Oceanografia e di Geofisica Sperimentale (OGS), Borgo Grotta Gigante 42 C, 34010 Trieste, Italy

*Corresponding author (e-mail: utinivella@ogs.trieste.it)

Abstract: The presence of gas hydrate and free gas within marine sediments deposited along the South Shetland margin, offshore the Antarctic Peninsula, was confirmed by low and high resolution geophysical data, acquired during three research cruises. Seismic data analysis has revealed the presence of a bottom-simulating reflector that is very strong and continuous in the eastern part of the margin. This area can be considered as a useful site to study the seismic characteristics of sediments containing gas hydrate, with a particular focus on the estimation of gas hydrate and free gas amounts in the pore space. Pre-stack depth migration and tomographic inversion were performed to produce a regional velocity field of gas-phase bearing sediments and to obtain information about the average thickness of gas hydrate and free gas layers. Using these data and theoretical models, the gas hydrate and free gas concentrations can be estimated. Moreover, the common image gather semblance analysis revealed the presence of detailed features, such as layers with small thickness characterized by low velocity alternating with high velocity layers, below and above the bottom-simulating reflector. These layers are associated with free gas trapped within the hydrate stability zone and deeper sediments. Thus, the use of the detailed and the regional velocity field analysis is important to give a more reliable estimate of gas content in the marine sediments.

Gas hydrates in marine environments have been mostly detected from analysis of seismic reflection profiles, where they produce remarkable bottom-simulating reflectors (BSRs; Cox 1983; Sloan 1998). Generally, the BSR is a very high-amplitude reflector that is associated with a phase reversal (Hyndman & Spence 1992; Max 2003). This phase reversal may indicate that sediments above the BSR are extensively filled with gas hydrates and lower sediments below it are filled with free gas in the pore space (Minshull et al. 1994; Sain et al. 2000). Several studies (e.g. Tinivella et al. 1998) revealed a seismic reflector below the BSR that can be associated with the base of the free gas zone, called BGR (base of the free gas reflector).

The scientific community is investing much effort in studying marine sediments containing gas hydrates to characterize the hydrate reservoir and to quantify the gas trapped within sediments from seismic data analysis (Chand & Minshull 2003; Zillmer 2006). Recently, the international community has considered CO_2 sequestration as a possible means of offsetting the emission of greenhouse gases into the atmosphere (Ledley et al. 1999). The CO_2 storage programme is a further reason to assess the feasibility of mapping and monitoring the reservoir by means of an efficient seismic analysis (Chadwick et al. 2002; Arts et al. 2004) and to obtain information about hydrate and free gas concentrations in a time-effective way.

The presence of a BSR, indicating a relevant gas hydrate reservoir on the South Shetland margin, was discovered during the Italian Antarctic cruise of 1989–1990, onboard R/V *OGS-Explora* (Lodolo et al. 1993). The South Shetland margin is a convergent plate boundary where oceanic lithosphere that is part of the Antarctic plate is subducting beneath the South Shetland micro-continental block (Fig. 1, left). Along the continental margin, a trench-accretionary prism–fore-arc basin sequence can be recognized (Maldonado et al. 1994; Kim et al. 1995). Active spreading along the Antarctic–Phoenix ridge stopped about 3.5 Ma ago (Larter & Barker 1991; Livermore et al. 2000), but subduction continued as a consequence of sinking and roll-back of the subducted slab coupled with extension in the Bransfield Strait marginal basin (Larter & Barker 1991; Kim et al. 1995; Jin et al. 2002; Dietrich et al. 2004). The part of the Antarctic plate involved in the subduction process (a remnant of the Phoenix plate) is laterally bordered by two main fracture zones, the Hero (to the south) and the Shackleton (to the north), respectively. These two fracture zones intersect the

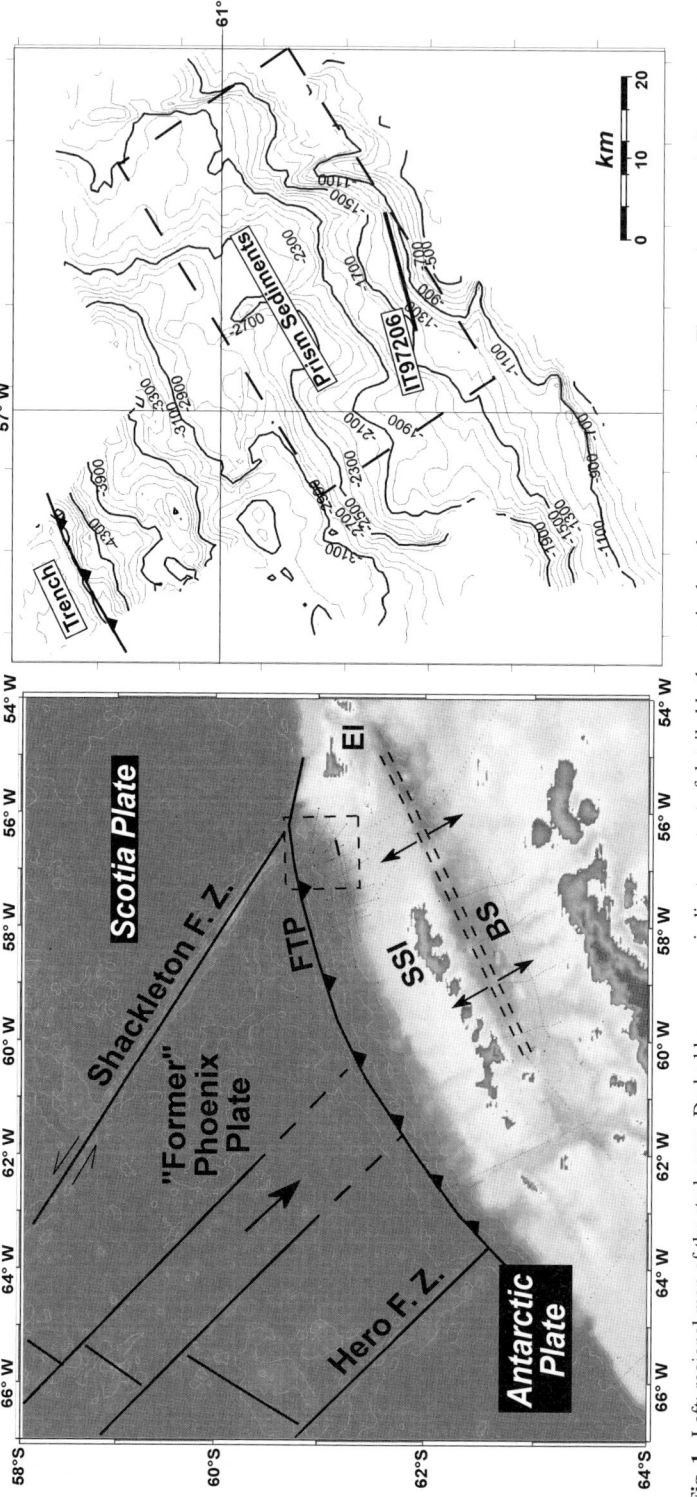

Fig. 1. Left: regional map of the study area. Dashed box on map indicates extent of detailed bathymetric data shown on the right part. The opposite arrows indicate the ongoing NW–SE extension across Bransfield Strait (after Lawver *et al.* 1996). B.S., Bransfield Strait; E.I., Elephant Island; F.T.P., Frontal Thrust Prism; S.S.I., South Shetland Islands. The subduction direction of 'Former' Phoenix Plate is indicated with solid arrow, while transform fault movements are indicated with double thin arrows. Right: new high-resolution bathymetric map and location of the seismic line analysed. The dashed box on the right panel delimits the area where the hydrate and free gas concentrations are estimated (see text). The depth labels are in metres and the contour interval is 100 m; the thick contours are every 500 m.

continental lithosphere and their landward projections are associated with structural and morphological variations in the overriding lithosphere (Grad et al. 1993; Kim et al. 1995; Jin et al. 2002; Loreto et al. 2006).

A narrow accretionary prism has developed along the continental margin, ranging from 20 to 40 km in width. The prism is characterized by a range of deformational features, deduced by seismic data interpretation (Lodolo et al. 2002): reverse and thrust faults mainly affect the frontal part of the prism; extensional faults further from the trench are oriented sub-parallel to the continental shelf; a strike–slip fault has been interpreted as being related to the Shackleton Fracture Zone. Small mid-slope basins are common within the prism, often bounded by extensional faults that locally reach the seafloor (Lodolo et al. 2002). A clear BSR was interpreted within prism sediments west of Elephant Island that is a continuous high-amplitude reflection in the area of interest (Fig. 1, right).

This seismic dataset was used in the past to extract detailed velocity information of the shallow structures by using conventional tomographic inversion (Tinivella et al. 1998) and jointly tomographic inversion and pre-stack depth migration tool (Tinivella & Accaino 2000; Tinivella et al. 2002). Here, we present a method to obtain a regional seismic velocity field and information about hydrate and free gas presence in the marine sediments, by using a method that is an improvement of the standard analysis of the pre-stack depth migration output (the common image gather, CIG). The velocity field is obtained with a layer stripping approach and tomographic inversion of the reflections observed in the CIGs (Accaino et al. 2005). We can identify two main advantages in the method used in this paper. Firstly, we perform the pre-stack depth migration once for each reflector, decreasing drastically the CPU time. Secondly, we increase the lateral resolution of the final velocity field with respect to a conventional CIG analysis, because we introduce the possibility of having lateral velocity variation in each CIG. By using this approach, the vertical resolution is reduced and the human and CPU time optimized. Finally, the average seismic velocity and geometrical information can be used to roughly estimate the gas reservoir potential in the study area.

Seismic data

A strong BSR was identified on multichannel seismic reflection profiles acquired during the Austral summers 1989/1990 (Lodolo et al. 1993) and 1996/1997 (Tinivella & Accaino 2000) on the South Shetland margin. The second cruise used an energy source of two generator-injector guns with a total volume of 4 litres firing every 25 m. The streamer was 3000 m long with a hydrophone group interval of 25 m, and the sampling interval was 1 ms. To better describe the area where the BSR is very strong and continuous, another cruise was carried out to acquire detailed bathymetric data (12 kHz was the acoustic frequency used), sub-bottom profile data and seismic data with a short hydrophone streamer (600 m) during the Austral summers 2003/2004.

The average seismic velocity in the gas hydrate and free gas zones can be determined using the pre-stack depth migration and tomographic approach iteratively (Accaino et al. 2005). We applied this method to the first 16.5 km of the seismic line IT97206, acquired during the second cruise. A location map incorporating the bathymetry data obtained on the third cruise is shown in Figure 1. The multibeam data were collected with the Reson multibeam echo sounding system (Reson Seabat 8150) on R/V OGS Explora over an area of about 5500 km^2 (Fig. 1, left). The multibeam bathymetric imagery was calibrated using the velocity profile in the water column reconstructed from the conducibility–temperature–density data, acquired during the Leg. Figure 1 was obtained using the processing software PDS2000, considering a cell grid size equal to 200 × 200 m. The seismic data used for the velocity analysis was re-sampled to 2 ms and a bandpass filter (10–80 Hz) was applied after the application of a geometrical spreading correction to the raw data. We report only the results after the first 3 km of the line, as the fold of coverage was insufficient on the earlier part.

Seismic velocity field

Our main target was to determine the seismic velocity field in order to obtain a seismic image in depth and an estimate of gas hydrate and free gas concentration in the pore space of marine sediments. For this purpose, we decided to use depth migration and the tomographic algorithm in the pre-stack domain to determine, iteratively and with a layer stripping approach, both the velocity field and the seismic image in depth. The inversion was performed using a tomographic software (CAT3D) and a modified version of the minimum time ray tracing based on Fermat's principle (Böhm et al. 1999). An iterative procedure was used for the inversion, based on the simultaneous iterative reconstruction procedure algorithm (van der Sluis & van der Vorst 1987; Stewart 1991). The ray tracing algorithm started from an initial hypothesis for its path and converged to a final geometry through an iterative procedure using the analytical solution of the Snell's law (Böhm et al. 1999). The method implies the discretization of a continuum into a certain number of layers, separated by curved interfaces defined by bi-cubic splines and subdivided in pixel. Details of the methodology are given in Accaino et al. (2005); here we summarize only the main principles.

The first step involved performing a pre-stack depth migration using a constant velocity field (in this case, a velocity of 1500 m s^{-1}). The pre-stack migration was performed using the free software Seismic Unix (Cohen & Stockwell 2001). After the migration, we picked the first sub-seafloor reflection v. offset in the CIGs. In the first step of the inversion procedure, we changed the velocity (including the possibility of having a lateral velocity variation inside the CIGs). The inversion was performed considering simultaneously all the selected events in all CIGs. The second step of the inversion was computation of an updated initial depth using the updated velocity field in each CIG. Then, these two steps of the inversion (the update of the velocity field and the layer depth) were performed iteratively until the differences in initial depths between two inversion steps were lower than a defined threshold. The final depth of the interface was given by interpolating the final depths of the reflection points at zero offset. For the inversion procedure we used the Cat3D software (Böhm et al. 1999).

The inversion of the second reflector was started by performing the pre-stack depth migration using the calculated velocity field in the upper layer, and a constant velocity in the next layer. Then, the same procedure was applied to evaluate the velocity and the geometry of the new layer. Note that the velocity and the depth were obtained by using a layer-stripping approach, i.e. layers were constrained (both for velocity and depth) by iterative inversion in depth order working downwards from the seafloor. We consider that this method has two main advantages. Firstly, the pre-stack depth migration was carried out once for each reflector, drastically decreasing the CPU time. Secondly, optimal lateral resolution of the final velocity field was achieved by modelling lateral velocity variation in each CIG. Moreover, to minimize the human time required to pick the reflections in the CIG domain, we used a semi-automatic picking tool available in the commercial software RADEX$^{\copyright}$.

Our choice of horizons to define layers for the inversion was based on consideration of the continuity of reflections along the seismic line. For this reason, we picked the seafloor (hereafter called Hor. 1), a shallow horizon (called Hor. 2, that is the base of the layer described above as the first layer) and two other reflectors (Hor. 3, corresponding to the base of layer 2, and Hor. 4, the base of layer 3), which locally corresponded to the BSR (the base of the gas hydrate layer) and the BGR (the base of the free gas layer) respectively. Below Hor. 4 (i.e. in the layer 4), we assumed a velocity at the top of the last layer of 2000 m s^{-1} and a vertical velocity gradient of 12 (m s^{-1}) m^{-1}: these values were chosen to obtain a satisfactory seismic image in depth. The final velocity model and the final pre-stack depth migration are shown in Figure 2, in which the main selected reflectors are indicated. Note that the final velocity model was vertically and horizontally smoothed to improve the migration and to attenuate lateral velocity variations.

In order to verify the reliability of the final velocity field, we picked the four horizons in the CIG domain to evaluate the flatness of the inverted reflections. The reliability of the inversion is confirmed by the low difference between horizons picked at offset −150 and horizons picked at −2500 m; the differences are shown in Figure 3 in terms of error, where the error is defined by:

$$error = (z_{-150} - z_{-2500})/z_{-150} \quad (1)$$

where z is the selected depth. The average absolute errors of the four horizons were equal to 0.29, 0.77, 0.51 and 0.78% respectively, with an average error equal to 0.59%; locally, the error increased because of the quality of the data and/or the lack of the lateral continuity of the reflections.

The final velocity field shows a first sub-seafloor layer characterized by velocities within a range of 1600–1800 m s^{-1} that increases strongly to about 2250 m s^{-1} at Hor. 2, corresponding to typical velocities of gas hydrate-bearing sediments. The velocity in layer 2 is quite uniform except across CIG 5000 and between CIGs 9900 and 12000 (Fig. 2, top), where the velocity increases to 2500 m s^{-1}. Layer 3 is characterized by strong lateral velocity variation; in particular we found low velocities (about 1400 m s^{-1}) and zones where velocities (2200–2500 m s^{-1}) are typical of the hydrate-bearing layer. Our schematic structural interpretation of the pre-stack depth migrated section (described below) suggests that some structural features, such as faults and folds, are localized around velocity anomalies in layers 2 and 3 (Fig. 2).

Pre-stack depth migration section

The final pre-stack depth migration of part of line IT97206 is characterized by higher signal-to-noise ratio. Between CIGs 3000 and 10000 (Fig. 2, bottom), a well-defined sedimentary basin has developed above the eastward dipping deeper strata, confirmed by onlapping seismo-stratigraphic

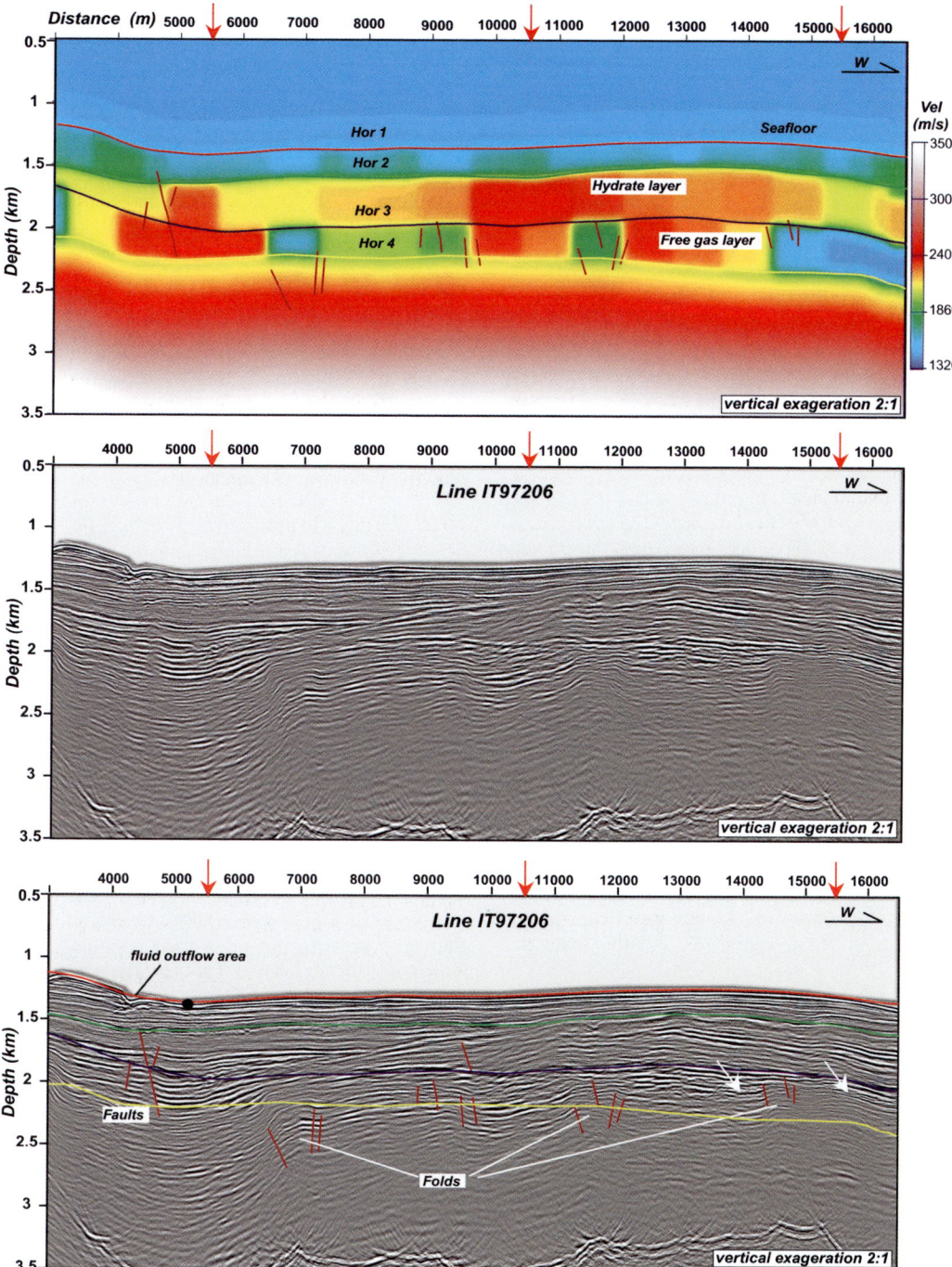

Fig. 2. Top: smoothed velocity field of seismic line IT97206 after tomographic inversion of CIGs. The distance in metres corresponds to the CIG number. Red arrows correspond to the location of CIGs and CDPs selected for detailed analysis (see text). Middle: pre-stack depth migration seismic line. Bottom: interpreted pre-stack depth migration of seismic line. Colour horizons correspond to the inverted horizons: Hor. 1 (red); Hor. 2 (green); Hor. 3 (blue); Hor. 4 (yellow). Interpreted faults are marked in red. The solid black circle indicates the depocentre. White arrows indicate low velocity layers with small thickness.

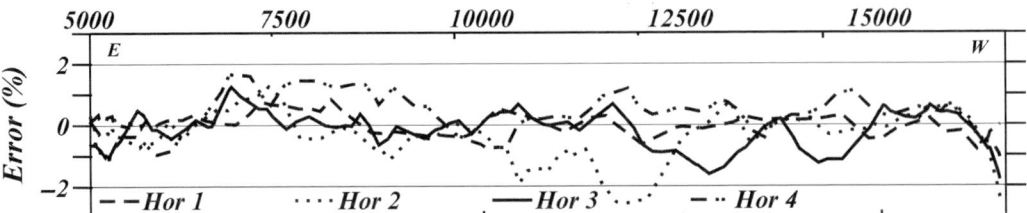

Fig. 3. Percentage differences between the picked horizons at offset equal to −150 and −2500 m (equation (1)). The distance in metres corresponds to the CIG number. Hor. 1, broken line; Hor. 2, dotted line; Hor. 3, solid line; Hor. 4, dashed–dotted line.

configurations and depocentre location (at CIG 5300). Sedimentary strata in the basin are locally interrupted and slightly shifted between 1600 and 2250 m depth (around CIG 4700). This discontinuity is interpreted as a fault that cross cuts the entire sedimentary basin and reaches the sea bed, as evidenced by a small incision in the seafloor. This fault could be an important pathway for fluid escape (Saffer et al. 2000; Accaino & Tinivella 2005).

Within deeper sediments, a broad anticline (between CIGs 7000 and 16500) is further deformed by three little folds (Fig. 2, bottom), with a wavelength of about 2 km. Between CIGs 8000 and 16500, a clear BSR is present that locally cross-cuts the sedimentary layers, and was accurately picked as Hor. 3 during tomographic inversion to produce the velocity field. Towards the eastern end of the section the BSR becomes less evident within the sedimentary basin, probably due to both the stratigraphic configuration of layers (sub-parallel to the seafloor) and a weaker acoustic contrast between the hydrate-bearing and gas-saturated sediments.

Detailed velocity analysis

The pre-stack depth migration (Fig. 2, bottom) indicates that Hor. 3 and Hor. 4 are only locally associated with the BSR and the BGR respectively. Moreover, locally, we did not detect a velocity inversion between layers 2 and 3, even if Hor. 3 does correspond to the BSR. To understand the cause of the lack of velocity inversion, we decided to perform detailed velocity analyses at three selected locations (see arrows in Fig. 2) using both common depth point (CDP) and CIG analyses.

Residual move-out of CDPs

We carried out semblance analyses of unprocessed CDPs to obtain stacking velocities, and then we carried out a residual NMO analysis (Yilmaz 2001) using the commercial software GeoDepth©. GeoDepth© progresses the residual moveout correction in one step by applying velocity analysis directly, following the equation:

$$\Delta t = \sqrt{t_0^2 + \left(\frac{1}{V_c^2} - \frac{1}{V_0^2}\right)x_{\text{ref}}^2} - t_0 \quad (2)$$

where t_0 is the reflection time at zero offset, x_{ref} the reference offset, which is the offset where the time residual is measured (in our case the maximum offset is 3150 m), Δt the time residual, defined as the time difference between the location of the reflection at zero offset and the location of the x_{ref}, V_0 the original stacking velocity and V_c the new corrected velocity. Interval velocities were calculated from the corrected stacking velocity using the Dix equation (Dix 1955). We show the detailed residual NMO analysis at three selected locations along the seismic profile, indicated by red arrows in Figure 2. In Figure 4 the NMO corrected CDPs, the interval velocities and the semblances are shown against two-way travel time; the r-parameter represents the time velocity error as obtained by semblance analysis. Note that the semblance energy is quite focused close to the r-parameter zero line, validating the results of the velocity analyses.

All the interval velocity profiles are characterized by alternating high and low velocity layers in both the gas hydrate stability region and below the BSR. Low interval velocities (less than 1500 m s^{-1}) are indicated in a few layers with small thicknesses (about 50–100 m thick). Moreover below the BSR, layers characterized by low velocity are observed at several depths, indicating that the concentration of free gas is variable both along the seismic section and in depth. High velocities are present locally in layer 3.

The interval velocity uncertainty calculated from stacking velocities is a function of the time thickness

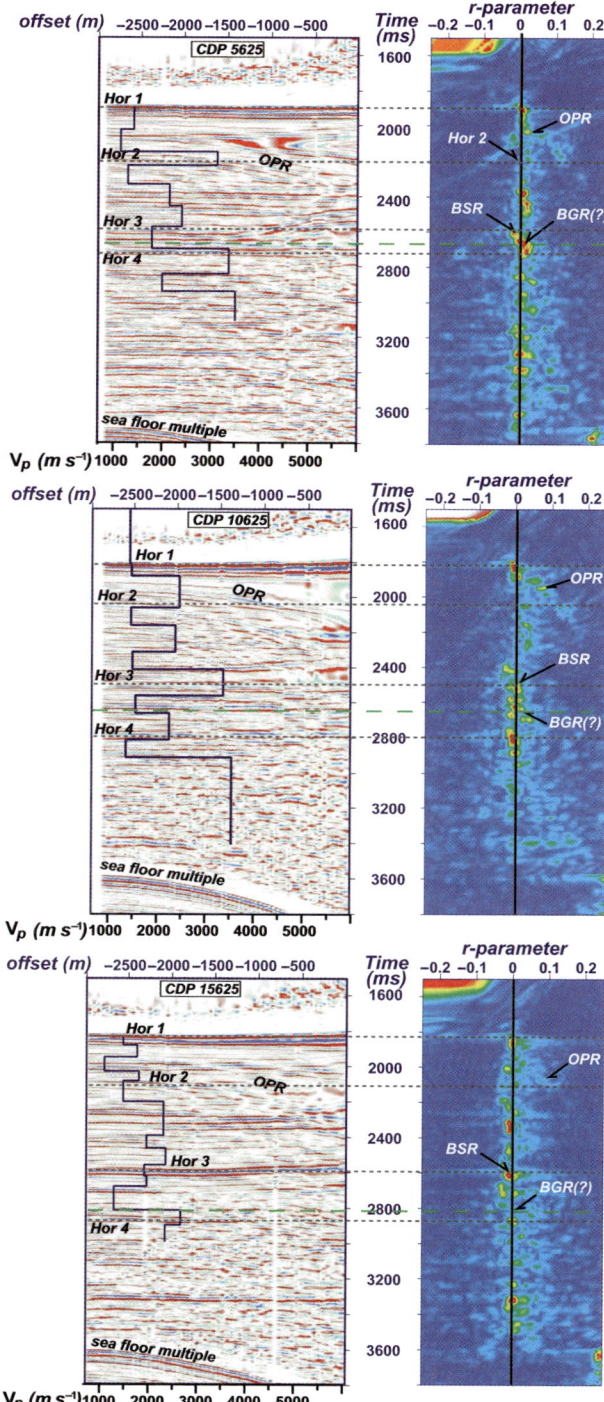

Fig. 4. CDPs after residual NMO correction (left) and residual semblance analysis (right) at the three selected location (see arrows in Fig. 2). The *r*-parameter represents the time velocity error as obtained by semblance analysis. The blue profiles are the interval velocities after residual move-out analyses. The main horizons are also indicated with dashed grey lines. The BSR and the BGR (dashed green lines) are indicated. OPR, out-of-plane reflection.

and reflection time of the investigated layer (Hajnal & Sereda 1981). Supposing that the error in stacking velocity is equal to 10 m s^{-1}, the average interval velocity error for the layer with small thickness assumes a value of about 450 m s^{-1}. This error is lower than the interval velocity variation for almost all layers (Fig. 4), confirming the reliability of the presence of the low velocity layers and small thickness. Moreover, the top and bottom of each layer is associated with a reflection caused by acoustic impedance change.

The CDP semblance analyses show some intervals with very low apparent velocity in the first layer. These low velocities are associated with out-of-plane reflections ('sideswipe'; labelled OPR in Fig. 4), probably due to the irregular morphology of the investigated strata. Note that there is strong normal moveout stretching at large offsets on the CDPs in Figure 4 after the residual move-out correction.

Residual move-out of CIGs

It is well known that pre-stack depth migration provides information about the quality of the velocity field (Yilmaz 2001). In fact, when an incorrect velocity is used to migrate multichannel data, the imaged depths in a CIG will differ from each other. In this situation, residual move-out is observed in migrated data; for this reason, residual move-out analysis is used to update the migration velocity (i.e. Liu 1995). In particular, if the residual semblance energy has mainly positive values, it indicates that the correct migration velocity is lower than the migration velocity used; on the contrary, if the residual semblance indicates negative values, the correct velocity is higher.

We carried out residual semblance analyses on our final CIGs using the free software Seismic Unix (Cohen & Stockwell 2001) as an additional check on the accuracy of the velocity field used to perform the migration and the local velocity deviations associated with free gas zones. We obtained a three-dimensional semblance cube, where the x-, y- and z-axes refer to the distance, the r-parameter associated with the residual semblance (the r-parameter represents the depth velocity error; Liu 1995) and the depth, respectively. Observing the three-dimensional cube, we were able to extract information about local velocity deviations associated with layers, characterized by small thickness, with low or high velocities with respect to the migration velocity. In particular, we used these analyses to detect the real base of the free gas zone and the local presence of the free gas in the hydrate stability zone.

To visualize the result, three vertical panels are shown in Figure 5, corresponding to three values of the r-parameter: −0.01 (top), 0.0 (middle) and 0.01 (bottom). The images highlight the deviations of the actual seismic velocities from the migration velocity field. In particular, the upper panel furnishes information on the presence of high-velocity zones with respect to the migration velocity, while the bottom panel indicates the areas where the actual velocity is lower than the migration velocity. Therefore, the regions that have been migrated with a wrong velocity can be detected by comparing the three panels.

Within layer 2 a prominent reflector (see white arrow in Fig. 5) may be associated with the base of an internal free gas layer, as suggested by the high concentration of energy in the positive r-parameter panel; this means that inside this low-velocity layer (about 50–100 m thick) the true velocity is lower than the migration velocity. Within layer 3, the residual semblance energy is higher in the positive panel (see black arrows in the lower r-parameter panel). These regions indicate the presence of low-velocity layers, as also detected by detailed velocity analysis (see Fig. 4). The low-velocity layers characterized by small thickness can be associated with reflections evident in the pre-stack depth migration (see white arrows in Fig. 2). The free gas layer thickness below the BSR probably ranges between about 100 and 250 m, confirming the results of previous studies in the area, in which a variable thickness of the free gas zone was interpreted (i.e. Tinivella et al. 2002).

To obtain better definition of the base of the free gas reflector between the CIGs 12000 and 15000, we produced eight depth slices of the three-dimensional cube (Fig. 6). The slices were extracted every 50 m, starting from 1900 m depth. The first two slices correspond approximatively to the depth of Hor. 3 (BSR), as confirmed by the strong semblance energy. The deeper slices highlight energy concentrated at positive r-parameter values (see arrows), which we interpret as being associated with the BGR. In this case, the depth of this reflector is between 2100 and 2150 m. Horizon 4 corresponds approximatively to the last slice (2250 m depth) in which the energy is well focused around the zero (or small negative values) of the r-parameter, indicating that the deeper velocity migration is correct or, locally, slightly underestimated.

Figure 7 shows three selected CIGs at the same location of the CDP analyses (see red arrows in Fig. 2) after the migration performed with constant velocity (1500 m s^{-1}; left), the final migration (middle) and the residual semblance analyses (right). The flatness of the main reflections confirms the accuracy of the final velocity field. To facilitate comparison of these results with the detailed CDP velocity analyses, we have overlaid both the tomographic velocities (green profiles) and the interval

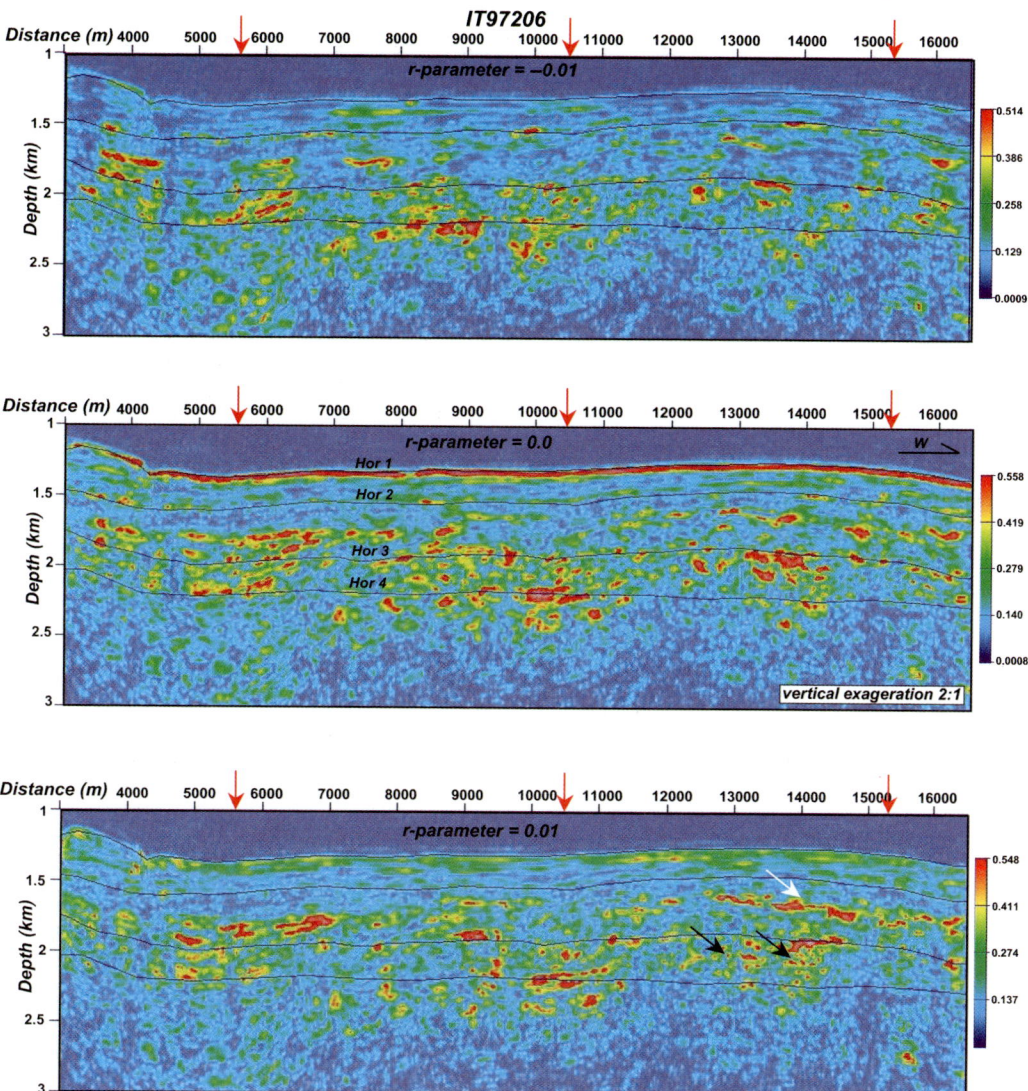

Fig. 5. Semblance analysis of CIGs for three values of r-parameter: -0.1 (top); 0.0 (middle); 0.1 (bottom). The distance in metres corresponds to the CIG number. The r-parameter represents the depth velocity error as obtained by semblance analysis. Black lines indicate the four main horizons. White and black arrows are referred to low velocity layers above and below the BSR respectively. The three red arrows correspond to the location of CIGs and CDPs selected for detailed analysis (see text).

velocities extracted from the CDP analysis (see above; blue profiles) on the final migration CIGs. Note that the shallow apparent low velocity events (labelled OPR in Fig. 7 and described above) are detected by the semblance analyses.

The velocity in the layer 2 is an average between low and high velocities, produced probably by an alternation of free gas and gas hydrate layers. Similar velocity trends above BSRs have been observed by several authors (e.g. Mienert & Posewang 1999; Bünz et al. 2005). Comparing the two sets of velocity profiles, the local absence of a velocity inversion between layers 2 and 3 can be understood. In fact, layer 3 is about 300 m thick, and within it there are several free gas layers characterized by small thicknesses separated by water-saturated sediments, as clearly observed at CIGs 10625 and 15625 (Fig. 7). Therefore, the average

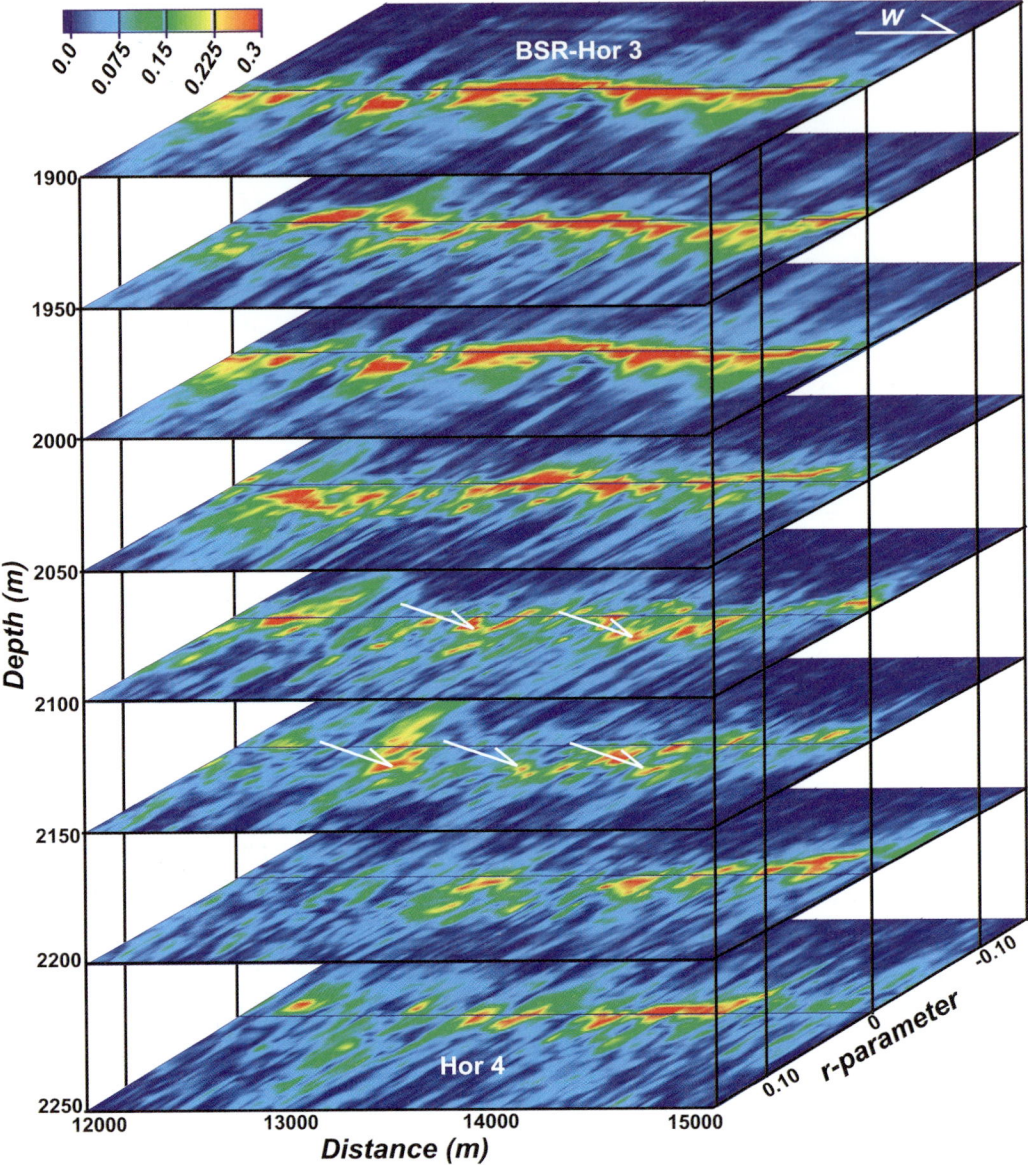

Fig. 6. Depth slices of three-dimensional semblance analysis of the CIGs. The distance in metres corresponds to the CIG number. The first two slices correspond to Hor. 3 (BSR), while the last one to Hor. 4. The white arrows indicate the possible location of the BGR or low-velocity events.

velocity can mask the presence of free gas. The semblance analyses highlight energy concentrations related to layers with seismic velocities lower than the migration velocity field within layer 3, validating the detailed velocity analyses on CDPs. These low-velocity events could correspond to the real BGR (Fig. 7). Finally, the dashed lines in Figure 7 show that the positive values of the r-parameter, which correspond to lower migration velocity with respect to actual velocity, are at the same depth of the low-velocity layers detected by the residual NMO analysis. Thus, even if the r-parameter value cannot be directly used to quantify the velocity error, in this case, we can associate these lower velocities with the free-gas bearing sediment layers.

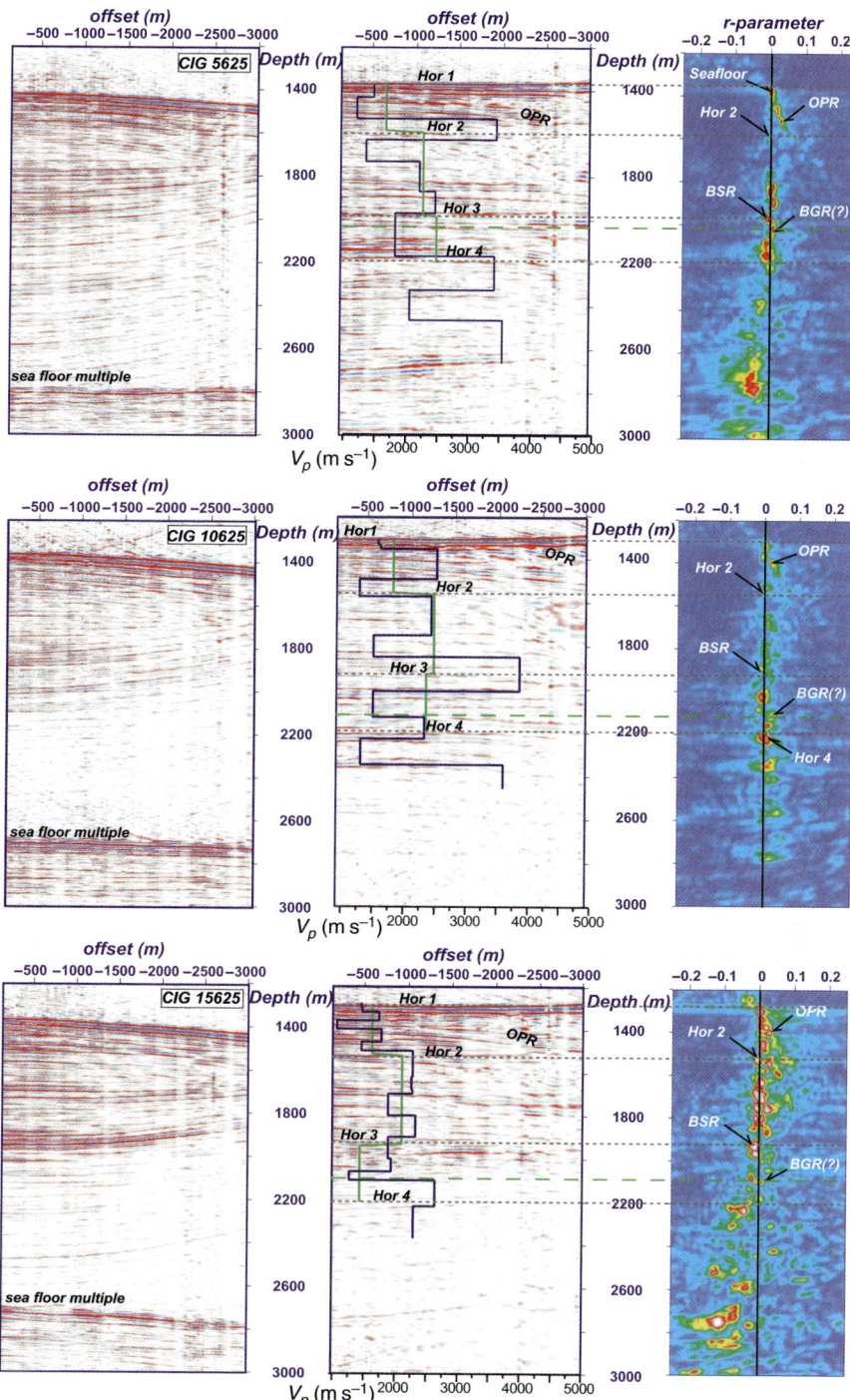

Fig. 7. Top, CIG 5625; middle, CIG 10625; bottom, CIG 15625. Left panels: CIGs after the pre-stack depth migration performed with a constant velocity field equal to 1500 m s^{-1}. Centre panels: CIGs after final pre-stack depth migration. Green profiles: tomographic velocities. Blue profiles: velocity after residual move-out analysis. The main horizons are also indicated with dashed grey lines. The BSR and the BGR (dashed green lines) are indicated. OPR, out-of-plane reflection. Right panels: semblance analysis of the selected CIGs, in which the main horizons are indicated.

Gas phase concentration

The velocity field can be translated in terms of gas hydrate and free gas concentration; for this purpose, we used the methodology described in Tinivella (1999) and tested in this area to quantify the gas hydrate and the free gas in the pore space (i.e. Tinivella et al. 2002). The compressional (V_p) wave velocity is expressed as (see explanation of symbols in Table 1):

$$V_p = \left\{ \left[\left(\frac{1}{C_m} + \frac{4}{3}\mu \right) \right. \right.$$

$$\left. + \frac{\frac{\phi_{eff}}{k}\frac{\rho_m}{\rho_f} + \left(1 - \beta - 2 \cdot \frac{\phi_{eff}}{k}\right) \cdot (1-\beta)}{(1 - \phi_{eff} - \beta)C_b + \phi_{eff}C_f} \right]$$

$$\left. \times \frac{1}{\rho_m\left(1 - \frac{\phi_{eff}}{k}\frac{\rho_f}{\rho_m}\right)} \right\}^{1/2} \quad (3)$$

The reference curves, i.e. the physical parameters v. depth for water-saturated sediments, are the Hamilton curves (Hamilton 1976, 1979); we used the average Poisson ratio for all sediments equal to 0.435, obtained by OBS data analysis in the same area (Tinivella & Accaino 2000). Local analysis suggests that free gas is uniformly distributed in the pore space (Tinivella & Accaino 2000); nevertheless, to determine the range of possible estimation for both gas hydrate and free gas amounts, we evaluated the concentration in two cases: (1) gas hydrate and free gas uniformly distributed and (2) gas hydrate randomly and free gas patchily distributed in the pore space. The compressibility (C_{tot}) is obtained in the uniform case by

$$C_{tot} = s_1 C_1 + s_2 C_2 \quad (4)$$

and in the random case by

$$C_{tot}^{-1} = s_1 C_1^{-1} + s_2 C_2^{-1} \quad (5)$$

where the index denotes the type of solid or fluid and the symbol s is the saturation of each compound.

In the case of the free gas patchily distributed, the fluid compressibility is given by

$$C_f^{-1} = \frac{4}{3}\mu \left\{ \left[\frac{S_w}{1 + (4/3\mu \cdot C_w)^{-1}} \right. \right.$$

$$\left. \left. + \frac{S_g}{1 + (4/3\mu \cdot C_g)^{-1}} \right]^{-1} - 1 \right\} \quad (6)$$

where the symbols are described in Table 1. We recall that the patchy distribution signifies that all water is concentrated in fully saturated patches, and gas is concentrated in patches without water (e.g. Dvorkin et al. 1999). The gas phase sections are shown in Figure 8. Note that the hydrate concentration ranges from 22%, in the random case, to 27% by volume, in the uniform case. The variation of the free gas concentration in the two hypotheses varies from 0.5 to 11%. It is interesting to note that the concentration trend for gas hydrate is opposite to that for the free gas: the estimated gas hydrate concentration is higher in the uniformly distributed case, whereas the estimated free gas concentration is higher in the patchily distributed case. Moreover, gas hydrate is shown to be widely distributed within layer 2, whereas free gas is present only in localized areas below Hor. 3 and is characterized by strong variation of concentration. As indicated in the detailed velocity analyses reported above, the free gas layer thickness and its velocities are overestimated by the inversion procedure and the real free gas concentration is higher than the evaluated concentration shown in Figure 8.

Note that the concentration estimations are affected by percentage errors that could be equal to about ± 20 and $\pm 7\%$ in the case of gas hydrate and free gas, respectively, as established by previous sensitivity tests performed in the same area using the same reference curves (Tinivella et al. 2002). This high value is related to the fact that no drilling data are available and the reference curves are extrapolated using seismic data where the BSR is not present.

Discussion and conclusion

The velocity analysis revealed the presence of three main layers characterizing the first kilometre of sediments below the seafloor. The first layer is characterized by an average velocity of 1735 m s^{-1} with an average thickness of 250 m, as revealed by both tomographic and semblance analyses. This velocity is higher than the normal compacted accretionary prism sediment velocity (Hamilton 1979). We suggest two possible explanations; the first one is that this high velocity could be related to low gas hydrate concentrations in the pore space. In this case, the velocity indicates an average hydrate concentration by volume ranging from $5.8 \pm 1.2\%$ (in the case of random hydrate distribution) to $6.9 \pm 1.4\%$ (in the case of uniform distribution). The second hypothesis, about the nature of this anomalous shallow high velocity, is related to the presence of biogenic silica (Opal-A) in low concentration. In fact, offshore from part of the Antarctic Peninsula to the south of our study area, Ocean Drilling Program Leg 178 results revealed high velocities and biogenic silica concentrations within the upper few hundred metres of sediment drift

Table 1. *List of parameters in equation (3) (after Tinivella 1999)*

ϕ	Porosity
ϕ_s	Solid proportion
ϕ_h	Gas hydrate proportion
ϕ_w	Water proportion
ϕ_g	Free gas proportion
$\phi_s + \phi_w + \phi_g = 1$	
$\phi_s + \phi_h + \phi_w = 1$	
$C_h = \phi_h/(\phi_h + \phi_w)$	Gas hydrate concentration
$s_s = \phi_s/(\phi_s + \phi_h)$	Grain saturation
$s_h = \phi_h/(\phi_h + \phi_s)$	Gas hydrate saturation
$s_w = \phi_w/(\phi_w + \phi_g)$	Water saturation
$s_g = \phi_g/(\phi_w + \phi_g)$	Free gas saturation
$\phi_{eff} = (1 - C_h)\phi$	Effective porosity
C_s	Grain compressibility
C_h	Gas hydrate compressibility
C_w	Water compressibility
C_g	Free gas compressibility
C_b	Hill average compressibility of the solid phase
C_f	Compressibility of the fluid phase
C_p	Pore compressibility
$C_m = (1 - \phi_{eff})C_b + \phi_{eff} C_p$	Compressibility of the matrix
$\beta = C_b/C_m$	
ρ_s	Grain density
ρ_h	Gas hydrate density
ρ_w	Water density
ρ_g	Gas density
$\rho_b = s_s\rho_s + s_h\rho_h$	Density of the solid phase
$\rho_f = s_w\rho_w + s_g\rho_g$	Density of the fluid phase
$\rho_b = (1 - \phi_{eff})\rho_b + \phi_{eff}\rho_f$	Average density
μ_{sm0}	Solid matrix shear modulus (no cementation)
μ_{smKT}	Kuster & Toksöz's (1974) shear modulus
$\mu_{sm} = (\mu_{smKT} - \mu_{sm0})[\phi_h/(1 - \phi_s)]^{3.8} + \mu_{sm0}$	Solid matrix shear modulus (percolation theory; Leclaire 1992)
μ_s	Grain rigidity
μ_h	Gas hydrate rigidity
$\mu = (\phi_s + \phi_h)(\phi_s s_s/\mu_{sm} + s_h/\mu_h)^{-1}$	Average rigidity of the skeleton
k	Coupling factor

deposits (Lonsdale 1990; Volpi *et al.* 2003). In this hypothesis, our average velocity equates to an average silica concentration of about 10%.

The second layer is characterized by high average velocity (2220 m s^{-1}) and an average thickness of 345 m. The bottom of this layer in the western part of the section corresponds to the BSR, while in the sedimentary basin the presence of the BSR is not clear because of the seismo-stratigraphy and structural features (faults) affecting sediments; in particular, the faults may act as conduits for gas to escape (Fig. 2). This probably explains why the calculated gas hydrate concentration is lower compared with the western area (Fig. 8). The top of this layer is a continuous reflector, characterized by normal phase, and it simulates the seafloor. If we assume that marine sediments in the first layer do not contain hydrate in the pore space, Hor. 2 could be locally interpreted as the top of the hydrate layer, as already observed in other areas (Posewang & Mienert 1999). This layer is characterized by an average hydrate concentration that ranges from 14.8 ± 3.0% (in the random distribution of gas hydrate case) to 17.7 ± 3.5% of volume (in the uniform distribution case). The hydrate concentration could be less than our estimation if significant concentrations of biogenic silica are present. On the other hand, the presence of a low velocity layer, detected by the detailed velocity analyses (Figs 4 & 7), decreases the average velocity in the whole second layer, causing an underestimate of the average hydrate concentration. This low-velocity layer may be associated with sediments with free gas in the pore space, as discussed, or, locally, different lithology.

The layer 3 is characterized by a low average velocity (1660 m s^{-1}) and an average thickness of 300 m. This low velocity can be interpreted as being due to the presence of free gas in the pore space with an average concentration ranging

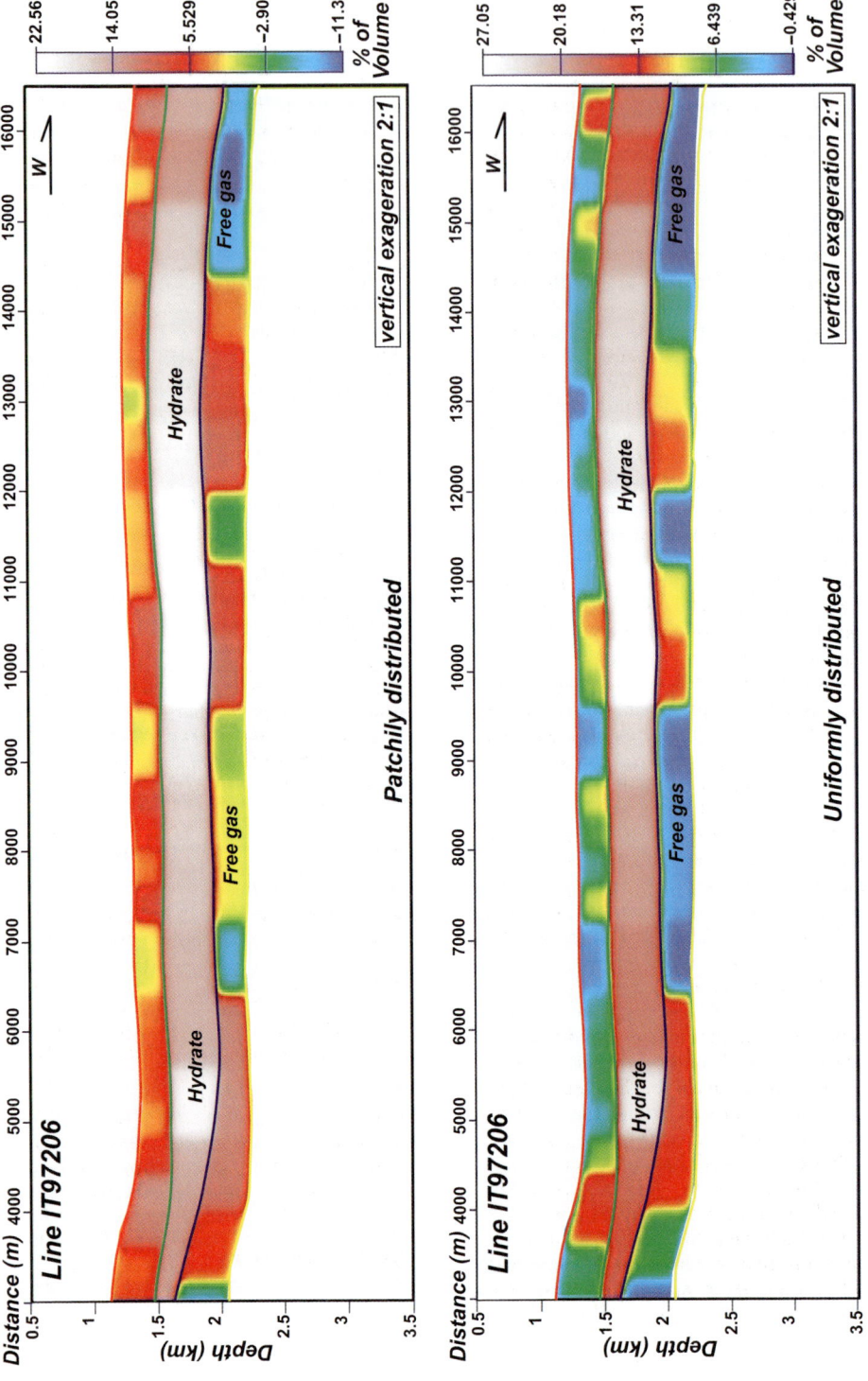

Fig. 8. Gas phase concentration obtained from smoothed velocity field considering random distribution of gas hydrate and patchy distribution of free gas (top) and uniform distribution of both gas phases (bottom). The positive and negative values are referred to gas hydrate and free gas concentrations, respectively. Note that the negative values above the BSR indicate the free gas presence and the positive values below the BSR indicate the gas hydrate presence. The four inverted horizons are indicated.

from $0.3 \pm 0.02\%$ (in the uniform hypothesis) to $8.9 \pm 0.6\%$ (in the patchy hypothesis). Also in this layer, biogenic silica may be present in the sediments, because the Opal A/C transition occurs at greater depth than the BSR produced by the hydrate/free gas transition. In this case, the free gas concentration is underestimated. Moreover, the detailed velocity analyses indicate that the free gas layer is thinner than layer 3 (about 100 m; see Fig. 6). Therefore, the concentration is underestimated because the average velocity includes layers without free gas in the pore space. Finally, Figure 7 indicates that free gas is present in one or more layers with small thickness below the BSR between gas free sediments.

Supposing the first hypothesis for the layer 1 to be valid, we can assume that in this area the gas hydrate layer is 595 m thick with a concentration of $5.8 \pm 1.2\%$ in the first 250 m and $14.8 \pm 3.0\%$ in the lower part. On the other hand, following the second hypothesis, the gas hydrate layer is 345 m thick with a minimum concentration of $4.8 \pm 1.0\%$ (considering that about 10% of biogenic silica is present). Moreover, we can assume that the free gas layer is 100 m thick with a minimum concentration of $0.3 \pm 0.02\%$ of volume. Previous interpretation of the entire seismic dataset available from the South Shetland margin (Lodolo et al. 2002) indicated that the area characterized by a continuous BSR, over which the new bathymetric data were acquired, is about 1170 km^2, as indicated in the dashed box in Figure 1. We can extrapolate that about 76.7 ± 15.6 km^3 of gas hydrate volume if the gas hydrate fills sediments from the sea bottom and 19.4 ± 4.0 km^3 of gas hydrate volume if biogenic silica is present in the pore space above the BSR. Finally, 0.35 ± 0.02 km^3 of free gas volume are trapped in the marine sediments below the BSR.

In conclusion, the tomographic analysis of CIGs can be considered as a useful tool to determine the velocity field at a regional scale and the seismic image in depth, reducing the human time with respect to other detailed inversion procedures without loss of precision. In fact, our velocity inversion is in agreement with other detailed analyses performed in the study area (i.e. Tinivella et al. 2002). Moreover, this procedure can be used to provide indications about layers characterized by anomalous physical properties with respect to the surrounding sediments. Thus, by jointly using the tomographic velocity model, the residual semblance analyses of the CIGs and the theoretical models, we can obtain information about the hydrate and free gas thickness and their relative amounts. This work underlines that several features can over- or underestimate the gas phase amount, such as the presence of biogenic silica or structural elements that can favour local fluid escape or accumulation.

These features can be detected by depth migration combined with local detailed analysis of CDPs and CIGs. Finally, the velocity model and the related gas-phase sections showed that there is probably more gas hydrate to the west than to the east, and that gas is concentrated in different parts of the profile than where hydrate is concentrated. This observation confirms that geological structures and processes control the gas and hydrate distribution, as observed along other margins by several authors (Ashi et al. 2002; Bünz et al. 2005).

We are very grateful to Graham Westbrook for useful scientific suggestions and to Serguei Bouriac for technical support in the use of Radex software. Thanks to Neslihan Ocakoğlu for preliminary review of the manuscript. An insightful review by Robert Larter and Tim Minshull helped to improve the manuscript. This work is partially supported by PNRA.

References

ACCAINO, F. & TINIVELLA, U. 2005. Gas hydrates and active fluid outflow offshore the South Shetland Margin. *5th ICGH Trondheim.*

ACCAINO, F., BOHM, G. & TINIVELLA, U. 2005. Tomographic inversion of common image gathers. *First Break*, **23**, 39–44.

ARTS, R., EIKEN, O., CHADWICK, R. A., ZWEIGEL, P., VAN DER MEER, L. & ZINSZNER, B. 2004. Monitoring of CO_2 injected at Sleipner using time-lapse seismic data. *Energy*, **29**, 1383–1392.

ASHI, J., TOKUYAMA, H. & TAIRA, A. 2002. Distribution of methane hydrate BSRs and its implication for the prism growth in the Nankai Trough. *Marine Geology*, **187**, 177–191.

BÖHM, G., ROSSI, G. & VESNAVER, A. 1999. Minimum time ray-tracing for 3-D irregular grids. *Journal of Seismic Exploration*, **8**, 117–121.

BÜNZ, S., MIENERT, J., BRYN, P. & BERG, K. 2005. Fluid flow impact on slope failure from 3D seismic data: A case study in the Storegga Slide. *Basin Research*, **17**, 109–122.

CHADWICK, R. A., ZWEIGEL, P., GREGERSEN, U., KIRBY, G. A., HOLLOWAY, G. A. S. & JOHANNSSEN, P. N. 2002. Geological characterisation of CO_2 storage sites: Lessons from Sleipner, Northern North Sea. *6th International Conference on Greenhouse Gas Control Technology*, Kyoto, Japan, expanded abstract.

CHAND, S. & MINSHULL, T. A. 2003. Seismic constraints on the effects of gas hydrate on sediment physical properties and fluid flow: A review. *Geofluids*, **3**, 1–15.

COHEN, J. K. & STOCKWELL, J. W. JR. 2001. *CWP/SU: Seismic Unix Release 35: A Free Package for Seismic Research and Processing.* Center for Wave Phenomena, Colorado School of Mines.

COX, J. L. 1983. *Natural Gas Hydrates: Properties, Occurrence and Recovery.* Butterworth, London.

DIETRICH, R., RÜLKE, A. ET AL. 2004. Plate kinematics and deformation status of the Antarctic Peninsula based on GPS. *Global and Planetary Change*, **42**, 313–321.

DIX, C. H. 1955. Seismic velocities from surface measurements. *Geophysics*, **20**, 68–86.

DVORKIN, J., MOOS, D., PACKWOOD, J. L. & NUR, A. M. 1999. Identifying patchy saturation from well logs. *Geophysics*, **64**, 1756–1759.

GRAD, M., GUTERCH, A. & JANIK, T. 1993. Seismic structure of the lithosphere across the zone of subducted Drake plate under the Antarctic plate, West Antarctica. *Geophysical Journal International*, **115**, 586–600.

HAJANL, Z. & SEREDA, I. T. 1981. Maximum uncertainty of interval velocity estimates. *Geophysics*, **46**, 1543–1547.

HAMILTON, E. L. 1976. Variations of density and porosity with depth in deep-sea sediments. *Journal of Sedimentary Petrology*, **46**, 280–300.

HAMILTON, E. L. 1979. V_p/V_s and Poisson's ratios in marine sediments and rocks. *Journal of the Acoustic Society of America*, **66**, 1093–1101.

HYNDMAN, R. D. & SPENCE, G. D. 1992. A seismic study of methane hydrate marine bottom simulating reflectors. *Journal of Geophysical Research*, **97**, 6683–6698.

JIN, Y. K., LARTER, R. D., KIM, Y., NAM, S. H. & KIM, K. J. 2002. Post-subduction margin structures along Boyd Strait, Antarctic Peninsula. *Tectonophysics*, **346**, 187–200.

KIM, Y., KIM, H.-S., LARTER, R. D., CAMERLENGHI, A., GAMBÔA, L. A. P. & RUDOWSKI, S. 1995. Tectonic deformation in the upper crust and sediments at the South Shetland Trench. *In*: COOPER, A. K., BARKER, P. T. & BRANCOLINI, G. (eds) *Geology and Seismic Stratigraphy of the Atlantic Margin*. Antarctic Research Series, **68**. AGU, Washington, DC, 157–166.

KUSTER, G. T. & TOKSÖZ, M. N. 1974. Velocity and attenuation of seismic waves in two-phase media: Part I. Theoretical formulations. *Geophysics*, **39**, 587–606.

LARTER, R. D. & BARKER, P. F. 1991. Effects of ridge crest-trench interaction on Antarctic-Phoenix spreading: Forces on a young subducting plate. *Journal of Geophysical Research*, **96**, 19583–19607.

LAWVER, L. A., SLOAN, B. J. *ET AL*. 1996. Distributed active extension in Bransfield basin, Antarctic Peninsula: Evidence from multibeam bathymetry. *GSA Today*, **6**(11), 1–6.

LECLAIRE, P. 1992. *Propagation acoustique dans les milieux poreoux soumis au gel – modélisation et expérience*. Thèse de Doctorat en Physique, Université Paris 7, Paris.

LEDLEY, T. S., SUNDQUIST, E. T., SCHWARTZ, S. E., HALL, D. K., FELLOWS, J. D. & KILLEEN, T. L. 1999. Climate Change and Greenhouse Gases. *EOS, Transactions of the American Geophysical Union*, **80**, 453–458.

LIU, Z. 1995. *Migration velocity analysis*. PhD thesis, Colorado School of Mines, CWP **168**.

LIVERMORE, R., BALANYÁ, J. C. *ET AL*. 2000. Autopsy on a dead spreading center: The Phoenix Ridge, Drake Passage, Antarctica. *Geology*, **28**, 607–610.

LODOLO, E., CAMERLENGHI, A. & BRANCOLINI, G. 1993. A bottom simulating reflector on the South Shetland margin, Antarctic Peninsula. *Antarctic Science*, **5**, 201–210.

LODOLO, E., CAMERLENGHI, A., MADRUSSANI, G., TINIVELLA, U. & ROSSI, G. 2002. Assessment of gas hydrate and free gas distribution on the South Shetland margin (Antarctica) based on multichannel seismic reflection data. *Geophysical Journal International*, **148**, 103–119.

LONSDALE, M. J. 1990. The relationship between silica diagenesis, methane, and seismic reflections on the south orkney microcontinent. *In*: BARKER, P. R., KENNETT, J. P. *ET AL*. (eds) *Proceedings of ODP, Scientific Results*, **113**. Ocean Drilling Program, College Station, TX; doi: 10.2973/odp.proc.sr.113.177.1990.

LORETO, M. F., DELLA VEDOVA, B., ACCAINO, F., TINIVELLA, U. & ACCETTELLA, D. 2006. Shallow geological structures of the South Shetland trench, Antarctic Peninsula (SLAPPSS project). *Ofioliti*, **31**, 135–143.

MALDONADO, A., LARTER, R. D. & ALDAYA, F. 1994. Forearc tectonic evolution of the South Shetland Margin, Antarctic Peninsula. *Tectonics*, **13**, 1345–1370.

MAX, M. D. 2003. *Natural Gas Hydrate in Oceanic and Permafrost Environments*. Kluwer Academic, Dordrecht.

MIENERT, J. & POSEWANG, J. 1999. Evidence of shallow- and deep-water gas hydrate destabilizations in North Atlantic polar continental margin sediments. *Geo-Maine Letters*, **19**, 143–149.

MINSHULL, T. A., SINGH, S. C. & WESTBROOK, G. K. 1994. Seismic velocity structure at a gas hydrate reflector, offshore western Columbia, from full waveform inversion. *Journal of Geophysical Research*, **99**, 4715–4734.

POSEWANG, J. & MIENERT, J. 1999. The enigma of double BSRs: Indicators for changes in the hydrate stability field? *Geo-Maine Letters*, **19**, 157–163.

SAFFER, D. M., SILVER, E. A., FISHER, A. T., TOBIN, H. & MORAN, K. 2000. Inferred pore pressures at the Costa Rica subduction zone: Implications for dewatering processes. *Earth and Planetary Science Letters*, **177**, 193–207.

SAIN, K., MINSHULL, T. A., SINGH, S. C. & HOBBS, R. W. 2000. Evidence for a thick free gas layer beneath the bottom simulating reflector in the Makran accretionary prism. *Marine Geology*, **164**, 37–51.

SLOAN, E. D. JR. 1998. *Clathrates of Natural Gases*. Marcel Dekker, New York.

STEWART, R. R. 1991. *Exploration Seismic Tomography: Fundamentals*. Course Note Series, **3**. Society of Exploration Geophysicists, Tulsa, OK.

TINIVELLA, U. 1999. A method for estimating gas hydrate and free gas concentrations in marine sediments. *Bollettino di Geofisica Teorica ed Applicata*, **40**(1), 19–30.

TINIVELLA, U. & ACCAINO, F. 2000. Compressional velocity structure and Poisson's ratio in marine sediments with gas hydrate and free gas by inversion of reflected and refracted seismic data (South Shetland Islands, Antarctica). *Marine Geology*, **164**, 13–27.

TINIVELLA, U., LODOLO, E., CAMERLENGHI, A. & BOEHM, G. 1998. Seismic tomography study of a bottom simulating reflector off the South Shetland Islands (Antarctica). *In*: HENNIET, J.-P. & MIENERT, J. (eds) *Gas Hydrate: Relevance to World Margin*

Stability and Climate Change. Geological Society, London, Special Publications, **137**, 141–151.

TINIVELLA, U., ACCAINO, F. & CAMERLENGHI, A. 2002. Gas hydrate and free gas distribution from inversion of seismic data on the South Shetland margin (Antarctica). *Marine Geophysical Research*, **23**, 109–123.

VAN DER SLUIS, A. & VAN DER VORST, H. A. 1987. Numerical solutions of large, sparse linear systems arising from tomographic problems. *In*: NOLET, G. (ed.) *Seismic Tomography*. Reidel, Dordrecht, 49–84.

VOLPI, V., CAMERLENGHI, A., HILLENBRAND, C.-D., REBESCO, M. & IVALDI, R. 2003. Effects of biogenic silica on sediment compaction and slope stability on the Pacific margin of the Antarctic Peninsula. *Basin Research*, **15**, 339–363.

YILMAZ, O. 2001. *Seismic Data Analysis: Processing, Inversion and Interpretation of Seismic Data*. Series: Investigation in Geophysics. Society of Exploration Geophysicists, Tulsa, OK, **10**.

ZILLMER, M. 2006. A method for determining gas-hydrate or free-gas saturation of porous media from seismic measurements. *Geophysics*, **71**, 21–32.

Mimicking natural systems: methane hydrate formation–decomposition in depleted sediments

M. W. EATON[1], K. W. JONES[2] & DEVINDER MAHAJAN[1,3]*

[1]*Materials Science and Engineering, Stony Brook University, Stony Brook, NY 11794, USA*
[2]*Environmental Sciences Department, Brookhaven National Laboratory, Upton, NY 11973, USA*
[3]*Energy Sciences and Technology, Brookhaven National Laboratory, Upton, NY 11973, USA*
*Corresponding author (e-mail: dmahajan@bnl.gov)

Abstract: We have initiated a systematic study of sediment–hydrate interaction under subsurface-mimic conditions to initially focus on marine hydrates. A major obstacle to studying natural hydrate systems has been the absence of a sophisticated mimic apparatus in which the hydrate formation phenomenon can be reproduced with precision. We have designed and constructed a bench-top unit, namely flexible integrated study of hydrates (FISH), for this purpose. The unit is fully instrumented to precisely record temperatures, pressures and changes in gas volume during absorption/evolution. The Labview software allows rapid and continuous data collection during the hydrate formation/dissociation cycle. In our integrated approach, several host sediments collected from Blake Ridge, a well-researched hydrate site, were characterized using the computed microtomography technique at Beamline X-26A of the National Synchrotron Light Source at Brookhaven National Laboratory. The characterized depleted sediments were then used to study the hydrate formation/decomposition kinetics under various pressures in the FISH unit. We report two hydrate formation methods: one under continuous methane gas-flow conditions (dynamic mode) and the other in which hydrates are formed from the dissolved gas phase by diffusion (static mode). Also reported is a depressurization method, namely the step-down pressure method, to yield gas evolution data. Data from such runs with host sediment from the deepest site (667 metres) is presented. During hydrate formation, the data reveals a temperature signature that is consistent with an exothermic hydrate formation event. In the decomposition cycle, data at various pressures was analysed to yield curves with similar slopes, suggesting a zero-order dependence. The capabilities of the FISH unit and the implications of these runs in establishing a database of sediment–hydrate kinetics and pore saturation are discussed.

Gas hydrates belong to a general class of inclusion compounds commonly known as clathrates: a compound of molecular cage structure made of host molecules encapsulating guest molecules. Common clathrate compounds of interest are those formed from CO_2-H_2O and CH_4-H_2O mixtures, the former for application in carbon sequestration and the latter for methane extraction (Sloan 1998), although CH_4 exchange with CO_2 within hydrates has been suggested as a method to extract CH_4 (Mahajan & Taylor 2006). Natural gas hydrate is a naturally occurring ice-like solid which is made from water molecules assembled via hydrogen bonding to form polyhedral cavities as the cage (host) in which other molecules (mostly methane) can exist as the guest. The physical appearance of the natural gas hydrate is close to pure frozen water, i.e. the ice phase. At standard pressure and temperature, a methane hydrate molecule contains approximately 164 volumes of methane for each volume of water.

Until recently, methane hydrates, known to scientists for almost 200 years, have remained a scientific curiosity. It was not until the 1930s that it was realized that methane hydrate was responsible for plugging natural gas pipelines, particularly those located in cold environments. For the next 40 years, a small body of researchers investigated the physics of various clathrates, including the construction of the first predictive models of their formation. A prime focus of this work continues to be the development of chemical additives and other methods to inhibit hydrate formation (Sloan 1998, 2004).

Methane hydrates have been discovered in the subsurface in permafrost regions, but most occur in oceanic sediments hundreds of metres below the seafloor where water depths are greater than 500 m. The amount of methane gas that is tied up as hydrates beneath the seafloor and in permafrost is several orders of magnitude higher than all other known conventional sources of methane. According to the US Geological Survey the estimated global natural gas hydrate reserves is in the range of 100 000 to about 300 000 000 trillion cubic feet (Collett 1995, 2002). This estimate, when compared with the 13 000 trillion cubic feet of conventional natural gas reserves, demonstrates the vastly more abundant natural gas hydrates around the globe. The Department of Energy (DOE) now estimates about 1% recovery of methane from the known methane hydrate reserves within the US would be enough (over 2000 trillion cubic feet) to satisfy US consumption for the next several decades (Boswell 2007). The potential of developing this indigenous resource has recently led hydrate research and development, with resource characterization (Sloan 1998; Collett 2002), mining and transport by economical conversion to liquid fuels attracting increased attention (Wegrzyn et al. 1999), with the ultimate goal of developing an environmentally benign method.

The characteristic instability of methane hydrates under ambient temperature and pressure conditions has been the subject of much scientific interest (Sloan 1998; Goel et al. 2001). Furthermore, the system complexity also stems from occurrence of hydrates in sediments. Knowledge critical to simulating hydrate formation/dissociation in sediments includes: (a) phase-diagram and kinetic data for hydrate formation; (b) thermo-physical properties of methane hydrate and sediments; (c) structure properties and permeability of sediments; and (d) advanced numerical methods for transport phenomena in porous media. Numerical modelling to obtain quantitative information on methane hydrate formation/dissociation in marine sediment conditions continues to progress (Moridis et al. 2005; Zheng et al. 2009).

The occurrence of marine hydrates within sediment matrix has serious implications with respect to site stability with respect to its surroundings during methane extraction. Other environmental consequences such as the impact of fresh water production (a by-product of methane extraction) on species that are normally found in the vicinity of hydrate mounds (ice worms and other microorganisms) pose a serious scientific challenge in developing a suitable production method (Mahajan & Somasundaran 2006).

The focus of the study presented in this paper was to initiate data collection in a sophisticated laboratory unit that could mimic seafloor conditions that sustain hydrates in the natural environment. Relevant to this aspect are the identification of pure gas hydrate structure classification and the type of hydrate structure in sediments. Of the 130 compounds (guest molecules) that are known to form clathrate hydrates, those involving natural gas compounds form one of the three classified structures, sI, sII and sH, with hydrate crystal lattice structure of body-centred cubic, diamond cubic and hexagonal, respectively. The basic building blocks in all hydrate structures are five polyhedra formed by hydrogen-bonded water molecules, but these structures differ in the size and the number of cavities per unit cell. Simple (pure) hydrates of CH_4 exhibit structure I (sI) in which CH_4 always occupies the small cavities (two 5^{12} cavities). In this structure, CH_4 (guest) with diameter of 4.36 Å fits into the cavity because the molecular diameter to cavity diameter ratio is 0.855. However, pure CH_4 is stabilized in sI only by the additional stability, though small, that is conferred when the other available $5^{12}6^2$ cavities are filled. It is also known that the hydrate structure is sensitive to other gas impurities. A good example is the structure transition of pure CH_4 hydrate from sI to sII in the presence of C_3H_8, a gas commonly present in CH_4 deposits. Thus, it is noteworthy that molecular size determines structure and equilibrium pressure (Sloan 1998). Recent analysis shows that the sI form is dominant in hydrate samples from the Mallik 2L-38 well (Winters et al. 2004).

It is now known that gas hydrate saturation is strongly influence by the presence of host sediments. There are a number of rock physics models in the literature that attempt to quantify the relationship between sediment elastic properties and gas hydrate saturations. The cementation models of Dvorkin & Nur (1996) treat the grains as randomly packed spheres where the gas hydrates occur at the contact point (model 1) or grow around the grains (model 2). These models predict large increases in elastic properties with only a small amount of gas hydrate but stay relatively flat as the concentration of gas hydrate increases further. Models 3 and 4 are variations of the cementation models, but consider the gas hydrate as either a component of the load-bearing matrix or as filling the pores (Dvorkin et al. 1999; Helgerud et al. 1999). A pore-space hydrate fills intergranular porosities of sands and sandstones, and is expected to be interconnected in their pore systems, which clearly contrasts with nodule and disseminated types (model 6). A pore-filling hydrate is small-sized, ranging up to 10 mm. However, it is considered to decompose continuously and effectively when produced. Model 5 is an inclusion-type model that treats gas hydrate and grains as the matrix and inclusions respectively, solving for elastic moduli of the system by iteratively solving either the inclusion-type or self-consistent type equations. Models 1–5 all consider gas hydrate as homogeneously distributed in the sediments. However, evidence of gas hydrate coring within the DSDP, ODP (Booth et al. 1998) and Mallik 2L-38 gas hydrate projects (Dallimore et al. 1999) reveals that hydrates often exist as pure aggregation (massive bodies, nodules, layers) and disseminate as fracture fillings in the shallow shaly sediments. A layered hydrate and massive hydrate are extensively continuous horizontally and concordant to strata, whose thickness should exceed 100 mm.

Higher gas hydrate concentrations create an increase in the rock elastic properties. Gas hydrate acts as an insulator and in water-saturated sediments lowers the electrical conductivity of sediments. Archie's equation (equation (1)) is used to relate electrical conductivity of hydrate formation to water saturation, S_w (Pearson et al. 1986) (it is actually the resistivity index, RI, a ratio of rock conductivities when the rock is fully and partially saturated), and n, the empirical saturation exponent.

$$S_w^n = 1/RI = (S_w < 1)/(S_w = 1) \qquad (1)$$

The resistivity data from gas hydrate research wells ODP Leg 164 site 994 (Paull et al. 1996) and Mallik 2L-38 (Dallimore et al. 1999) was used to calculate the fraction of the total pore space occupied by gas hydrate; n in both studies was chosen to be 1.9386. The saturation exponent is controlled by the distribution of the conductive and non-conductive phase in the pore space and thus depends on wetting properties, saturation history and the rock microstructure.

Gas hydrates forming in conventional gas reservoirs are likely to be pore-space gas-hydrates. Little expansion of the frozen texture would be expected if gas hydrates were formed at reasonable to considerable depth. On the other hand, gas hydrates formed under shallow seals could grow and expand since they would not have to overcome excessive overburden stress conditions. Subsequently when buried to greater depth, ice or gas-hydrate supported texture would develop. Gas from gas hydrates within framework-supported textures is likely to be produced without any changes in texture, implying stable gas productivity. On the other hand, production from gas hydrates within an ice or gas-hydrate supported texture would be unstable, owing to possible changes occurring in the texture with production. That is, the connecting pores of the gas-hydrate or ice supported texture could collapse and reduce the permeability during gas extraction from the formation. Therefore, knowledge of properties such as porosity and permeability of sediments is crucial to help establish the type of hydrate formation under given mimic conditions.

The foregoing discussion presents an argument for a need to establish sediment–hydrate interaction under conditions that are relevant to naturally occurring hydrate systems. There is a paucity of such data, although recent studies by Winters et al. (2004) and Kneafsey et al. (2007) are noteworthy. The purpose of this work, then, is to preliminarily investigate the performance of the new unit, which can mimic seafloor conditions in the formation and dissociation of pure methane hydrates formed in once hydrate-bearing, now depleted, sediments. The thermodynamics of pure, laboratory hydrates are well established, and much work has been done to determine what effect, if any, the presence of porous media has on such values (Handa & Stupin 1992; Clarke et al. 1999; Clennell et al. 1999; Anderson et al. 2001; Sheshadri et al. 2001). While an initial attempt has been made at modelling the kinetics, the assumptions used grossly oversimplify the problem, and because of the current ongoing work (Hong et al. 2003; Moridis et al. 2005; Gerami & Pooladi-Darvish 2007) this work should serve as a means for addition to and refinement of the current models, not an alternative.

Materials and methods

Unit description

The main component of the hydrate formation/ decomposition unit is a high-pressure reactor fabricated from 316 stainless steel. Originally a Jurgeson boiler liquid-level gauge, the unit has been retrofitted with temperature and pressure gauges, in addition to an inlet gas sparger. The sparger is constructed of glass wool sandwiched between two 50 μm stainless steel sieves, and serves to divide the incoming gas into fine bubbles. Not only does this eliminate channelling effects, that is, the formation of 'least resistance' paths that cause the gas to bypass the bulk of the sediment, but it also serves to more closely approximate natural formation conditions (i.e. approaching a single-phase solution of methane dissolved in water). The 200 ml vessel has rectangular viewing windows constructed from borosilicate (rated to 20.7 MPa) and is immersed in a temperature-controlled bath consisting of an equal volume mixture of ethylene glycol and water, for operating temperatures from 253 to 313 K. Experimental gas (Scott Specialty Gas methane, 99.999%) and deionized water (DI) are brought into contact in the reactor in a countercurrent fashion: water, when added during the run, enters from the top, while the gas enters from the bottom to help the liquid clay sediments achieve a uniform system.

Hydrate formation method

As detailed in a previous work (Eaton et al. 2007), the hydrates in the cell can be formed in several different ways. In a 'dynamic mode', hydrates are formed in a pre-pressurized cell by decreasing the temperature through the phase boundary. This method is typical of other laboratory studies using powdered ice and subsequently raising the temperature to allow for conversion. In a 'static mode', the temperature is held constant, and gas is slowly added to the cell. This method is more representative of the seafloor conditions where not only is the temperature relatively constant, but the methane gas/water does not agitate the sediment and the methane hydrate formation phenomenon probably occurs mostly by diffusion.

The sediment to water weight ratio was approximately 1:3, and the gas entered at a flow-controlled 215 ml min^{-1}. This gas flowrate was chosen not only because it approaches the lower limit of the attached Brooks Instruments mass flow controller, but also because previous experiments (Eaton et al. 2007) have shown that 'saturation' of the system occurs at flowrates between 75 and 220 ml min^{-1}. At higher flowrates, the gas-to-sediment-water exposure time is too limited, retarding hydrate growth and gas consumption. At flowrates below 75 ml min^{-1}, the total amount of gas consumed does not appear to increase with decreasing flowrates, suggesting no further benefit to inlet restriction. Temperature measurements of the gas and sediment phases were obtained via thermocouples inserted at the top and bottom of the cell (± 0.1 K), respectively, and results were recorded every second using the LabView software. Likewise, cell pressure (± 10.3 kPa) was recorded via a transducer located at the top of the unit, and measurements were fed into the LabView software. Figure 1 shows the cell inside the water-bath housing. Note that the instrument located halfway down the cell is an acoustic probe, and was not used in these experiments.

Three sediment samples, gathered from Blake Ridge (cruise, ODP Leg 164; latitude, 31°48.210'N; longitude, 75°31.343' W; hole/core, 995A-80X-1; water depth, 2278.5 m) were obtained from the Unites States Geological Survey (USGS), Woods Hole, Massachusetts. Previous measurements of stress history and geotechnical properties (water content, 39.3% M_w/M solids; porosity, 51.0%; maximum past stress, 2730 kPa) have been reported by Winters et al. (2004). The three samples ranged in sub-seafloor depths from 1 to 667 m. All sediment-based work presented here was performed on the 667 m sample.

Formation was initiated by the introduction of gas into the system until the desired pressure was achieved (6.21, 8.27 or 10.34 MPa). Formation pressures were arbitrarily chosen, but varied to ascertain the effect of formation pressure on formation and decomposition kinetics/thermodynamics. Once formation ended, as indicated by the cessation of gas consumption, hydrates were held at a constant temperature and pressure for 48 h, in order to allow for equilibration and possible further conversion. Figure 2 shows a typical temperature/pressure v. time trace for the initial formation event. Note that during the 48 h equilibration period, data collection continued, although no significant events were noted for any of the runs.

Hydrate decomposition

Decomposition was done in a step-down manner. That is, a backpressure regulator (BPR) attached to the cell outlet was adjusted so that the containing pressure was approximately 200 psi lower than the current system pressure. After releasing the cell pressure through communication with the BPR and allowing the cell to thermally re-equilibrate, pressure on the BPR was again lowered. This was done until the cell reached atmospheric pressure. As the decomposition progressed, vented gas was measured and amounts were digitally recorded. A pressure trace and vented gas rates, can be seen in Figures 3 and 4, respectively.

Results

Preliminary results obtained from formation and dissociation runs in the Blake Ridge sediments are presented. During formation, it was observed that the temperature of the sediment in response to exothermic hydrate formation monotonically increased with formation pressure (Table 1). This type of pressure 'spike' can be seen in Figure 2a–c, which show formation at 1500, 1200 and 900 psi (10.34, 8.27 and 6.21 MPa) respectively. This is most likely due to the inherent increased rate of formation at increasingly higher pressure. For comparison, Figure 2d is a temperature v. time trace for a system without any sediment. The loss of sediment mass was compensated for by the addition of extra

Fig. 1. The seafloor mimic FISH unit at Brookhaven National Laboratory. Only the hydrate forming vessel is shown here.

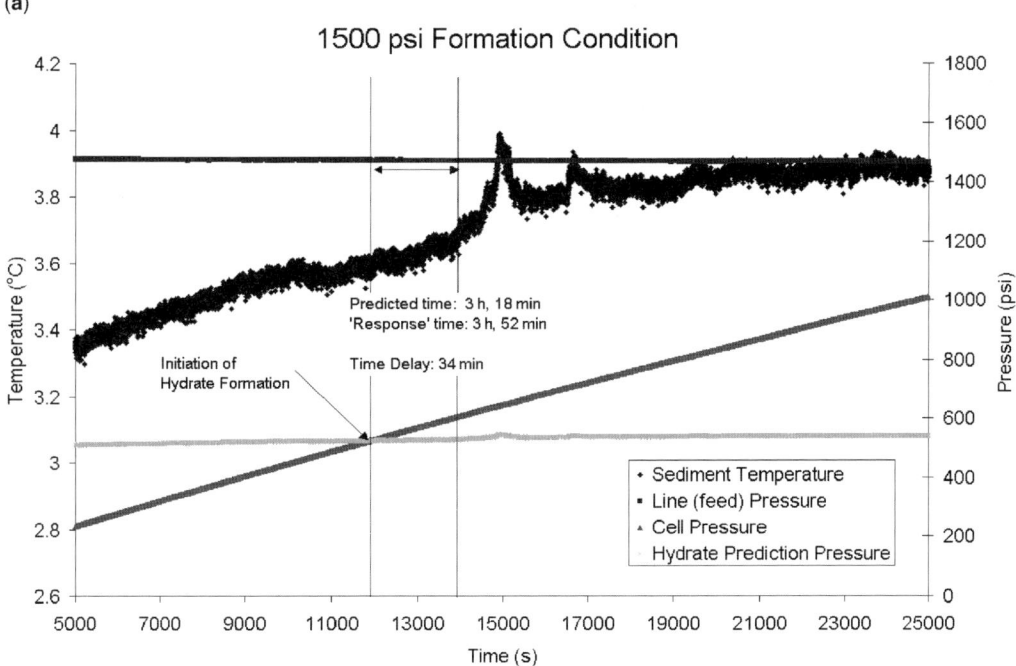

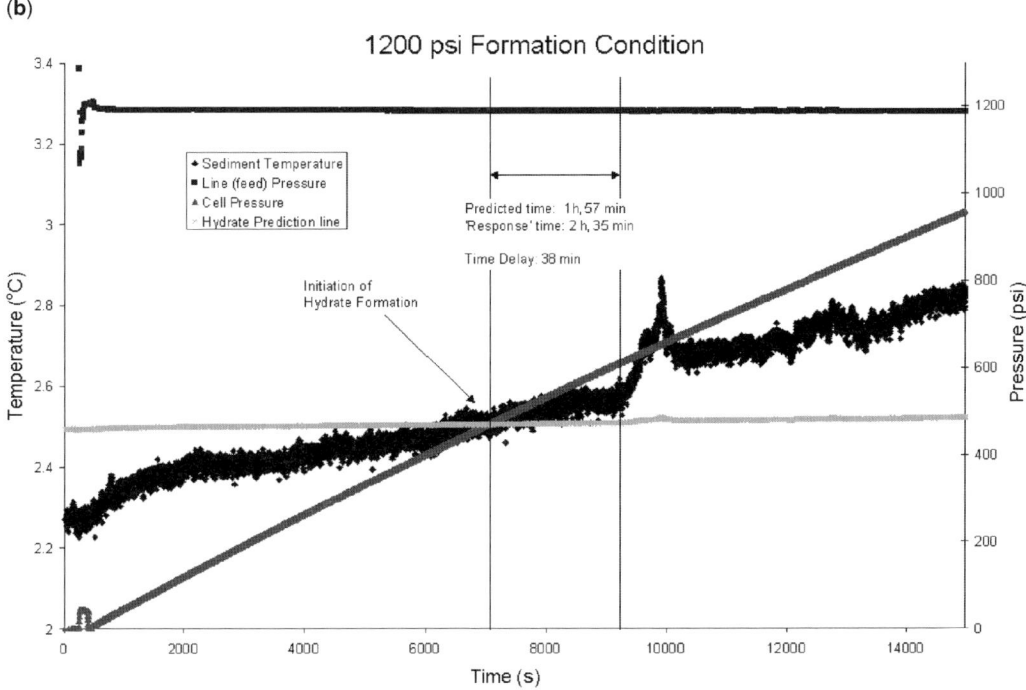

Fig. 2. (a) Temperature v. run time trace for a 1500 psi (10.34 MPa) formation condition. (b) Temperature v. run time trace for a 1200 psi (8.27 Mpa) formation condition. (c) Temperature v. run time trace for a 900 psi (6.21 MPa) formation condition. (d) Temperature v. run time trace for a 1500 psi (10.34 MPa) formation condition in pure water.

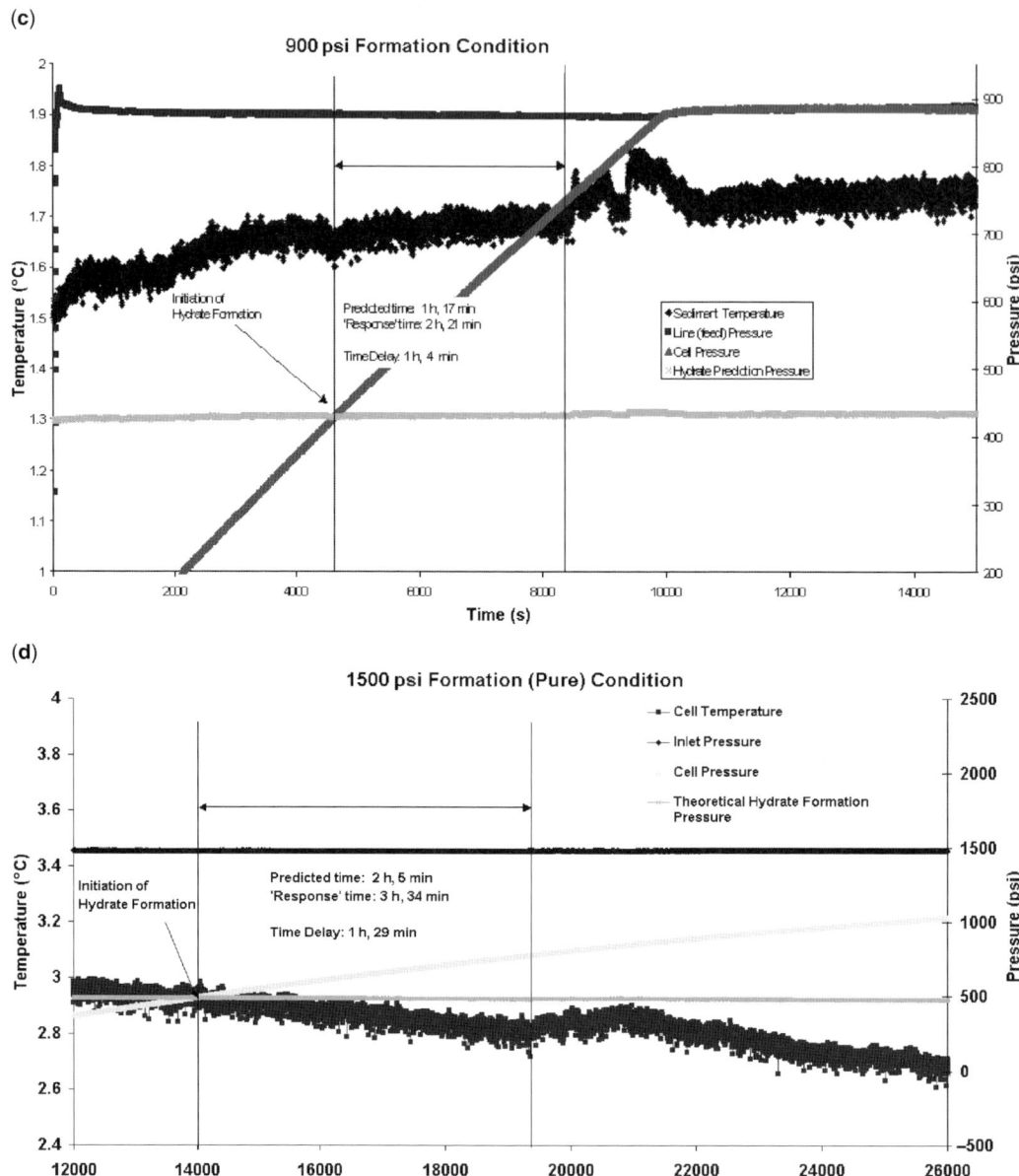

Fig. 2. (*Continued*).

DI water. Note the far lower temperature spike. We are unsure why the lower spike occurs. It could be due to the five-fold increase in thermal conductivity of hydrate in sediment (Asher 1987), which conducts the heat of formation more quickly to the thermocouple, or there may be a catalytic effect from the sediment which we have not considered.

As described by current kinetic models, the rate of dissociation (or, conversely, formation) is influenced strongly by the difference between the equilibrium pressure (for a given temperature) and the system pressure. Because of the higher driving force in a 10.34 MPa system as compared with the 6.21 MPa system, rates are expected to be higher,

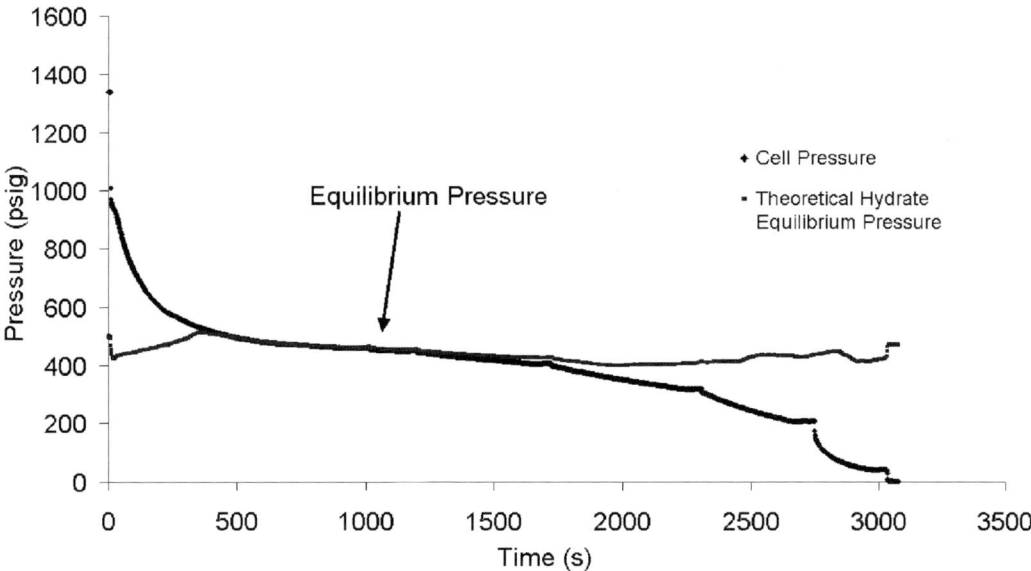

Fig. 3. Dissociation profile of pressure v. time.

and therefore the rate of heat accumulation and momentary temperature elevation is increased. Another phenomenon that occurred was the decrease in time necessary to witness the temperature spike. Theoretical spike appearance should coincide with the cell pressure crossing the hydrate phase boundary for a given temperature. Using accepted hydrate thermodynamic predictions (Ballard 2002) in conjunction with the rate of change in cell pressure with time, this 'spike time' was calculated. Actual temperature spike time was recorded at the initiation of the temperature rise outside of the

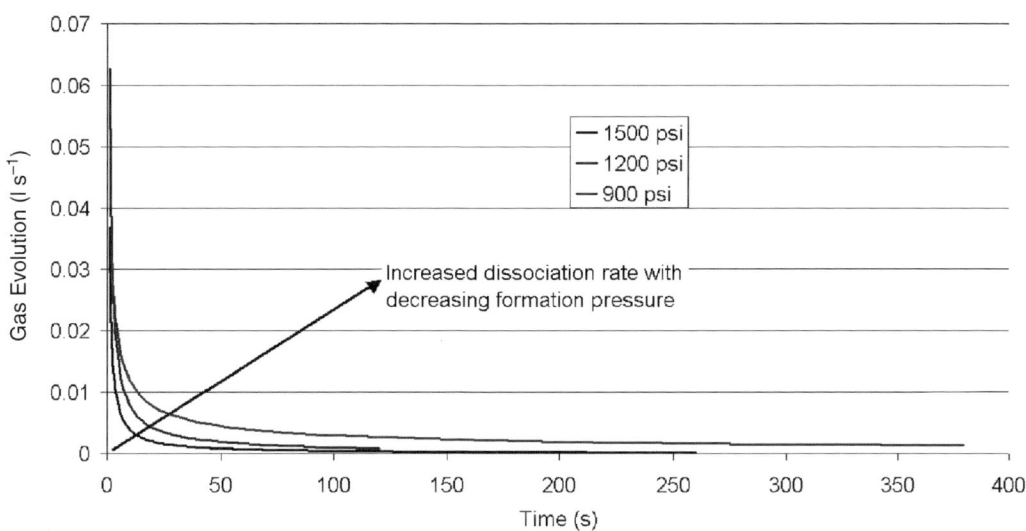

Fig. 4. Dissociation profile of pressure v. time.

Table 1. *Analysis of temperature spike magnitude and time of appearance v. formation pressure*

Sediment	System P (psi)	System P (MPa)	Bath (°C)	Estimated t (min)	Actual spike t (min)	Δt	T_{spike} (°C)
Y	900	6.2	1.8	70	141	1:11	0.18
Y	1200	8.27	2.6	117	155	0:38	0.32
Y	1500	10.34	3.5	198	232	0:34	0.4

error on the thermocouple measurement. However, it can be seen in Table 1 that there was always a delay (up to 70 min at the lowest formation pressures) when observing the temperature rise. It is not clear if this is a function of the hydrate–sediment morphology influencing heat transfer properties of the hydrate–sediment matrix, or if it has a more fundamental explanation. Transfer of these data to current hydrate modellers will hopefully shed light on this issue.

As noted, decomposition was performed in a step-down manner. This novel approach was used to investigate peculiarities in the decomposition process that would otherwise get lost or be blunted by the effect of venting a high-pressure hydrate directly to atmospheric pressure. From Figure 3, a number of interesting things can be seen. One is that a plateau in pressure occurred in the hydrate depressurization process. This, as demonstrated by many researchers before, is indicative of hydrate dissociation. At the phase boundary between hydrate and a gas–water mixture, the hydrate dissociates at constant pressure (the equilibrium pressure for a given temperature). As shown, this pressure almost exactly coincides with that of the predicted hydrate equilibrium pressure (difference of approximately 0.2%). This finding is unusual, as Blake Ridge hydrates have demonstrated significant shifts from predicted equilibrium conditions. More precisely, the bottom simulating reflector, indicative of the base of the hydrate formation for the Blake Ridge, is shallower than it should be (Matsumoto *et al.* 1996). There are a few explanations for this discrepancy, with the most feasible being the use of desalinated water and excess non-pore space hydrate contribution. Because this experiment uses de-ionized water, and excess salinity has been found around the Blake Ridge hydrate zone, it is conceivable that it is the salinity that is affecting equilibrium, which our system does not mimic. Further work with more complex systems is necessary to verify this assumption. Lastly, our system uses a 3:1 by mass water to sediment ratio. This ratio, while ensuring that the sediment is fully saturated with water, may cause a slight deviation from 'natural condition' results by hydrating a significant portion of non-sediment-bound water. Consequently, properties measured for the system as a whole (thermodynamic equilibrium, kinetics, etc.) may tend towards those observed in pure systems, explaining the observation of formation/decomposition pressures close to those of pure H_2O–CH_4.

It should be noted that several 'steps' (drops in BPR pressure) took place during the plateau section of Figure 3. However, hydrate decomposition maintained pressure until all hydrated gas was exhausted. In the ensuing pressure step-downs, the asymptotic decrease to the BPR pressure is indicative of simple gas venting, without pressure supplementation by hydrate dissociation.

In addition to cell pressure, gas evolution rate was also measured. As noted above, venting behaviour for pressures above the equilibrium pressure and below the equilibrium pressure (post-hydrate exhaustion) demonstrate no unusual (or hydrate influenced) effects. However, for pressures at or slightly below equilibrium pressure ($500 < t < 1550$ s), hydrate dissociation sustained a low-level but continuous output of gas. Figure 4 shows the gas production rate for dissociation of hydrates formed at the three different pressures. From these rates, a *very* simple gas phase rate constant, k, could be calculated (0.015 ± 0.003 s^{-1}). It is important to note that this rate constant does not take into account surface area of the hydrate, fugacity or other thermodynamic quantities (unlike current models), but merely calculates a rate law based upon the rate of produced gas. The calculated values of k do not appear to depend on the formation pressure (within experimental error), which agrees with current kinetic models that suggest that dissociation rate depends only on the fugacity of the system, not its history. Additionally, the kinetic rate constant was calculated after 45 s of venting, when outlet gas was assumed to be derived entirely from dissociated hydrate, and not from gas in the head-space or free gas trapped in the sediment–hydrate matrix.

It is unknown at this time why the apparent increase in venting rate occurs with decreased formation pressure at early times (despite the same system pressure), although it is quite possible that, owing to the increased hydrate saturation at higher pressures, and therefore increased amount of pore-space filling, trapped gas simply takes longer to emerge from the pore space.

Conclusions and future work

A major obstacle to studying natural hydrate systems has been the absence of a sophisticated apparatus in which the hydrate formation phenomenon can be reproduced with precision. Our newly designed unit, flexible integrated study of hydrates (FISH), capable of temperature, pressure and precision flow control, has begun to overcome most of these obstacles and allowed us to acquire data for use to supplement and refine hydrate formation and dissociation knowledge. The unit is undergoing further modifications to enhance its capability for measuring multiple properties during hydrate–sediment interactions.

For our study, we used depleted sediments from Blake Ridge, a well-researched hydrate site, as hosts to mimic natural environment that sustains hydrate occurrence under a narrow range of temperature and pressure conditions. We have shown that the computed microtomography technique at the National Synchrotron Light Source at Brookhaven National Laboratory is excellent for conducting microscopic/spectroscopic analyses of depleted sediments to yield associated porosity, tortuosity and mineral properties values.

We have presented preliminary data in this paper to discern the role of sediments in hydrate formation/dissociation kinetics under *in situ* conditions. Early results agree with both intuition and current kinetic models. A sensitivity analysis revealed a kinetic rate constant independent, within error, of formation pressure. While some confounding factors need to be addressed before a definitive conclusion is reached, studies have revealed that hydrates formed in Blake Ridge sediments, shown in nature to exhibit a shift in equilibrium, are apparently not susceptible to such a shift using pure water and pure methane gas. We are also analysing the dissociation data in detail to extract pore-filling values for the system under investigation.

Although the hydrate-forming gas used in the experiments presented in this paper was essentially pure methane yielding structure I hydrate, the use of natural sediment (from a natural environment) dictates using an 'impure' system later, through the addition of higher hydrocarbons. While Blake Ridge hydrates have shown an abundance of biogenic methane, and necessarily a lower concentration of higher hydrocarbons (C2, C3, etc.), sites such as the Gulf of Mexico often contain sII and in some cases, sH (Sassen & MacDonald 1994), which by their very nature form and behave differently than do sI hydrates.

Future work for this system includes a larger span of temperature and pressure formation conditions, investigation of the effects of sediment character (varies with depth) on hydrate formation/decomposition behaviour, the use of simulated/real seawater, and the use of multicomponent gas mixtures to more closely approximate nature in our versatile FISH unit. It should be stressed again that work from this apparatus will be used to supplement, improve and test kinetic models, not create them. The use of computed microtomography (CMT) to establish porosity and permeability differences provides the means to establish and analyse the hydrate 'life cycle' under *in situ* conditions. Our goal is to determine basic kinetic and thermodynamic data, and a morphological response to varying grain size and gas composition – possibly lending insight to the types of proposed hydrate formations. Natural gas hydrates represent a huge untapped energy source in both the permafrost and ocean floor environments and data regarding the behaviour of hydrates in such locales is in high demand – understanding their behaviour may allow for efficient, economic and environmentally compatible extraction techniques.

This work was supported by the U.S. Department of Energy under Contract No. DE-AC02-98CH10886 through National Energy Technology Laboratory. DM also thanks Stony Brook University for a start-up grant. The authors wish to thank Dr. William Winters, United States Geological Survey, Woods Hole, Massachusetts, for providing sediment samples from the Ocean Drilling Program upon which this study was based.

References

ANDERSON, R., BURGASS, R. W., TOHIDI, B. & ØSTERGAARD, K. K. 2001. Experimental measurement of gas hydrate stability zones in porous media. *European Association of Geoscientists and Engineers 63rd Conference & Technical Exhibition*, Amsterdam, 567.

ASHER, G. B. 1987. *Development of a Computerized Thermal Conductivity Measurement System Utilizing the Transient Needle Probe Technique – An Application to Hydrates in Porous Media*. Colorado School of Mines.

BALLARD, A. 2002. *A non-ideal hydrate solid solution model for a multi-phase equilibria program*. Ph.D. Thesis, Colorado School of Mines, Golden, CO.

BOOTH, J. S., CLENNELL, B., PECHER, I. A., WINTERS, W. J., RELLE, M. K. & DILLON, W. P. 1998. Laboratory investigation of gas hydrate genesis in sediments – modes of occurrence, volumes and growth patterns. *In*: *Gas Hydrates in Nature – Results from Geophysical and Geochemical Studies*. European Geophysical Society, XXIII, General Assembly, Nice.

BOSWELL, R. 2007. Resource potential of methane hydrate coming into focus. *Journal of Petroleum Science and Engineering*, **56**, 9–13.

CLARKE, M. A., POOLADI-DARVISH, M. & BISHNOI, P. R. 1999. A method to predict equilibrium conditions of gas hydrate formation in porous media.

Industrial and Engineering Chemistry Research, **38**, 2485–2490.

CLENNELL, B. M., HOVLAND, M., BOOTH, J. S., HENRY, P. & WINTERS, W. J. 1999. Formation of natural gas hydrates in marine sediments 1. Conceptual model of gas hydrate growth conditioned by host sediment properties. *Journal of Geophysical Research*, **104**, 22985–23003.

COLLETT, T. S. 1995. Gas hydrate resources of the United States. *In*: GAUTIER, D. L., DOLTON, G. L., TAKAHASHI, K. I. & VARNES, K. L. (eds) *National Assessment of United States Oil and Gas Resources on CD-ROM*. United States Geological Survey Digital Data Series, **30**, 1.

COLLETT, T. S. 2002. Energy resource potential of marine gas hydrates. *American Association of Petroleum Geologists Bulletin*, **86**, 1971–1992.

DALLIMORE, S. R., UCHIDA, T. & COLLETT, T. S. 1999. *Scientific Results from JAPEX/JNOC/GSC Mallik 2L-38 Gas Hydrate Research Well, Mackenzie Delta, Northwest Territories, Canada*. Geological Survey of Canada Bulletins **544**, February.

DVORKIN, J. & NUR, A. 1996. Elasticity of high-porosity sandstones: Theory for two North Sea datasets. *Geophysics*, **61**, 1363–1370.

DVORKIN, J., PRASAD, M., SAKAI, A. & LAVOIE, D. 1999. Elasticity of marine sediments. *Geophysical Research Letters*, **26**, 1781–1784.

EATON, M., MAHAJAN, D. & FLOOD, R. 2007. A novel high-pressure apparatus to study hydrate–sediment interactions. *Journal of Petroleum Science and Engineering*, **56**, 101–107.

GERAMI, S. & POOLADI-DARVISH, M. 2007. Predicting gas generation by depressurization of gas hydrates where the sharp-interface assumption is not valid. *Journal of Petroleum Science and Engineering*, **56**, 146–164.

GOEL, N., WIGGINS, M. & SHAH, S. 2001. Analytical modeling of gas recovery from *in situ* hydrates dissociation. *Journal of Petroleum Science and Engineering*, **29**, 115–127.

HANDA, P. Y. & STUPIN, D. 1992. Thermodynamic properties and dissociation characteristics of methane and propane hydrate in 70-Å-radius silica gel pores. *Journal of Physical Chemistry*, **96**, 8599–8603.

HELGERUD, M., DVORKIN, J., NUR, A., SAKAI, A. & COLLETT, T. 1999. Elastic wave velocity in marine sediments with gas hydrates: Effective medium cooling. *Geophysical Research Letters*, **26**, 2021–2024.

HONG, H., POOLADI-DARVISH, M. & BISHNOI, P. R. 2003. Analytical modeling of gas production from hydrates in porous media. *Journal of Canadian Petroleum Technology*, **42**, 45–56.

KNEAFSEY, T. J., TOMUTSA, L., MORIDIS, G. J., SEOL, Y., FREIFELD, B. M., TAYLOR, C. E. & GUPTA, A. 2007. Methane hydrate formation and dissociation in a partially saturated core-scale sand sample.

Journal of Petroleum Science and Engineering, **56**, 108–126.

MAHAJAN, D. & SOMASUNDARAN, P. (ORGANIZERS). 2006. *An International Workshop on Science and Technology Issues in Methane Hydrate R&D*, Kauai, HI, 5–9 March. Available at: www.eci.org.

MAHAJAN, D. & TAYLOR, C. E. (eds). 2006. A special volume on gas hydrates and clathrates. *Journal of Petroleum Science and Engineering*, **56**, 1–8.

MATSUMOTO, R., WATANABE, Y. *ET AL*. 1996. Distribution and occurrence of marine gas hydrates. Preliminary results of ODP Leg 164. Blake Ridge drilling. *Chishitsugaku Zasshi*, **102**, 932.

MORIDIS, G. J., SEOL, Y. & KNEAFSEY, T. 2005. Studies of reaction kinetics of methane hydrate dissociation in porous media. *Proceedings of the 5th International Conference on Gas Hydrates*, Trondheim, 13–16 June.

PAULL, C. K., BUELOW, W., USSLER, W. & BOROWSKI, W. S. 1996. Increased continental-margin slumping frequency during sea-level lowstands above gas hydrate-bearing sediments. *Geology*, **24**, 143–146.

PEARSON, C., MURPHY, J. & HERMES, R. 1986. Acoustic and resistivity measurements on rock samples containing tetrahydrofuran hydrates. *Journal of Geophysical Research*, **91**, 14132–14138.

SASSEN, R. & MACDONALD, I. R. 1994. Evidence of structure H hydrate, Gulf of Mexico continental slope. *Organic Geochemistry*, **22**, 1029–1032.

SESHADRI, K., WILDER, J. W. & SMITH, D. H. 2001. Measurements of equilibrium pressures and temperatures for propane hydrate in silica gels with different pore-size distributions. *Journal of Physical Chemistry*, **105**, 2627–2631.

SLOAN, E. D. 1998. *Clathrate Hydrates of Natural Gases*, 2nd edn. Marcel Dekker, New York.

SLOAN, E. D. 2004. Introductory overview: Hydrate knowledge development. *American Mineralogist*, **89**, 1155–1161.

TAYLOR, C. E., LEKSE, J. & ENGLISH, N. 2004. NETL's methane hydrate research. *Proceedings of AAPG Hedberg Conference Gas Hydrates: Energy Resource Potential and Associated Geologic Hazards*, 12–16 September 2004, Vancouver.

WEGRZYN, J. E., MAHAJAN, D. & GUREVICH, M. 1999. Catalytic routes to transportation fuels utilizing natural gas hydrates. *Catalysis Today*, **50**, 97–108.

WINTERS, W. J., WAITE, W. F. & PECHER, I. A. 2004. Comparison of methane gas hydrate formation on physical properties of fine- and coarse-grained sediments. *American Association of Petroleum Geologists Hedberg Conference*, 12–16 September, Vancouver.

ZHENG, L., ZHANG, H., ZHANG, M., KERKAR, P. & MAHAJAN, D. 2009. Modeling methane hydrate formation in marine sediments. *American Association of Petroleum Geologists Bulletin*, in press.

Effects of solid surfaces on hydrate kinetics and stability

B. KVAMME*, A. GRAUE, T. BUANES, T. KUZNETSOVA & G. ERSLAND

Department of Physics and Technology, University of Bergen, Allégaten 55, N-5007 Bergen, Norway

*Corresponding author (e-mail: Bjorn.Kvamme@ift.uib.no)

Abstract: Reservoirs of clathrate hydrates of natural gases (hydrates), found worldwide and containing huge amounts of bound natural gases (mostly methane), represent potentially vast and yet untapped energy resources. Since CO_2-containing hydrates are considerably more stable thermodynamically than methane hydrates, if we find a way to replace the original hydrate-bound hydrocarbons with the CO_2, two goals can be accomplished at the same time: safe storage of carbon dioxide in hydrate reservoirs, and *in situ* release of hydrocarbon gas. We have applied the techniques of magnetic resonance imaging as a tool to visualize the conversion of CH_4 hydrate within Bentheim sandstone matrix into the CO_2 hydrate. Corresponding model systems have been simulated using the phase field theory approach. Our theoretical studies indicate that the kinetic behaviour of the systems closely resembles that of CO_2 transport through an aqueous solution. We have interpreted this to mean that the hydrate and the matrix mineral surfaces are separated by liquid-containing channels. These channels will serve as escape routes for released natural gas, as well as distribution channels for injected CO_2.

The storage of CO_2 in reservoirs has, through the trial injection of production CO_2 from the Sleipner oil-and-gas field into the Utsira formation, already been established as a feasible alternative for reducing CO_2 emissions into atmosphere. Seismic profiles of the Utsira formation before and after the injection show the contours of CO_2 plumes trapped beneath distinct clay layers. The safety of this storage option has been extensively studied using different reservoir simulation tools (see Xu *et al.* 2004 and references therein, for representative examples). The northern parts of the North Sea, as well as the Barents Sea, have regions of seafloor temperatures that lie just above and even below zero Celsius. This means that pressure and temperature conditions in natural-gas reservoirs present under the seafloor may well be within the CO_2 hydrate stability zone (provided that their surroundings are saturated with the respect to the aqueous CO_2–hydrate equilibrium). The properties of the fluid phase in contact with the matrix pore walls will largely be determined by interaction of its molecules with the mineral surfaces. If layers of molecules are adsorbed onto the mineral surfaces, they will form a separate phase. Which phases potentially present in the porous media will be stable at these conditions will be decided by the coupled processes driving the system towards local and global minimum free energy, in accordance with the Gibbs phase rule. The efficiency of the hydrate sealing will therefore depend on whether or not CO_2 hydrate formed during aquifer storage will stick to the mineral surfaces.

Another type of potential CO_2 storage can be provided by reservoirs that already contain natural gas hydrates. CO_2 hydrate is significantly more stable than natural gas hydrates. Injection of CO_2 into hydrate reservoirs will therefore naturally convert the *in situ* hydrate into CO_2 hydrate while at the same time releasing the trapped natural gas. The Storegga slide is probably the most extensively investigated one among the Norwegian gas-hydrate reservoirs. The second-largest gas field on the Norwegian shelf, the Ormen Lange, is located inside the Storegga area. The annual natural gas production from this field amounts to 2.60×10^{10} standard m³. The Storegga slide has been a target of several international studies, with the recent estimates (Bünz *et al.* 2003) indicating that the hydrate region encompasses approximately 4000 km² of layered hydrate. The average thickness of hydrate layers varies between 40 and 50 m. Thus the volume of natural gas resources present in this area in the form of gas hydrates could be as high as 1.80×10^{12} standard m³, or 70 times the current annual production from the Ormen Lange. On the global basis, the energy associated with hydrocarbons trapped in hydrate reservoirs may exceed all known resources of coal and conventional hydrocarbon reservoirs by a factor of two. This is the motivation for a rather extensive hydrate programme in Japan,

From: LONG, D., LOVELL, M. A., REES, J. G. & ROCHELLE, C. A. (eds) *Sediment-Hosted Gas Hydrates: New Insights on Natural and Synthetic Systems.* The Geological Society, London, Special Publications, **319**, 131–144.
DOI: 10.1144/SP319.11   0305-8719/09/$15.00 © The Geological Society of London 2009.

MH21 (http://www.mh21japan.gr.jp/english/), which focuses specifically on exploitation of hydrate resources in the Nankai Trough.

Techniques and approaches

Nuclear magnetic relaxation properties are different depending on whether hydrogen is a part of liquid water or solid hydrate. This makes it possible to use the techniques of magnetic resonance imaging (MRI) to visualize and follow the kinetics of hydrate formation (Kvamme et al. 2004). The hydrogen in CH_4 will be visible using MRI for a gas/fluid phase in bulk, but inside the hydrate cavities the relaxation time for the hydrogen in CH_4 will be too short for detection and will thus appear as invisible. This means that MRI can also be used for imaging the reformation of CH_4 hydrate into CO_2 hydrate through injection of CO_2, since the released methane will be visible with MRI. The resolution of the MRI apparatus (Kvamme et al. 2004) is on the order of 100 μm. Additional information from the experiments (Kvamme et al. 2004) are obtained through mass balance and monitoring the fluids supplied to the core as well as fluids extracted from the experimental cell. Material balance calculations provide mass transfer profiles and kinetic rates in addition to macroscopic imaging of porous media effects on phase transitions and flow patterns. Limitations of the resolution are the lack of ability to track the detailed mechanisms of the phase transitions, along with the corresponding profiles for changes in energies and free energies.

Implicit dynamic coupling of mass transport, heat transport and thermodynamically governed phase transitions makes it difficult to extrapolate results obtained at given experimental temperatures and pressures to other conditions without invoking more fundamental theoretical concepts. Therefore, we supplemented our experimental efforts with computer simulations that followed the phase field theory (PFT) approach (Gránásy et al. 2003, 2004a, b; Kvamme et al. 2004; Svandal et al. 2006a, b) and utilized thermodynamic properties derived from molecular dynamics (Kvamme & Tanaka 1995; Svandal et al. 2006a, b) and equations of state for fluid-phase CO_2 and CH_4 (Soave 1972). Interface properties needed by the PFT can also be estimated along the lines discussed by Kvamme et al. (2005a, b, 2007). For the systems under consideration (hydrates, aqueous solutions, fluids inside a porous rock matrix), heat transport will be roughly two orders of magnitude faster than mass transport. Thus one may safely ignore the heat transport dynamics for these systems (isothermal approximation) without adversely affecting the final kinetic results. Another advantage of the PFT approach lies in the fact that the solution of its coupled differential equations allows for assessment of the relative significance of different phase transition contributions at the actual thermodynamic conditions and growth morphologies. As such,

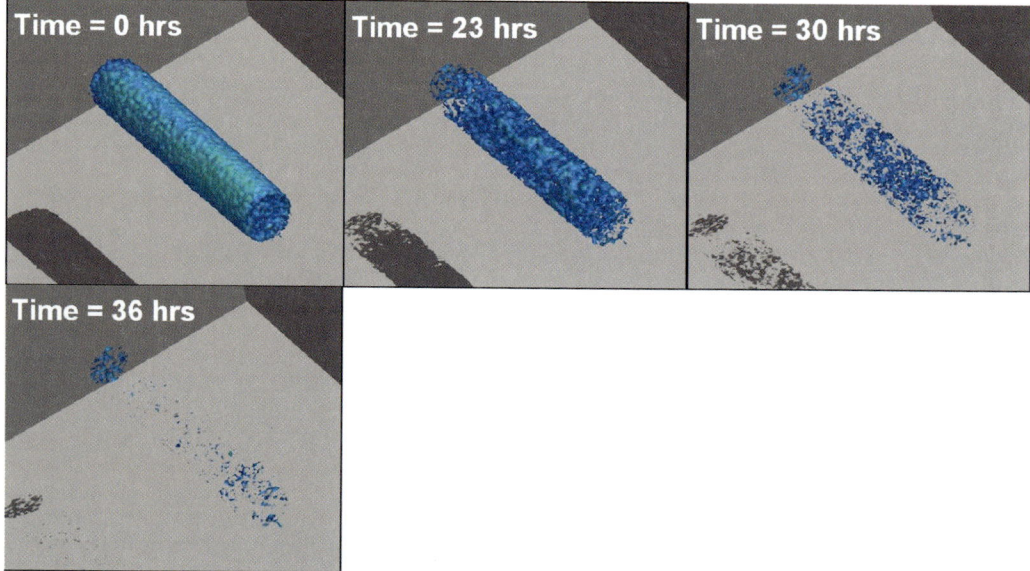

Fig. 1. MRI visualization of CH_4 hydrate formation in a sandstone core plug. Details on the properties of the plug are given elsewhere (Kvamme et al. 2004). CH_4 is supplied from the right-hand side into the plug at 83.75 bar (1200 psig) and 4 °C.

PFT simulations can also serve as a basis for development of simplified kinetic correlations.

Experimental results

MRI was applied to visualize hydrate phase transition kinetics in a porous Bentheim sandstone. Figure 1 shows snapshots of CH_4 hydrate formation in the plug. The consumption of CH_4 was supplied from the right-hand side of the plug. The plug was initially saturated with brine and methane; the water saturation was approximately 52% (12.15 cm^3, 0.092 mol), using a brine with salinity of 0.1 wt% NaCl. The remaining pore volume was filled with methane at a pressure of 83.75 bar (1200 psig) at a temperature of 4 °C, yielding a gas saturation of 48% PV.

Unlike reported results from bulk experiments on heterogeneous hydrate formation (Makogan 1997) on the interface between aqueous phase and hydrate former, there was no observed induction time for the hydrate growth in this porous system, as illustrated in Figure 2. This is also in accordance with our previous results (Kvamme et al. 2004). For comparison we therefore conducted a corresponding experiment without the presence of porous material.

Two half cylinders of polyoxymethylene (POM) were separated by a spacer that left two connected compartments, as shown in Figure 3. The lower half section was filled with water and the remaining part of the compartments was pressurized with CH_4 to 83.75 bar (1200 psig) and kept at 4 °C. Some snapshots from the progress of hydrate formation are given in Figure 4. The corresponding hydrate fraction as function of time is plotted in Figure 5. We note two obvious differences between this system and the Bentheim sandstone system (Figs 1 & 2 as well as experiments reported by Kvamme et al. 2004). The type of solid material (and corresponding differences in wetting properties) confining the fluid/hydrate system is clearly different. Secondly there is a huge difference in material/fluid contact area. CH_4 is the primary wetting component towards POM. This fact, and the associated effect of capillary CH_4 transport along the walls and downwards into the compartment, may explain some of the observed breakthrough patterns (Fig. 4). However, as discussed in more detail in the next section, all regions of the initial film of hydrate are competing for the available molecules in order to obtain further growth. According to the first and second laws of thermodynamics, the

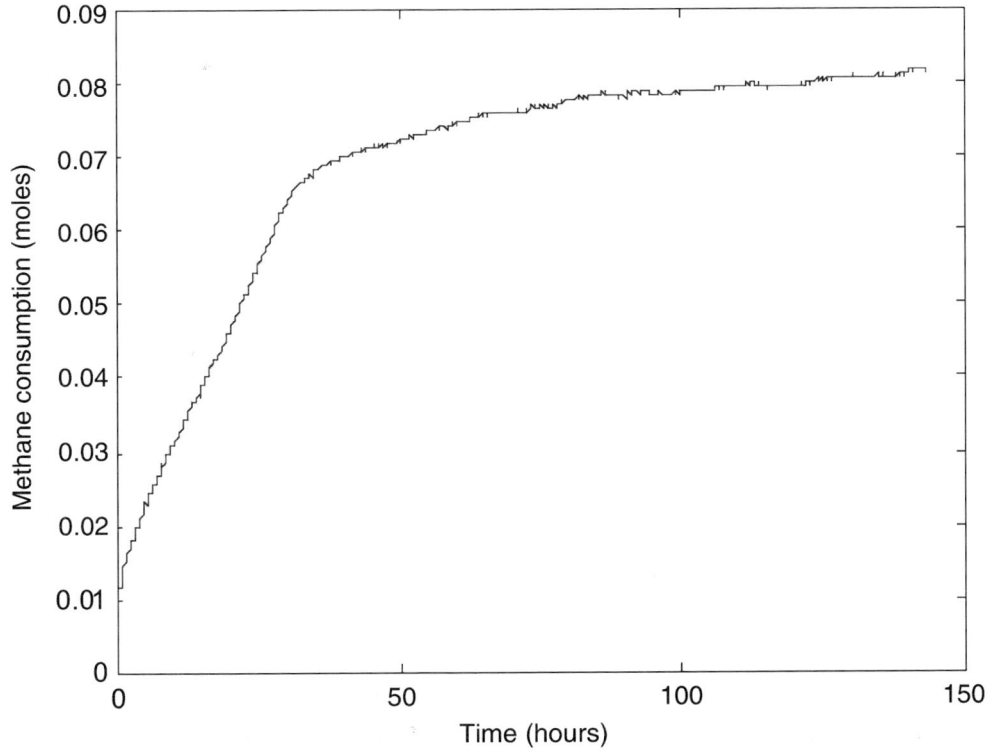

Fig. 2. Gas consumption as function of time for the experiment illustrated in Figure 1.

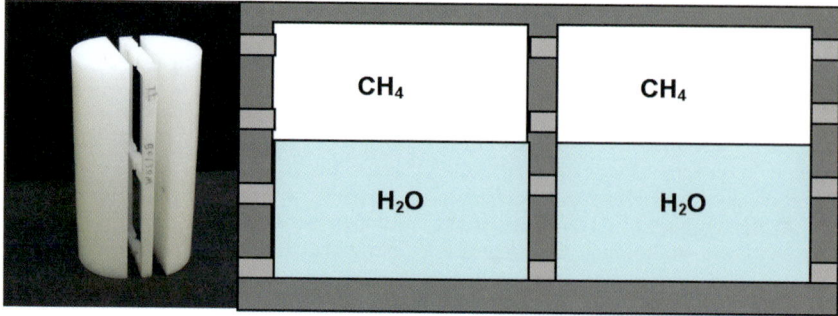

Fig. 3. Experimental setup for bulk hydrate formation experiment. Left: massive silicone rubber half cylinders and fluid/hydrate chamber. Right: schematic picture of fluid/hydrate compartments and connecting channels.

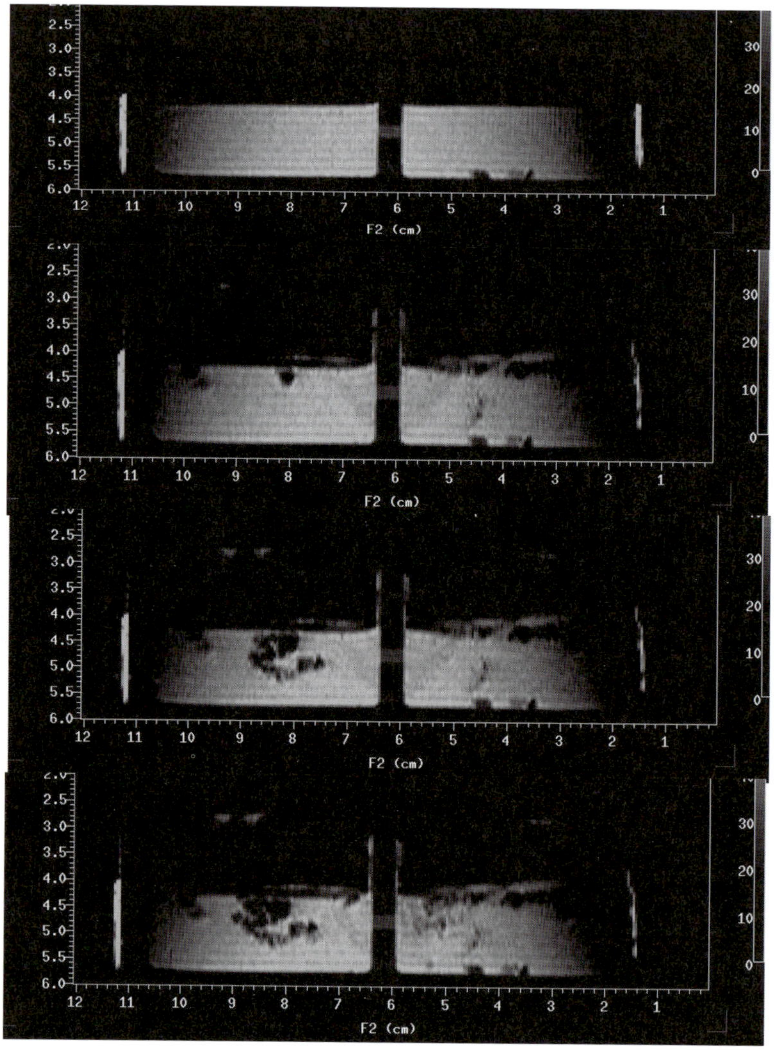

Fig. 4. MRI images of the hydrate formation. Times after start are (from top) 0, 113.9, 126.8 and 185.2 h.

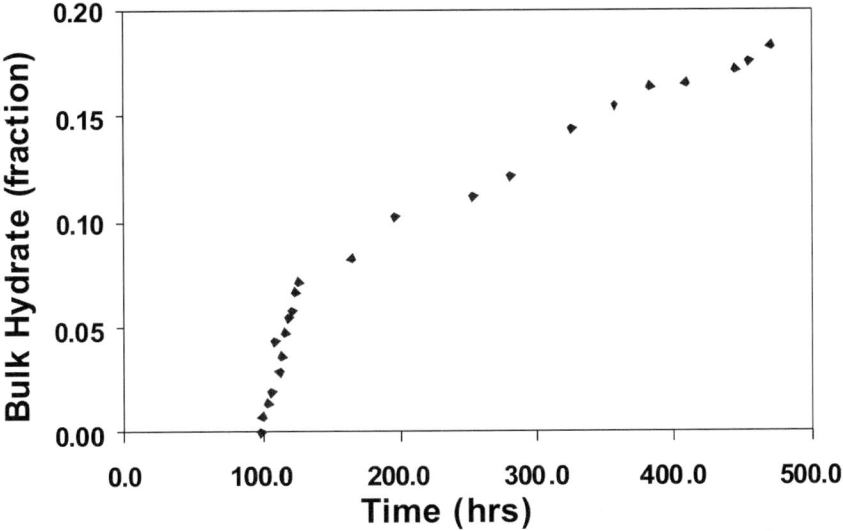

Fig. 5. Hydrate fraction as function of time for the bulk hydrate experiment.

regions of lowest free energy of the non-equilibrium hydrate film will gain further growth at the cost of less stable neighbouring regions when the direct supply of new hydrate formers is limited by the low transport rates through the solid hydrate membrane.

Theoretical modelling

Heterogeneous hydrate formation from a single hydrate former on the interface between the hydrate former phase and aqueous solution involves the coexistence of three phases that contain only two components, and thus only a single degree of freedom. The presence of solid walls will further reduce the number of degrees of freedom by one, since the density and the component distribution in the adsorbed layer close to the solid surface will differ from those of the surrounding phases and must be treated as a separate phase. This means that there are no degrees of freedom in this system. The second system is reformation of CH_4 hydrate though injection of CO_2. If this phase transition is a direct solid-state exchange the number of degrees of freedom is two and there is a theoretical possibility for the system to establish equilibrium but only if the mineral–water interactions do not result in the formation of a separate liquid water phase. In either case the combined first and second laws of thermodynamics will direct the dynamic paths of changes in the systems towards either equilibrium or local and global minimum free energy.

The conversion of CH_4 hydrate to CO_2 hydrate has been studied experimentally in a system where two half cylinders of Bentheim sandstone are divided by a silicone rubber spacer like the one used in the bulk experiment (see Fig. 3, left). CH_4 hydrate at 83.75 bar (1200 psig) and 4 °C is initially produced according to the procedures described by Kvamme et al. (2004). The remaining excess of methane in the spacer is then flushed with CO_2 and the system is kept at constant pressure and constant temperature. The resulting experimental conversion rate from CH_4 hydrate into CO_2 is plotted with circles in Figure 11.

If the conversion is a solid-state conversion it is to be expected that the system could be modelled as a two-phase system with only two actively moving components, CH_4 and CO_2. The water molecules would as such just represent a solid cavity structure with specific impact on the thermodynamic properties CH_4 and CO_2 and vice versa.

In this work we use a version of phase field theory which includes three fields: the phase ϕ, the molar CO_2 concentration c and the microscopic orientation, θ. Note that for historical reasons $\phi = 0$ corresponds to solid and $\phi = 1$ to liquid in the scope of PFT. From the free energy functional:

$$F = \int d^3 r \left\{ \frac{1}{2} \varepsilon_\phi^2 T (\nabla \phi)^2 + \frac{1}{2} \varepsilon_c^2 T (\nabla c)^2 \right. $$
$$\left. + f_{\text{ori}}(|\nabla \phi|) + f(\phi, c) \right\} \quad (1)$$

$$f(\phi, c) = wTg(\phi) + f_L p(\phi) + f_S [1 - p(\phi)] \quad (2)$$

The thermodynamic properties of the hydrate, f_H, and fluid phase, f_L, are derived from molecular simulations as described by Svandal et al. (2006a). The functions $g(\phi)$ and $p(\phi)$ are not uniquely defined but constrained by the requirements of thermodynamic consistency. The following forms (Kvamme et al. 2004) have been adopted throughout this work: $g(\phi) = \frac{1}{4}\phi^2(1-\phi)^2$ and $p(\phi) = \phi^3(10 - 15\phi + 6\phi^2)$. If ε_c is set equal to zero or equal to ε_ϕ then there are two unknown parameters in addition to the orientation dependency of the free energy. These two parameters can be estimated from the interface properties (Cahn & Hilliard 1958):

$$d = \left(\frac{\varepsilon^2 T}{2}\right)^{1/2} \int_{0.1}^{0.9} d\xi \{\Delta f[\xi, c(\xi)]\}^{-1/2} \quad (3)$$

where d is the 10–90 interface thickness for a 10–90% confidence interval. Parameters ε and w appear in Δf [see equations (2) and (5)] in addition to the appearance of ε in the prefactor. Manipulations of equation (3) (Cahn & Hilliard 1958) gives a corresponding expression for the interface free energy of the solid–liquid interface:

$$\gamma_\infty = (\varepsilon^2 T)^{1/2} \int_0^1 d\xi \{\Delta f[\xi, c(\xi)]\}^{1/2} \quad (4)$$

where $\Delta f = f - f_0$, and

$$f_0 = f_L(c_L^{eq}) + \frac{\partial f_L}{\partial c}\bigg|_{c_L^{eq}}(c - c_L^{eq})$$

$$= f_S(c_S^{eq}) + \frac{\partial f_S}{\partial c}\bigg|_{c_S^{eq}}(c - c_S^{eq}) \quad (5)$$

Equation (5) is the common tangent equation for the equilibrium condition. With appropriate values for d and γ_∞ from experiments or theoretical studies/molecular simulations then equations (3) and (4) can be solved iteratively for ε and w. For details see Kvamme et al. (2004) and references therein. The orientational free energy contributions require typically one or two empirical parameters, depending on the complexity of the crystal morphology. Throughout this work we adopt the form:

$$\varepsilon_\phi = \varepsilon_{\phi_0}\left[1 + \frac{s_0}{2}\cos(n\vartheta - 2\pi\theta)\right] \quad (6)$$

$$\vartheta = \arctan[(\nabla\phi)_y/(\nabla\phi)_x]$$

which results in dendritic growth of hydrate from aqueous CO_2 solution (Gránásy et al. 2004c; Kvamme et al. 2004; Svandal et al. 2006b; Tegze et al. 2006). Subject to conservation of mass the equations of motion are derived (Gránásy et al. 2004c; Kvamme et al. 2004; Svandal et al. 2006b; Tegze et al. 2006). In the examples used in this work we set the mobility (Gránásy et al. 2004c; Kvamme et al. 2004; Svandal et al. 2006b; Tegze et al. 2006) of phase field and concentration field as identical and equal to an interpolation between liquid and solid diffusivity coefficients of CO_2 according to the local phase field. The latter value is not accurately known but is likely to be smaller than 10^{-12} m^2 s^{-1} (Radhakrishnan & Trout 2002), which is the value we have used. For liquid CO_2 we have used 1.56×10^{-9} m^2 s^{-1}.

$$\dot{\phi} = M_\phi \frac{\delta F}{\delta \phi} + \dot{\mathsf{s}}_\phi \quad (7)$$

$$\dot{c} = \nabla\left(M_c \nabla \frac{\delta F}{\delta c}\right) + \mathsf{s}_c \quad (8)$$

$$M_c = \frac{D_S + (D_L - D_S)p(\phi)}{RT} \quad (9)$$

$$\dot{\theta} = M_\theta \frac{\delta F}{\delta \theta} + \mathsf{s}_\theta \quad (10)$$

The PFT approach is a mean field theory and the no ice terms ζ_c and ζ_ϕ are added so as to mimic some of the effects of the dynamics across the boundary between the limits of the simulation cell and the infinite surroundings while at the same time normalizing so that the total average net effect on the system should be zero. The mobility of the concentration field is very close to the diffusivity, as reflected in equation (9), while the mobility of the phase field is more complex. This is the mobility related to the change in fluid structure, from fluid phase(s) ($\phi = 1$) over to solid structure ($\phi = 0$). Physically this implies changes in the librational modes from fluid state(s) with kinetic and translational modes and over to more restricted modes in a solid. Corresponding changes in free energy can be estimated from molecular simulations. For the solid states the free energies of librational modes have been sampled as described by Kvamme & Tanaka (1995) and also systemized into free energy changes due to guest inclusions. Liquid free energies can be calculated from a number of different approaches. For polar fluids like water we have good experiences with the use of Mezei's (1992) polynomial path approach for scaling of interactions (Kuznetsova & Kvamme 2001, 2002). Work on estimation of mobilities along these lines is initiated but in the absence of appropriate values of mobilities at this stage we use the same mobilities for the phase field and the concentration field.

The interface width needed in equation (3) is estimated using molecular dynamics simulations of model hydrate-aqueous fluid system. The SPC/E (Berendsen *et al.* 1987) model for water and the CO_2 model of Harris & Yung (1995) were used. The envelope of the density peaks, which may be loosely identified as the spatial variation of the amplitude of the dominant density wave (i.e. a constant times $\phi(z)$), was fitted to the following hyper-tangential functional form:

$$X(z) = A + 1/2 B\{1 + \tanh[(z-z_0)/(2^{3/2}\delta)]\} \quad (11)$$

where the interface thickness δ is related to the 5–95% interface thickness d (the distance on which the phase field changes between 0.1 and 0.9) as $d = 2^{5/2} \operatorname{atanh}(0.9)\delta$. Note that this interface profile is valid rigorously only when the chemical effects at the interface can be ignored. In practice, equation (11) appears to approximate the interfacial profiles reasonably well. The average value of d was estimated to be 0.85 ± 0.07 nm.

Figures 6 and 7 present a simulation snapshot (taken by the VMD software package; Humphrey *et al.* 1996) and the corresponding density profiles typical for a system involving hydrate and aqueous liquid. The interfacial zone where the liquid water is less structured than in hydrate but still affected by the hydrate's proximity is clearly identifiable.

Crystalline hydrates are characterized by several structural details specific to hydrates only (numbered 1–5 in Fig. 7). Hydrogen density has proved to be an especially useful hydrate marker, since its density profile in bulk hydrate features a distinctive triple-peak, with the central hydrogen peak coinciding with the crest of carbon density and the hydrate-oxygen valley (marker 3 in Fig. 7). This hydrogen signature is totally absent in either liquid water or aqueous CO_2 (consider the lopsided structure of marker 4 inside the interfacial zone). Thus one can define the interface thickness as the transition length along which the triple-peak feature decays from the one characteristic for bulk to non-existent (in combination with other criteria).

Another hydrate-structure marker is given by the positioning of water-oxygen valleys relative to the carbon peaks (markers 1 and 5). The regular succession of alternating high and low peaks evident in Figure 7 is a typical feature of structure I carbon dioxide hydrate when it is projected on a given plane; the water-oxygen valleys flank the high peaks and coincide with the low ones. The latter valleys vary in depth throughout the whole of the hydrate crystal; these long-range envelope oscillations can be clearly seen in coarse-grain density profiles.

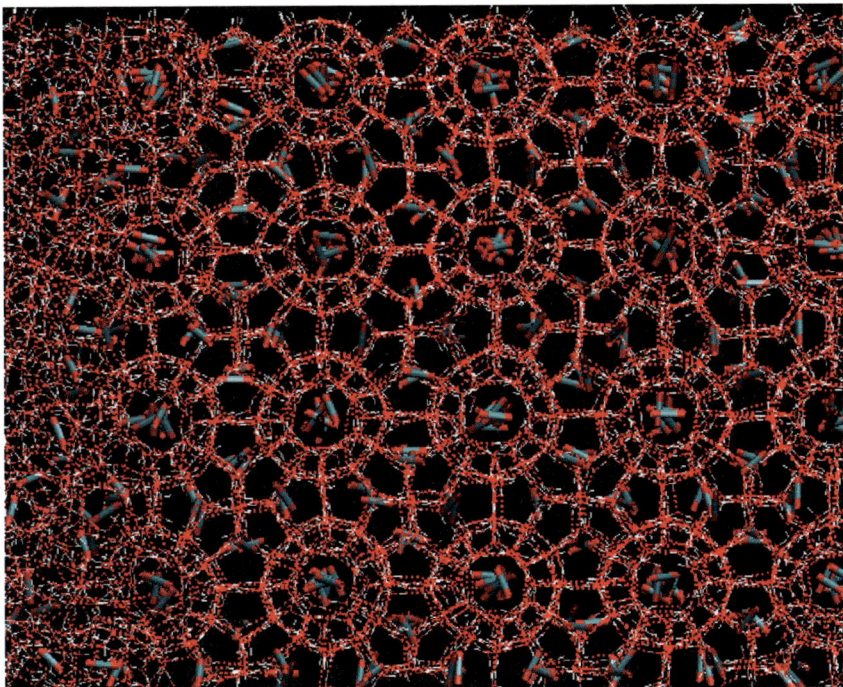

Fig. 6. Snapshot of molecular dynamics simulation featuring carbon dioxide hydrate in contact with aqueous CO_2. Hydrogen bonds are depicted by stippled lines.

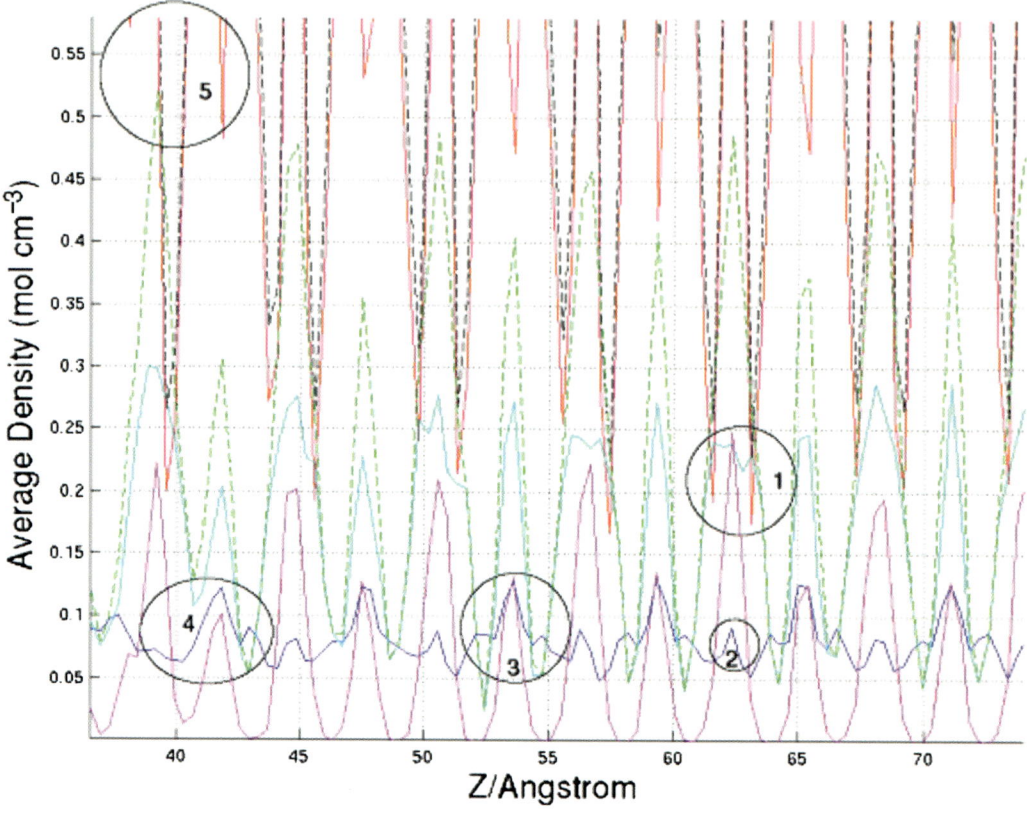

Fig. 7. Density profiles characterizing the system in Figure 6. Blue line, liquid/hydrate hydrogen; cyan, liquid/hydrate water oxygen; red, CO_2 oxygen; magenta, CO_2 carbon.

Work is in progress on estimation of interface free energy by means of Molecular Dynamics simulations, but for the present simulations we have used the liquid water/ice value of 29.1 mN m^{-1} (Hardy 1977) applied together with an interface thickness of 10 Å.

In Figure 8 we plot four snapshots from a PFT simulation of a CO_2 slab surrounded by saturated (mole-fraction CO_2 equal to 0.033) aqueous solution of CO_2 at 150 bar and 274 K. Yellow indicates hydrate. The initial hydrate cores on the left side have different orientational fields while the initial hydrate particles on the right-hand side have the same orientational field. A consequence of the first and second law is that neighbouring regions will compete for available CO_2 and regions of lower free energy will survive at the cost of the regions that have a higher free energy. The ultimate consequence of this can lead to penetration of the hydrate film and massive hydrate growth.

Within the simplified two-phase exchange model, the PFT is solved for different diffusivity coefficients ranging between 10^{-9} and 10^{-12} m^2 s^{-1}. As indicated earlier, the first of these corresponds to typical diffusivities in aqueous solution values, while the latter value is assumed to be solid hydrate value. Interpolation between the simulated results for different diffusivities gives the solid curve in Figure 11, corresponding to a value of the diffusivity coefficient of 1.7×10^{-9} m^2 s^{-1}, which is close to the value of diffusivity of CO_2 in aqueous solution (1.56×10^{-9} m^2 s^{-1} at 30 °C).

Solid surfaces without any thermodynamic properties will have an impact on the progress of the growth due to gradients in free energies resulting from the impact of equations (7)–(10) on the development of the free energy through the system. This is illustrated in Figure 9 where we plot some snapshots from a simulation of hydrate growth from saturated (with respect to fluid CO_2/aqueous solution) solution of CO_2 at 150 bar and 1 °C. The solid circles are region of no mass flux but the black regions have no thermodynamic properties.

Physical interactions between typical sandstone and surrounding molecules will typically imply the existence of a film of aqueous solution between

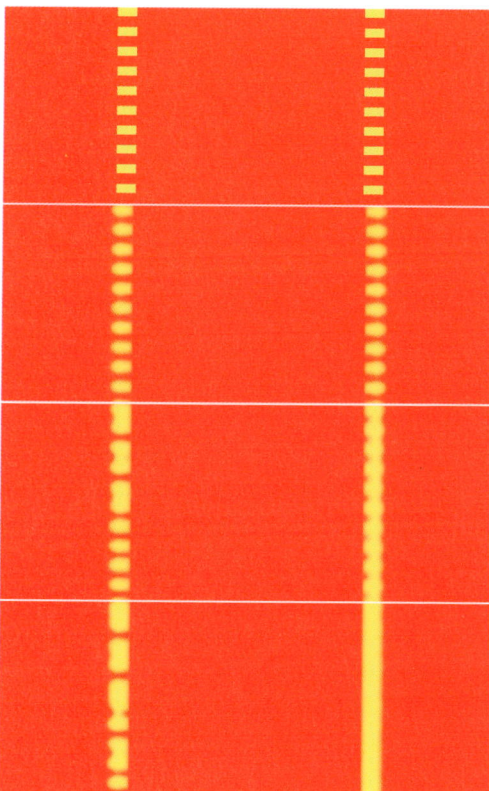

Fig. 8. Results from PFT simulation of an initially punctuated hydrate film on the interface between CO_2 (inner red section) and saturated aqueous solution of CO_2 at 150 bar and 274 K. The orientation field on the left side is set at random and different for the different hydrate particles. Hydrate particles on the right-hand side were fixed at the same value.

the mineral and the hydrate, as also seen from samples of *in situ* hydrate (Techmer *et al.* 2005). This is in contrast to clay minerals which may actually stick to the hydrate (Ordriozola *et al.* 2004; Titiloye & Skipper 2005).

Molecular dynamics simulation details and applied force fields

The molecular dynamics used the constant-temperature, constant-pressure algorithm from the MDynaMix package of Lyubartsev & Laaksonen (2000). The starting interfacial system was constructed from slabs of bulk water–carbon dioxide mixture and appropriately cleaved calcite crystal, thermalized initially at 298 K and set side by side, with the periodic boundary conditions applied. The resulting systems were subsequently equilibrated for several tens of picoseconds before the average collection began. The production time amounted to 200 ps. The time step was set to 10^{-16} s to allow for accurate integration of internal degrees of freedom. The system was kept at a constant temperature of 298 K. Electrostatic interactions were handled by the Ewald summation technique with a variable number of reciprocal vectors. A Linux-based message passing interface (MPI) was used to implement parallel computation on a cluster dual-processor machines. The number of processors ranged from 2 to 12. The cut-off radius for the Lennard–Jones potential and electrostatic forces was 12 Å. Six different systems were simulated; system details are stated in Table 1. The calcite crystal had either $(10\bar{1}4)$ (dominant) or $(10\bar{1}0)$ (next dominant) planes facing the water–CO_2 interface.

Density profiles of x–y averaged quantities of interest were generated by partitioning the simulation box into discrete bins in the z direction. The atomic density profile for atom of type i was thus obtained by

$$\langle \rho_i(z) \rangle = \frac{\langle m_i N_i(z) \rangle}{A \Delta z} \quad (12)$$

where N_i is the number of i-type atoms in a slab located between z and $z + \Delta z$, A is the cross-section area, $A = L_x L_y$, m_i is the atom's mass, Δz the slab's width. Varying the binning width in the 0.095–0.475 Å range (1800–360 bins) did not affect the profiles in any significant way.

The three-site F3C water model of Levitt *et al.* (1997) (with SPC/E charges) was used in the simulations. The model differs from the majority of water force fields in that its hydrogens possess a small amount of van der Waals interaction useful to offset the otherwise unshielded Coulombic attraction between positively charged water hydrogens and calcium anions.

For modelling calcite, we used a combination of two molecule models: a one-site model for the Ca^{2+} ion and a four-site model for the CO_3^{2-} ion. Both models were adopted from Hwang *et al.* (2001). The ζ, R_0, D_0 exp-6 parameters were fitted (Mayo *et al.* 1990) to yield accurate lattice parameters.

The water–CO_2 interactions used a three-site Lennard–Jones model carbon dioxide model of Panhuis *et al.* (1998), specifically parameterized to reproduce the behaviour of lone CO_2 molecules in water. As for CO_2–calcite and CO_2–CO_2 interactions, we modified the original five-site CO_2 model due to Tsuzuki & Tanabe (1999) by merging the van der Waals and Coulombic interaction sites to make it into a three-site one. The charges were also exchanged for those of Panhuis *et al.* (1998). This force field combination was

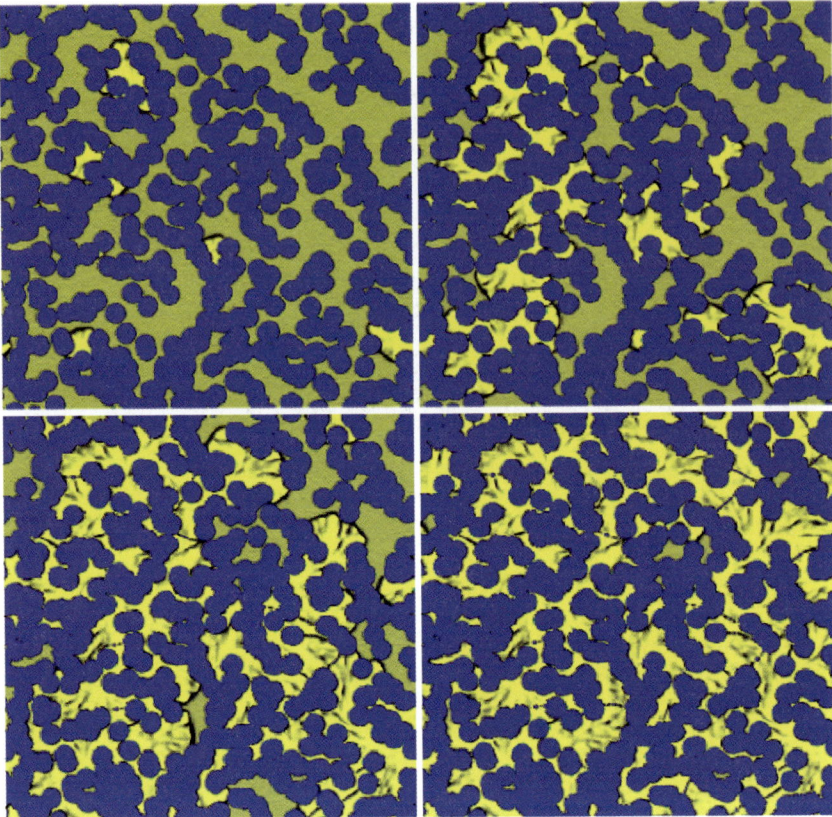

Fig. 9. PFT simulations of hydrate (yellow) growth in between solid particles (blue).

tested by comparing our water oxygen–CO_2 oxygen and water oxygen–CO_2 carbon radial distribution functions with those of Panhuis *et al.* (1998).

Interfacial system

The findings of this work are not directly comparable to either experimental or modelling results, since nobody (to our knowledge) has performed either experiments or numerical simulations involving calcite in contact with water–CO_2 mixture. This is why the validity of our approach was tested on calcite–water systems. Two calcite surfaces ($10\bar{1}4$ and $10\bar{1}0$) were investigated to see which surface had the lowest surface energy. Both experimental results and simulations agree that the ($10\bar{1}4$) surface is the most stable one under both dry and wet conditions (corresponds to the lowest surface energy). Our results for pure water and calcite and calcite in vacuum showed a good qualitative agreement with (wildly varying) results of previous numerical treatment of other authors, as well a good quantitative agreement with the work of Hwang *et al.* (2001). The best agreement was

Table 1. *Excess surface energies and potential energies for calcite–water–CO_2 systems*

System	Cell dimensions (Å)			Potential energy per 'particle' (kJ mol^{-1})	Excess surface energy (J m^{-2})
	x	y	z		
Pure water	24.4	25.9	29.0	-43.0770 ± 0.0009	—
Aqueous CO_2	17.1	20.0	29.3	-42.9560 ± 0.0018	—
($10\bar{1}0$) Calcite–aqueous CO_2	17.1	20.0	49.3	-544.1338 ± 0.0006	0.79(4)
($10\bar{1}4$) Calcite–aqueous CO_2	24.4	25.9	49.0	-662.2044 ± 0.0005	0.41(9)

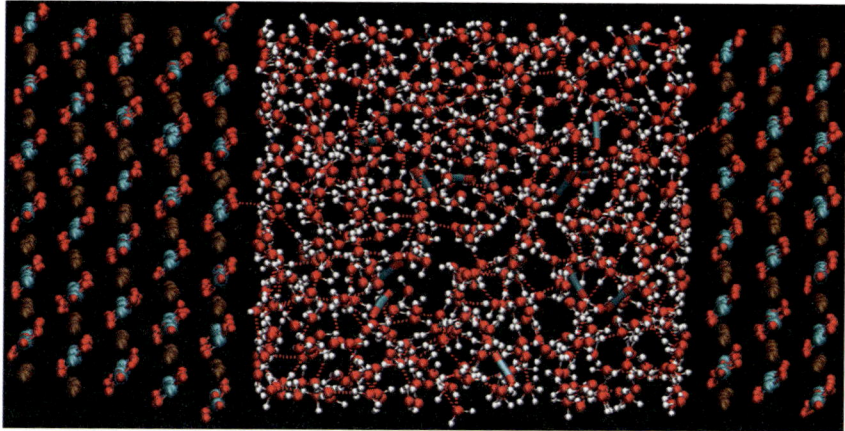

Fig. 10. Molecular simulation snapshot (taken by the VMD software package; Humphrey *et al.* 1996): calcite cleaved along the dominant (10$\bar{1}$4) plane in contact with water–carbon dioxide mixture. Carbon dioxide shows no preference for the mineral–liquid interface.

with Hwang *et al.* (2001) due to their model used and their approach was followed, though with modifications (cross-interactions). The differences between wet and dry surface energies were also consistent. If a mineral is water-wet, its excess surface energy should be significantly lower for the hydrated surface than for the dry surface. This can be stated as the second point of agreement (Fig. 10).

The next step was to investigate how adding CO_2 at 0.033 mf to the water would affect the excess surface energies of the two calcite-liquid surfaces (10$\bar{1}$4 and 10$\bar{1}$0) under study. Excess surface energy, γ^E, a good indicator of the relative stability of surfaces as well as the strength of hydration, is presented in Table 1.

$$\Delta E = E_{composite} - E_{bulk} - E_{water} \quad (13)$$

$$\gamma^E = \frac{\Delta E}{(2A_s N_A)} \quad (14)$$

where $E_{composite}$ is system's total potential energy, A_s is surface area and N_A is the Avogadro number.

The difference in surface energy, $\Delta\gamma^E$, between the most stable and the next most stable calcite surfaces amounted to

$$\Delta\gamma^E = \gamma^E_{(10\bar{1}0)} - \gamma^E_{(10\bar{1}4)} = (0.794 - 0.419)\,\mathrm{J\,m^{-2}}$$
$$= 0.375\,\mathrm{J\,m^{-2}}$$

The excess surface energies of surfaces wetted by the mixture of water and CO_2 retain the relative stability order, that is, the excess energy of (10$\bar{1}$4) surface is still by far the smallest one (0.794 J m^{-2}). It is lower by 0.375 J m^{-2} than in the (10$\bar{1}$0) surface (0.419 J m^{-2}), indicating that the (10$\bar{1}$4) surface is still the most stable one energetically. However, it does appear that the increase in excess surface energy associated with the addition of carbon dioxide is larger for the stable (10$\bar{1}$4) surface.

There are several instructive features to be observed in the density profiles. First, the average water densities have higher peaks at each side of the crystal. The water density in the immediate vicinity of (10$\bar{1}$0) calcite was low due to water molecules not fitting into the small surface ridges; the pronounced second and third water layers indicated water's affinity for calcite. Second, a closer look at the calcite crystal has shown that density peaks corresponding to different calcite components are aligned in the crystal bulk, but shifted relative to each other in the interfacial layers. In the layer closest to the interface, one can see that oxygen and calcium peaks clearly lie closer to the interface than that of carbon. This is probably due to positively charged calcium ion interacting significantly with the water oxygen, and carbonate oxygen attracted by the hydrogen. This is the origin of the excess energy. Two layers away from the interface, the calcium cation is still closer to the interface than the carbon, but the oxygen peak is shifted to the other side of the carbon peak. This is probably a consequence of the disorder in the first layer. Third, it appeared from the density distribution plots that carbon dioxide had no particular affinity for the calcite interface; it was spread evenly throughout the water. So one may assume that whatever affected the surface energy was water restructuring to accommodate the carbon dioxide. Carbon dioxide is a significantly larger molecule than water so it does not fit into the calcite surface as well as water does. Both VMD-generated views and density profiles suggest that carbon dioxide molecules are not attracted to the calcite interface. They do not cluster either, but tend to position themselves closer to each other than to the interface. This may be one of the reasons why the PFT simulations reproduce almost perfectly the experimentally measured conversion rates from CH_4 hydrate over to CO_2 hydrate (Fig. 11) with a mobility that is very close to diffusivity of CO_2 in aqueous solution. It should also be pointed out that, within the resolution of the MRI experiments, we cannot exclude the possibility of liquid water as part of the microscopic reformation mechanism. Work is still in progress on extended PFT simulations (Kvamme et al. 2005b).

Practical implications

The experiments, and the corresponding modelling, conducted in this work show that storage of CO_2 in hydrate reservoirs is a feasible alternative to the aquifer storage. Even without exploitation of the released hydrocarbons, the natural trapping capacity of shale layers in these reservoirs has been one of the factors contributing to the long-term preservation of natural-gas hydrates. Hydrate reservoirs should therefore be able to provide safe storage of CO_2 for very long time scales.

In contrast to the sandstone, there is theoretical evidence that hydrate is stable towards the surfaces of the dominant clay minerals. The combination of clay and hydrate-forming conditions will therefore provide ideal conditions for CO_2 storage in reservoirs.

An important challenge would be to address the issue of the size of the liquid channels separating hydrate from the minerals. Since the adsorbed layer of molecules on the mineral surfaces are separate phases with thermodynamic properties that can be derived from molecular dynamics simulations, this system can also be handled with the PFT approach using one phase field for the fluid/hydrate phase

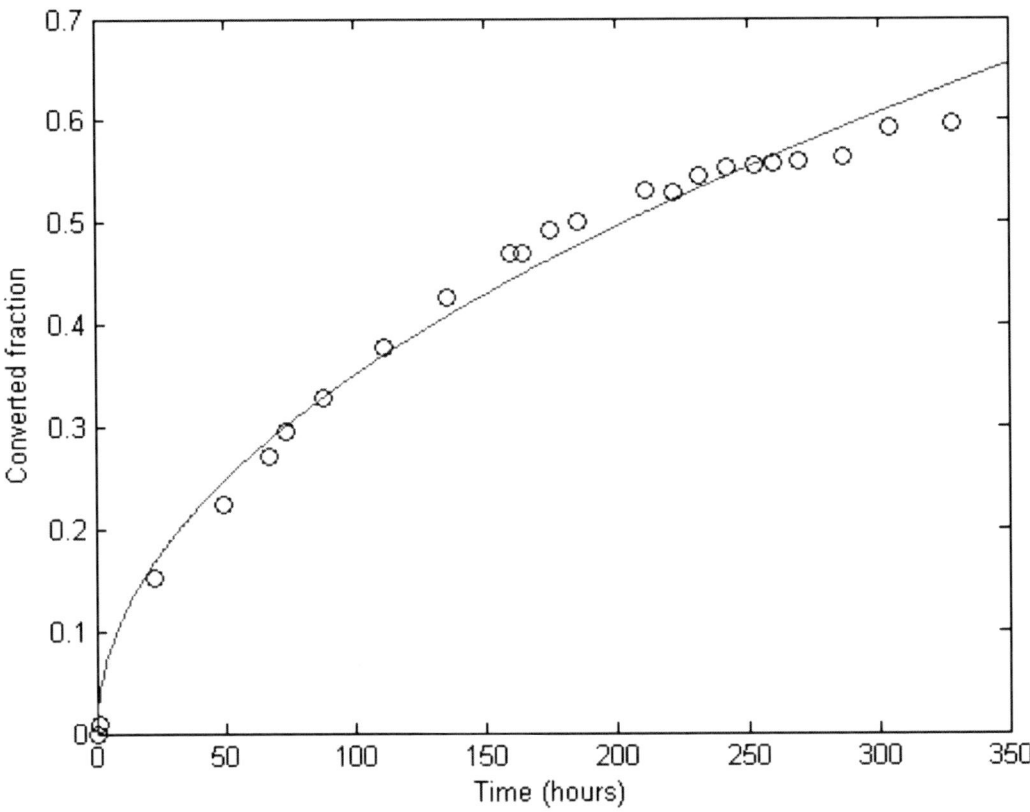

Fig. 11. Conversion of CH_4 hydrate over to CO_2 hydrate at 1200 psia and 4 °C. The porous media is described elsewhere (Kvamme et al. 2004). Circles are from MRI measurements while the curve represents interpolated results from a phase field approach with an efficient diffusivity coefficient for CO_2 equal to 1.7×10^{-9} m^2 s^{-1}.

transition and another phase field for the adsorbed phase/surrounding system. Since the PFT essentially explores the effects of the first and the second laws of thermodynamics, the integration of the system dynamics should be able to trace the logical interfaces for different mineral characteristics (clay/shale/sandstone) and also regions of fluid phases/channels. Practically this will imply an extension of the three phase (hydrate/aqueous/fluid) system described by Kvamme *et al.* (2005*b*).

Conclusion

We have applied the techniques of magnetic resonance imaging to demonstrate experimentally that the injection of liquid CO_2 can convert CH_4 hydrate within Bentheim sandstone into CO_2 hydrate while simultaneously releasing *in situ* CH_4. Phase field theory simulations of corresponding model systems indicate that the reformation kinetic rate is directly proportional to the kinetic rate characteristic for CO_2 transport through an aqueous solution. We explain this observation by the presence of aqueous channels separating the hydrate from the mineral surfaces. Under the actual storage conditions, these liquid channels may serve as transport routes for released natural gas, as well as distribution channels for injected CO_2. This means in practice that few natural gas hydrate reservoirs have zero permeability and that the small channels between minerals surfaces and hydrates may assist in necessary local distribution of fluids.

When it comes to the CO_2 storage in conventional aquifers whose prevailing temperatures and pressures include the regions of hydrate stability, the hydrate formation may play either positive or negative role. Formation of hydrate in sandstone may severely reduce its permeability, which can prove beneficial if the hydrate layer grows horizontally, since it will reduce the flux of CO_2 towards the seafloor. On the other hand, upward-oriented hydrate growth may inhibit the sideways spread of the CO_2 plume and thus inhibit distribution and gradual dissolution of CO_2 into the groundwater. Hydrate formation towards clay and shale is expected to provide long-term stability, and we therefore recommend searching for reservoirs that have existing clay/shale sealing layers as well as cold zones that promote hydrate formation.

We have also demonstrated a comprehensive theoretical approach enabling the evaluation of kinetics associated with formation, dissociation and reformation of hydrate in reservoirs. All parameters needed by the kinetic model can be estimated from molecular dynamics simulations of corresponding model systems, with the single exception being morphology-related effects, which typically needs one or two empirical parameters.

Our molecular dynamics computations involving excess surface energy of calcite in contact with pure water and aqueous carbon dioxide have indicated that the same thermodynamic concept could be applied to study the effect of mineral surfaces on the surrounding fluid phases.

Financial support from the Research Council of Norway, ConocoPhillips and Hydro is greatly appreciated. We acknowledge the invaluable contributions from James Howard and Jim Stevens at ConocoPhillips and Bernard Baldwin, Green Country Petrophysics LLC, in obtaining the experimental MRI results. The phase field theory simulations are based on extensions of a numerical coding provided by Laszlo Granasy.

References

BERENDSEN, H. J. C., GRIGERA, J. R. & STRAATSMA, T. P. 1987. The missing term in effective pair potentials. *Journal of Physical Chemistry*, **91**, 6269–6271.

BÜNZ, S., MIENERT, J. & BERNDT, C. 2003. Geological controls on the Storegga gas-hydrate system of the mid-Norwegian continental margin. *Earth and Planetary Science Letters*, **209**, 291–307.

CAHN, J. W. & HILLIARD, J. E. 1958. Free energy of a nonuniform system. I. Interfacial free energy. *Journal of Chemical Physics*, **28**, 258–267.

GRÁNÁSY, L., PUSZTAI, T., JUREK, Z., CONTI, M. & KVAMME, B. 2003. Phase field theory of nucleation in the hard sphere liquid. *Journal of Chemical Physics*, **119**, 10376–10382.

GRÁNÁSY, L., PUSZTAI, T., BÖRZSÖNYI, T., WARREN, J. A., KVAMME, B. & JAMES, P. F. 2004*a*. Nucleation and polycrystalline solidification in binary phase field theory. *Physics and Chemistry of Glasses*, **45**, 107–115.

GRÁNÁSY, L., PUSZTAI, T., TEGZE, G., KUZNETSOVA, T. & KVAMME, B. 2004*b*. Phase field theory of hydrate nucleation: Formation of CO_2 hydrate in aqueous solution. *In*: *Recent Advances in the Study of Gas Hydrates*. Kluwer Academic/Plenum, Dordrecht.

GRÁNÁSY, L., PUSZTAI, T. & WARREN, J. 2004*c*. Modelling polycrystalline solidification using phase field theory. *Journal of Physics: Condensed Matter*, **16**, 1205–1235.

HARDY, S. C. 1977. Wavelength of instabilities. *Philosophical Magazine*, **35**, 471.

HARRIS, J. G. & YUNG, K. H. 1995. Carbon dioxide's liquid–vapor coexistence curve and critical properties as predicted by a simple molecular model. *Journal of Physical Chemistry*, **99**, 12021.

HUMPHREY, W., DALKE, A. & SCHULTEN, K. 1996. VMD – visual molecular dynamics. *Journal of Molecular Graphics*, **14**, 33–38.

HWANG, S., BLANCO, M., DEMIRALP, E., CAGIN, T. & GODDARD, W. A. 2001. The MS-Q force field for clay minerals: Application to oil production. *Journal of Physical Chemistry B*, **105**, 4122–4127.

KUZNETSOVA, T. & KVAMME, B. 2001. Viabilty of atomistic potentials for thermodynamic properties of carbon dioxide at low temperatures. *Journal of Computational Chemistry*, **22**, 1772.

KUZNETSOVA, T. & KVAMME, B. 2002. Atomistic computer simulations for thermodynamic properties of carbon dioxide at low temperatures. *Energy Conversion Management*, **43**, 2601–2623.

KVAMME, B. & TANAKA, H. 1995. Thermodynamic stability of hydrates for ethylene, ethane and CO_2. *Journal of Physical Chemistry*, **99**, 7114.

KVAMME, B, GRAUE, A. ET AL. 2004. Towards understanding the kinetics of hydrate formation: Phase field theory of hydrate nucleation and magnetic resonance imaging. *Physical Chemistry Chemical Physics*, **6**, 2327–2334.

KVAMME, B., KUZNETSOVA, T. & UPPSTAD, D. 2005a. Modelling excess surface energy in dry and wetted calcite systems. *Lecture Series on Computer and Computational Sciences*, Vol. 1. VSP International Science Publishers, The Netherlands, 4pp.

KVAMME, B., SVANDAL, A., BUANES, T. & KUZNETSOVA, T. 2005b. Phase field approaches to the kinetic modeling of hydrate phase transitions. *In*: *Proceedings of Natural Gas Hydrates: Energy Resource Potential and Associated Geologic Hazards*, 12–16 September 2004, Vancouver.

KVAMME, B., KUZNETSOVA, T., HEBACH, A., OBERHOF, A. & LUNDE, E. 2007. Measurements and modeling of interfacial tension for water + CO_2 systems at elevated pressures. *Computational Material Sciences*, **38**, 506–513.

LEVITT, M., HIRSHBERG, M., SHARON, R., LADIG, K. E. & DAGETT, V. 1997.Calibration and testing of a water model for simulation of the molecular dynamics of proteins and nucleic acids in solution. *Journal of Physical Chemistry B*, **101**, 5051–5061.

LYUBARTSEV, A. P. & LAAKSONEN, A. M. 2000. DynaMix – a scalable portable parallel MD simulation package for arbitrary molecular mixtures. *Computer Physics Communications*, **128**, 565–589.

MAKOGAN, Y. F. 1997. *Hydrates of Hydrocarbons*, Pennwell, Tulsa, OK.

MAYO, S. L., OLAFSON, B. D. & GODDARD, W. A. III. 1990. DREIDING: A generic force field for molecular simulations. *Journal of Physical Chemistry*, **94**, 8897–8909.

MEZEI, M. 1992. Polynomial path for the calculation of liquid-state free-energies from computer-simulations tested on liquid water. *Journal of Computational Chemistry*, **13**, 651–656.

ODRIOZOLA, G., AGUILAR, J. F. & LOPEZ-LEMUS, J. 2004. Na-montmorillonite hydrates under ethane rich reservoirs: NPzzT and μPzzT simulations. *Journal of Chemical Physics*, **121**, 4266.

PANHUIS, M. I. H., PATTERSON, C. H. & LYNDEN-BELL, R. M. A. 1998. Molecular dynamics study of carbon dioxide in water: Diffusion, structure and thermodynamics. *Molecular Physics*, **94**, 963–972.

RADHAKRISHNAN, R. & TROUT, B. L. 2002. A new approach for studying nucleation phenomena using molecular simulation: Application to CO_2 hydrate clathrates. *Journal of Chemical Physics*, **177**, 1786–1796.

SOAVE, G. 1972. Equilibrium constants from a modified Redlich–Kwong equation of state. *Chemical Engineering Science*, **27**, 1197.

SVANDAL, A., KUZNETSOVA, T. & KVAMME, B. 2006a. Thermodynamic properties and phase transtions in the $H_2O/CO_2/CH_4$ system. *Physical Chemistry Chemical Physics*, **14**, 1707–1713.

SVANDAL, A., KVAMME, B., GRÀNÀSY, L. & PUSZTAI, T. 2006b. The phase field theory applied to CO_2 and CH_4 hydrate. *Journal of Crystal Growth*, **287**, 486–490.

TECHMER, K., HEINRICHS, T. & KUHS, W. F. 2005. Cryo-electron microscopic studies on the structures and composition of Mallik gas-hydrate-bearing samples. *In*: DALLIMORE, S. R. & COLLETT, T. S. (eds) *Scientific Results from the Mallik 2002 Gas Hydrate Production Research Well Program*, Mackenzie Delta, Northwest Territories, Canada. Geological Survey of Canada Bulletins, **585**.

TEGZE, G., PUSZTAI, T. ET AL. 2006. Multi-scale approach to CO_2-hydrate formation in aqueous solution: Phase field theory and molecular dynamics. Nucleation and growth. *Journal of Chemical Physics*, **124**, 234710.

TITILOYE, J. O. & SKIPPER, N. T. 2005. Monte Carlo and molecular dynamics simulations of methane in potassium montmorillonite clay hydrates at elevated pressures and temperatures. *Journal of Colloid and Interface Science*, **282**, 422–427.

TSUZUKI, S. & TANABE, K. 1999. Molecular dynamics simulations of fluid carbon dioxide using the model potential based on *ab initio* MO calculation. *Computational Material Science*, **14**, 220–226.

XU, T., APPS, J. A. & PRUESS, K. 2004. Numerical simulation of CO_2 disposal by mineral trapping in deep aquifers. *Applied Geochemistry*, **19**, 917–936.

Gas hydrate growth and dissociation in narrow pore networks: capillary inhibition and hysteresis phenomena

R. ANDERSON*, B. TOHIDI & J. B. W. WEBBER

Centre for Gas Hydrate Research, Institute of Petroleum Engineering, Heriot-Watt University, Edinburgh, EH14 4AS, UK

*Corresponding author (e-mail: ross.anderson@pet.hw.ac.uk)

Abstract: Marine sediments hosting gas hydrates are commonly fine-grained (silts, muds, clays) with very narrow mean pore diameters (~ 0.1 μm). This has led to speculation that capillary phenomena could play an important role in controlling hydrate distribution in the seafloor, and may be in part responsible for discrepancies between observed and predicted (from bulk phase equilibria) hydrate stability zone (HSZ) thicknesses. Numerous recent laboratory studies have confirmed a close relationship between hydrate inhibition and pore size, stability being reduced in narrow pores; however, to date the focus has been hydrate *dissociation* conditions in porous media, with capillary controls on the equally important process of hydrate *growth* being largely neglected. Here, we present experimental methane hydrate growth and dissociation conditions for synthetic mesoporous silicas over a range of pressure–temperature (PT) conditions (273–293 K, to 20 MPa) and pore size distributions. Results demonstrate that hydrate formation and decomposition in narrow pore networks is characterized by a distinct hysteresis: solid growth occurs at significantly lower temperatures (or higher pressures) than dissociation. Hysteresis takes the form of repeatable, irreversible closed primary growth and dissociation PT loops, within which various characteristic secondary 'scanning' curve pathways may be followed. Similar behaviour has recently been observed for ice–water systems in porous media, and is characteristic of liquid–vapour transitions in mesoporous materials. The causes of such hysteresis are still not fully understood; our results suggest pore blocking during hydrate growth as a primary cause.

Naturally occurring gas hydrates (or clathrate hydrates) in sediments may pose a hazard to deepwater drilling and production operations (Kvenvolden 1999; Milkov *et al.* 2000), have potential as a strategic low-carbon energy reserve (Kvenvolden 1999; Lee & Holder 2001), could provide a means for deep ocean CO_2 disposal through sequestration/storage (Hunter 1999; Brewer *et al.* 1999), and have long-term significance with respect to ocean margin stability, methane release to the atmosphere and global climate changes (Kvenvolden 1999; Dickens 2003).

Although our understanding of sediment-hosted gas hydrates has grown considerably in recent years, we still lack fundamental knowledge concerning the mechanisms of hydrate growth, accumulation and distribution within the subsurface. Clathrates have been recovered in shallow ocean floor sediment cores from numerous sites around the world (e.g. Ocean Drilling Program (ODP) Leg 164, Blake Ridge, offshore South Carolina (Paull *et al.* 2000), and Leg 204, Cascadia Margin, offshore Oregon (Tréhu & Shipboard Scientific Party 2003)). Sediments hosting gas hydrates are generally characterized by organic matter-rich fine-grained silts, muds and clays, with lesser coarser sandy layers present at some sites. Hydrates commonly display a wide range of growth habits, and are often patchily distributed within the host sediment according to texture (Booth *et al.* 1996). In fine-grained strata, hydrates are generally found in the form of segregated nodules, lenses, pellets or sheets. In contrast, where coarser layers are present, clathrates often form an interstitial pore fill between sediment grains. This variation in growth patterns according to sediment type suggests that host sediment properties may play an important role in controlling hydrate morphology and distribution within the subsurface (Clennell *et al.* 1999; Henry *et al.* 1999).

Further evidence for potential host sediment controls on hydrate equilibria comes from the predicted depth of the Base of the Hydrate Stability Zone (BHSZ) in seafloor sediments. While ODP coring has confirmed that the BHSZ commonly lies close to pressure and temperature conditions calculated from bulk (unconfined) phase equilibria, there are a number of sites where the thickness is notably less than predicted (e.g. Blake Ridge (Paull *et al.* 2000), and Cascadia Margin (Tréhu & Shipboard Scientific Party 2003)). The depth of the BHSZ is dependent on various factors, including gas concentration (gas concentration must exceed aqueous equilibrium solubility in the presence of hydrate), composition (the addition of CO_2, H_2S and higher thermogenic hydrocarbons such as ethane/propane increases hydrate stability), pore water salinity (dissolved salt reduces stability) and the local geothermal gradient. However, where these are relatively well established from drilling/coring (such as at the Blake Ridge and Cascadia Margin), additional factors must be sought to explain predicted/actual BHSZ discrepancies. One potential influence may be the host sediments themselves. The mechanisms by which sediment properties could alter hydrate stability and/or influence distribution are still relatively poorly understood (Max 2000); however, one potentially important factor which has received considerable attention in recent years is capillary inhibition (Clennell et al. 1999; Henry et al. 1999).

Phase behaviour in confined geometries

It is well established that the pressure–temperature (PT) conditions of first-order phase transitions (e.g. solid–liquid, liquid–vapour) may be significantly altered in confined geometries. In narrow pores, high-curvature phase interfaces can induce strong differential capillary pressures, altering the chemical potential of components relative to bulk (unconfined) conditions. For solid–liquid transitions, where pore sizes are sufficient for phases to retain the structural and physical properties of the bulk phase, solid melting temperatures are generally depressed as a function of pore radius in accordance with the Gibbs–Thomson equation (the constant pressure analogue of the constant temperature Kelvin equation for vapour pressure in mesoporous media; (Enüstün et al. 1978; Christensen 2001). For simple, single-component systems (e.g. ice–water), the common form of the equation relates the pore solid melting point depression, ΔT_p, from the bulk (unconfined) melting temperature, T_b, to the pore radius, r, through:

$$\Delta T_p = T_b \cdot \frac{F \gamma_{sl} \cos \theta}{r \rho_l \Delta H_{sl}} \quad (1)$$

where γ_{sl} is the solid–liquid interfacial free energy (often referred to as the surface or interfacial tension), F the shape factor of the interface (dependent on interface curvature), ρ_l the density of the liquid phase, ΔH_{sl} the latent heat (enthalpy) of fusion, and θ the contact angle between the solid phase and the pore wall (180° measured inside the solid phase if an unfrozen liquid layer is assumed, thus $\cos \theta = -1$). Where γ_{sl} and ΔH_{sl} are relatively constant over the PT conditions of interest, equation (1) dictates a linear relationship between $\Delta T_p/T_b$ and reciprocal pore radius, as confirmed experimentally for many organic and inorganic liquids (Rennie & Clifford 1977; Jackson & McKenna 1990, 1996; Christensen 2001). It should be noted that equation (1) assumes that the solid phase pressure (P_s) is equal to the bulk pressure (P_b), that is, $P_l < P_s = P_b$. Where the liquid phase (P_l) is at bulk pressure ($P_s > P_l = P_b$), the value ρ_l should be replaced by ρ_s, the density of the solid phase (Enüstün et al. 1978).

Although the thermodynamics of solid–liquid equilibria in small pores (particularly ice–water equilibria) has been the subject of investigation for over 100 years (Christensen 2001), only relatively recently did Handa & Stupin (1992) demonstrate that methane hydrate dissociation temperatures are depressed in narrow pores. Seafloor sediments hosting gas hydrates are commonly fine-grained (silts, muds, clays), with narrow mean pore diameters (~ 0.1 μm; Griffiths & Joshi 1989; Clennell et al. 1999). In light of this, it has previously been speculated that capillary phenomena could play an important role in controlling hydrate stability and distribution within sediments, and may be partly responsible for observed discrepancies between predicted and actual BHSZs (Ruppel 1997; Clennell et al. 1999; Henry et al. 1999).

In the most extensive theoretical analyses to date, Clennell et al. (1999) & Henry et al. (1999) (companion papers) developed a capillary-thermodynamic model for hydrate formation in the seafloor which attempted to account for the effect of pore size on equilibrium conditions. From model predictions, the authors could not confirm that capillary inhibition alone was responsible for observed discrepancies between predicted and actual BHSZs, although it was concluded that capillary phenomena did most likely play an important role in controlling hydrate phase behaviour and distribution, particularly in segregation and lens/nodule/layer formation. A lack of firm conclusions concerning the extent to which pore size affects the HSZ could in part be attributable to a lack of available values for hydrate–liquid (water) interfacial free energy (the authors used a value for ice–water interfacial free energy), and, significantly, an absence of reliable data relating pore size/geometry to hydrate growth/dissociation conditions with which to validate model predictions.

The potential role capillary effects may have in controlling hydrate growth and accumulation within sediments has led to considerable experimental and theoretical research into the phenomenon over the past 8 years. Work has focused primarily on (relatively) well characterized porous silicas (Uchida et al. 1999, 2002; Seshadri et al. 2001; Wilder et al. 2001a, b; Seo et al. 2002; Smith *et al.* 2002*a, b*, 2004; Wilder & Smith 2002; Zhang *et al.* 2002, 2003; Seo & Lee 2003; Aladko *et al.* 2004; Dicharry *et al.* 2005), and more recently on natural sands and clays (Uchida *et al.* 2004). Although there are a number of discrepancies between studies, particularly regarding experimental data interpretation (as discussed by Anderson *et al.* 2003*a*), the overall conclusion is that narrow pores have a significant and consistent inhibiting effect on hydrate stability. However, phase behaviour in porous media is highly complex, and there are many potentially important factors that have not yet been addressed. One significant, and particularly relevant, issue is that the focus to date has been the measurement and prediction of hydrate *dissociation* conditions in porous media, with the equally, if not more important process of hydrate *growth* being largely overlooked.

As suggested by Clennell *et al.* (1999), capillary theory predicts a considerable hysteresis may exist between solid growth and melting conditions in narrow pores. The hysteresis loops commonly associated with gas (e.g. nitrogen) adsorption/desorption in mesoporous materials and hydrocarbon reservoir rock drainage/imbibition curves are testament to the fact that such behaviour is a common characteristic of capillary pressure controlled phase transitions and fluid flow within porous media.

In Anderson *et al.* (2003*b*), we reported experimental CH_4, CO_2 and CH_4-CO_2 clathrate hydrate dissociation and ice melting data for mesoporous silica glasses. This data was used to estimate values for ice–water and hydrate–liquid (water) interfacial free energies through a modified version of equation (1), and subsequently employed to validate a capillary corrective function for hydrate thermodynamic models which allows the prediction of hydrate dissociation conditions for narrow cylindrical- or spherical-like pores (Llamedo *et al.* 2004). The added effect of pore water salinity was also investigated (Østergaard *et al.* 2002). Here, we report the results of a detailed experimental investigation of methane hydrate *growth and dissociation* conditions in synthetic mesoporous silica glasses. Data reveal an equilibrium hysteretic hydrate formation/decomposition behaviour not previously observed for clathrates in porous media. Through an analysis of experimental data, we will assess potential origins of the observed hysteresis phenomena, and then comment briefly on potential implications for seafloor hydrate systems.

Experimental equipment and methods

A specifically designed high-pressure (max. 41 MPa) set-up was used in experiments. The set-up, shown in Figure 1, consists of an equilibrium cell (75 cm³ volume) with removable sample cup, central PRT (platinum resistance thermometer), inlet/outlet valve, Quartzdyne pressure transducer and insulated coolant jacket.

The PRT was calibrated with a Prima 3040 precision thermometer, and measures cell temperature to ± 0.01 K with an estimated accuracy of ± 0.1 K. The transducer, via a computer interface, can measure system pressure to within $\pm 6.9 \times 10^{-6}$ MPa, and has a quoted accuracy of ± 0.008 MPa for the complete operating range of 0–138 MPa. System temperature was controlled by circulating fluid from a programmable cryostat (253–373 K) through the cell jacket, and could be kept stable to within ± 0.02 K. Cell temperature and pressure were continuously monitored and recorded using a computer.

Double-distilled water was used in all experiments. High-purity methane (99.995 mol%) was

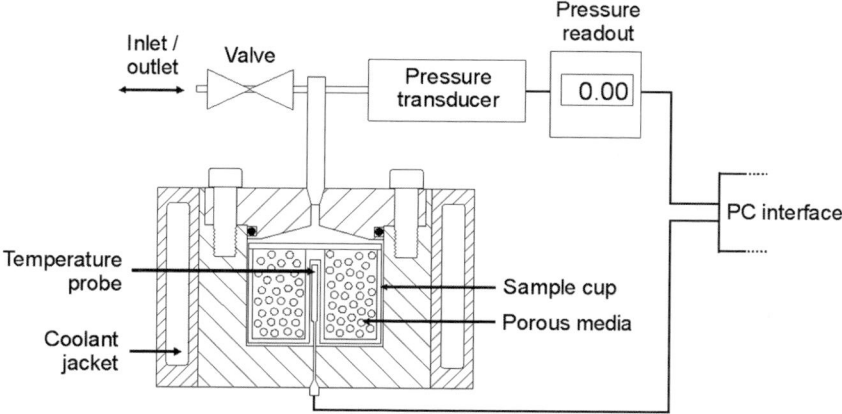

Fig. 1. Schematic illustration of the high-pressure set-up used in experiments.

supplied by Air Products. Porous silica samples, known as Controlled Pore Glass (CPG), were purchased from CPG Inc., USA (now Millipore, USA), and consist of 37–74 μm porous silica shards. Three samples, of 30.6, 15.8 and 9.2 nm nominal pore diameters, were used for experiments. Sample pore size distributions were previously characterized independently by NMR (nuclear magnetic resonance) cryoporometry (Anderson et al. 2003b; Dore et al. 2004).

Test procedures were as follows. CPG silicas were dried overnight in an oven, then saturated (water volume, V_w > pore volume, V_p) with a measured volume of distilled water. Prepared samples were placed in the cell, the cell cooled and water frozen (to minimize evaporation), then air evacuated. Temperature was subsequently raised again to the desired starting temperature (generally outside the bulk hydrate stability zone) before methane was injected to the initial starting pressure. To form hydrates in the first instance, the cell was cooled rapidly until growth commenced, as indicated by pressure–temperature relations. Subsequent to this, hydrate growth and dissociation PT pathways for sample hysteresis regions were determined by a stepped temperature cycling method based on the approach of Tohidi et al. (2000) & Anderson et al. (2003b). The method involves heating/cooling of the cell in steps (generally 0.2–0.5 K), with sufficient time being given (in this case 8–24 h) for the system to reach equilibrium (as indicated by stable pressure) following each step, which results in very reliable and highly repeatable (to within ± 0.1 K) measurements.

Results and discussion

Equilibrium methane hydrate growth and dissociation conditions were determined at various pressures for the three different CPG silica samples (30.6, 15.8 and 9.2 mean pore diameters). Figure 2 shows an example of typically observed clathrate growth and dissociation pressure–temperature pathways, in this case for the 30.6 nm sample. As system water volume exceeds CPG pore volume, gas hydrates form both within and outside the pore network; hydrates outwith the pores dissociate at the bulk (unconfined) methane hydrate + liquid + gas (H + L + G) phase boundary, with both pore

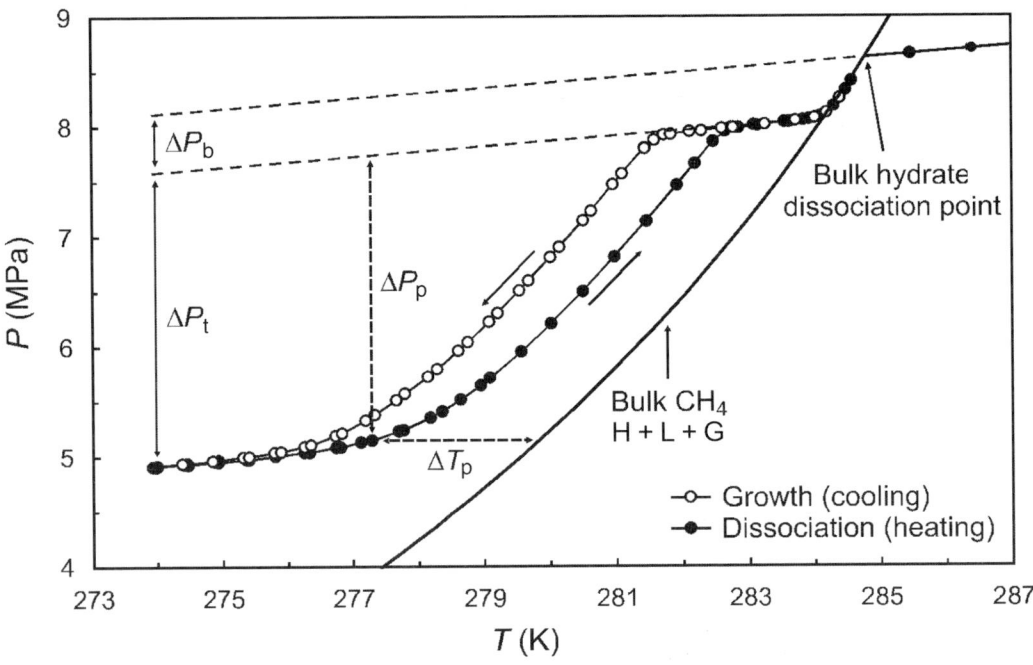

Fig. 2. Plot of primary growth and dissociation PT data for the 30.6 nm mean pore diameter CPG silica saturated with water. ΔP_t and ΔP_b are the total change in pressure associated with hydrate formation in the pores and the bulk respectively. ΔT_p and ΔP_p are the temperature depression of pore hydrate/growth dissociation conditions (from the bulk methane H + L + G phase boundary) and change in pressure associated with pore hydrate formation at any given recorded equilibrium PT condition on the heating/cooling curves respectively. Bulk CH_4 data: polynomial fit to Deaton & Frost (1946) and McLeod & Campbell (1961).

hydrate growth and dissociation conditions being depressed to significantly lower temperatures. The hysteresis between pore clathrate growth and dissociation conditions is distinct – hydrate formation occurs at temperatures significantly lower than decomposition, with irreversible (unidirectional) PT pathways forming a complete closed hysteresis loop. To our knowledge, this clear, repeatable (in the same closed system over 6 months), equilibrium PT hysteresis between growth and dissociation has not previously been reported for clathrate hydrates in porous media. Similar (although not so consistently repeatable) equilibrium hysteretic behaviour has been described for ice–water transitions in hardened cement pastes (Schulson et al. 2000; Swainson & Schulson 2001), however it is generally not reported in most literature studies of solid–liquid transitions in mesoporous materials.

In contrast to the repeatable, equilibrium hysteresis observed here, significant differences between measured freezing and melting temperatures due to stochastic heterogeneous nucleation phenomena have been reported for fluids confined to porous materials (Faivre et al. 1999; Morishige & Kawano 1999). In this case, hysteresis can be attributed to kinetic issues arising as a result of the supercooling generally required to initiate solid nucleation in the absence of a pre-existing crystalline phase. Here, we have eliminated the need for nucleation by ensuring clathrate is present in the bulk (outside the pores) when cooling to initiate pore hydrate growth. Theoretically, this means only progressive solid growth front propagation into media on cooling is required.

Characteristics of hysteresis loops

From Figures 2 and 3, we see that primary pore hydrate dissociation and growth patterns are characterized by a sigmoidal (with respect to linear liquid + gas only PT relationships) curves indicative of formation/decomposition across a Gaussian-like distribution of pores typical of Controlled Pore Glasses (Anderson et al. 2003a, b; Østergaard et al. 2002). Partial or complete dissociation curves for various synthetic (Vycor and sol–gel) mesoporous silicas have been reported previously by other workers (Uchida et al. 1999, 2002; Seshadri et al. 2001; Wilder et al. 2001a, b; Seo et al. 2002; Smith et al. 2002a, b, 2004; Wilder & Smith 2002; Zhang et al. 2002, 2003; Seo & Lee 2003; Dicharry et al. 2005), and show very similar characteristics.

For the purposes of interpretation, we can re-plot heating curve data in terms of the volume of pore hydrate formed relative to growth/dissociation temperature depression. Figure 4 shows a plot of

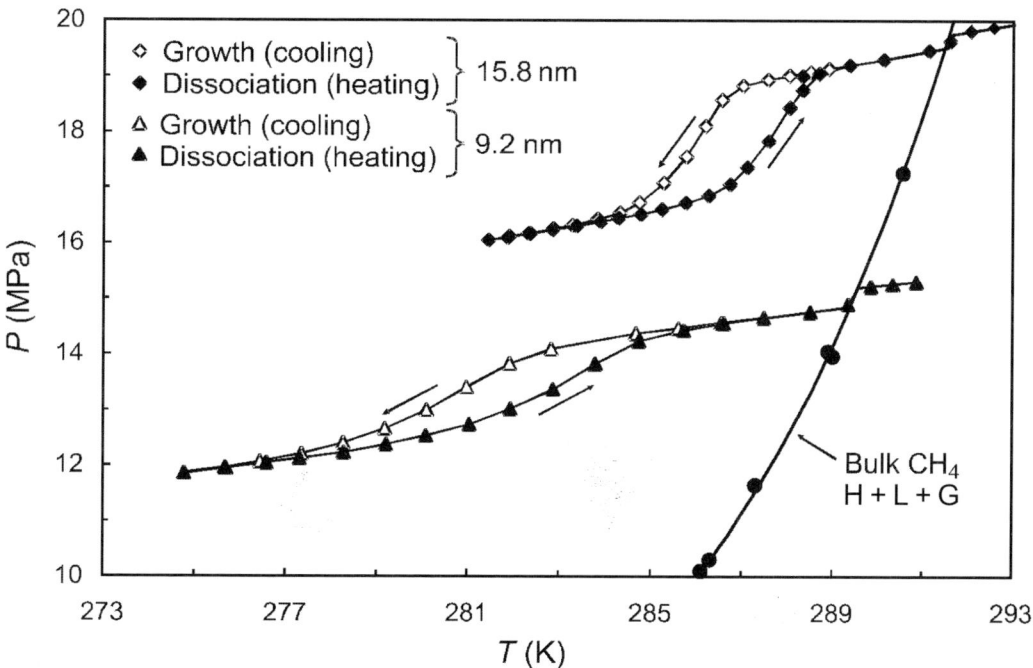

Fig. 3. Examples of primary methane hydrate growth and dissociation loop PT data for the 9.2 nm and 15.8 mean pore diameter CPG silicas. Bulk CH_4 data: polynomial fit to Deaton & Frost (1946) and McLeod & Campbell (1961).

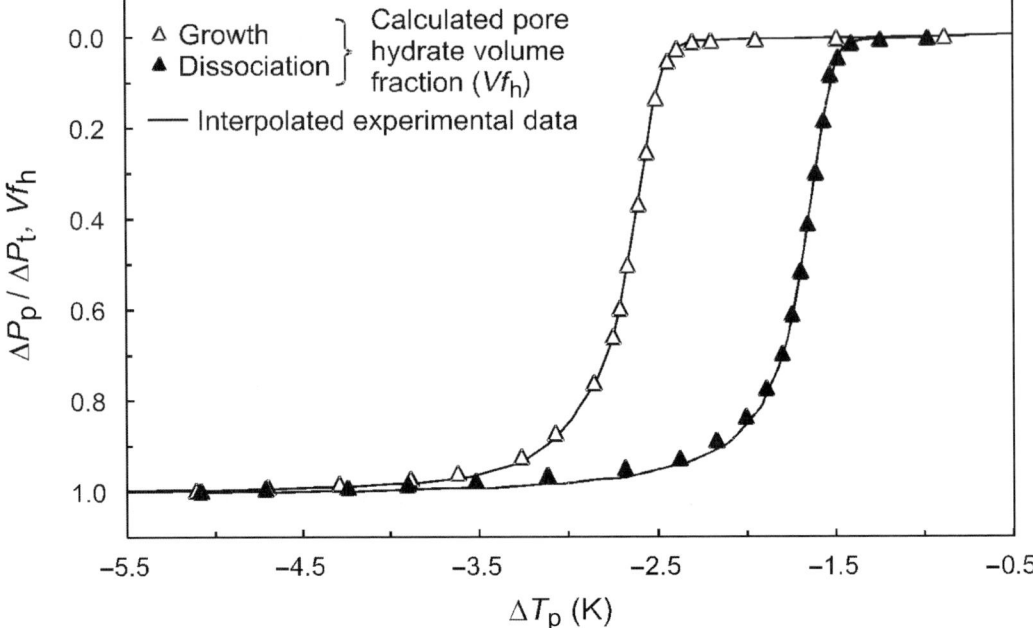

Fig. 4. Plot of ΔT_p v. $\Delta P_p/\Delta P_t$ (experimental data) and calculated volume fraction of pore hydrate (Vf_h) present for selected experimental points (30.6 nm sample). The fraction of total pressure change associated with pore hydrates essentially equals the volume fraction of pore hydrate formed at any point.

ΔT_p v. $\Delta P_p/\Delta P_t$ for the 30.6 nm sample where, as illustrated in Figure 2, ΔT_p is the temperature depression of hydrate growth/dissociation conditions from the bulk methane hydrate phase boundary, ΔP_p is the change in pressure associated with pore hydrate formation at any point and ΔP_t is the total change in pressure associated with pore hydrate formation. Also plotted for comparison is the calculated volume fraction of pore gas hydrate (Vf_h) present at each point. Hydrate volume fractions were calculated by standard iterative mass balance/volume methods assuming a methane hydration ratio of 1:6 (Handa 1986; Lievois et al. 1990; Ciscone et al. 2005). As can be seen, the faction (of total) pressure change associated with pore hydrates for each point is essentially equal to the volume fraction of pore hydrate present at that condition. Thus, in further analyses, we can consider that as a good approximation, $\Delta P_p/\Delta P_t = Vf_h$.

As our interest lies in the relationship between pore radius and hydrate growth/dissociation conditions, we could theoretically use equation (1) to convert ΔT_p to equivalent r, allowing the examination of data in terms of Vf_h v. r. However, this would require the assumption of specific pore/solid–solid interface shapes for both solid formation and melting conditions, as defined by the shape factor, F, in equation (1). To avoid this assumption, we can compare growth and dissociation in terms of acting capillary pressure, P_c, at ΔT_p. P_c can be calculated by substituting the right hand side of the Young–Laplace equation:

$$P_c = P_s - P_l = \frac{F\gamma_{sl}\cos\theta}{r} \quad (2)$$

where P_s is the pressure of the solid (hydrate) phase and P_l the pressure of the liquid phase, into equation (1) and rearranging to yield:

$$P_c = \frac{\Delta T_p}{T_b} \cdot \rho_l \Delta H_{sl} \quad (3)$$

Figure 5 presents pore hydrate volume fractions as a function of calculated capillary pressure during growth and dissociation for all the three CPG silicas. As can be seen, hysteresis loops for all samples show very similar characteristics, with primary growth (increasing P_c) and dissociation (decreasing P_c) PT pathways being of sigmoidal form, consistent with Gaussian-like pore size distributions. As would be expected, capillary pressures increase as a function of pore diameter, the 9.2 nm mean pore diameter sample having the highest capillary pressure range/greatest degree of hydrate inhibition.

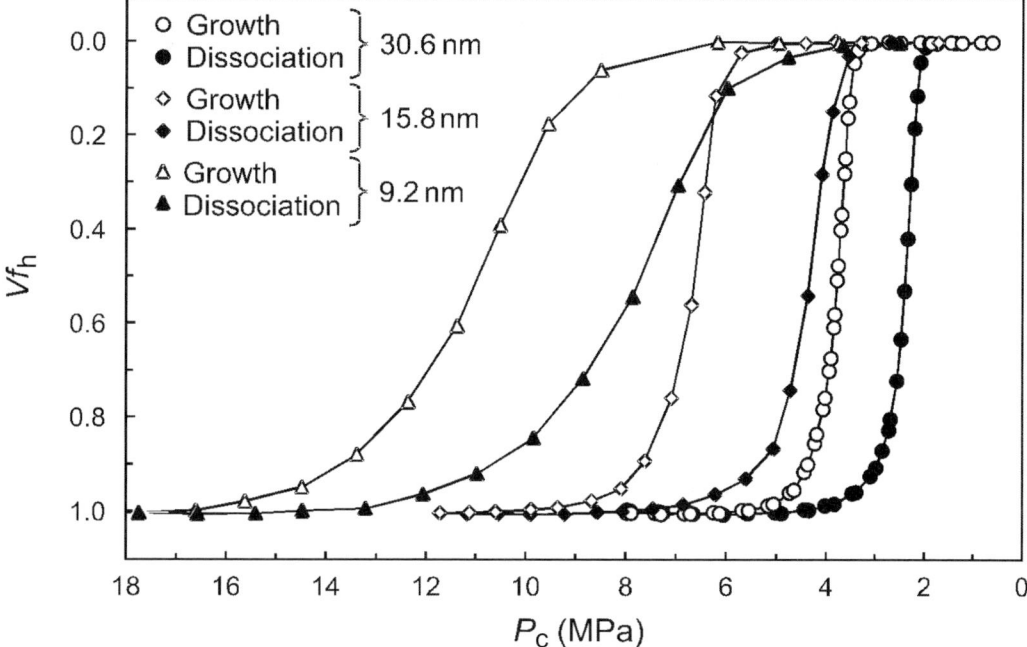

Fig. 5. Volume fraction pore hydrate versus capillary pressure for primary growth and dissociation loop data for all CPG silica samples studied.

By initiating cooling from any point on the primary dissociation curve, or conversely, by heating from any point on the primary growth curve, a variety of secondary characteristic 'scanning' growth/dissociation PT pathways may be followed, as illustrated in Figures 6 and 7 for the 30.6 nm sample. We adopt the term 'scanning' because it is generally used to describe similar curves in gas adsorption/desorption studies of mesoporous materials (Mason 1982, 1988). As for primary growth/dissociation pathways, scanning curves are irreversible, leading to an infinite number of possible, but consistent and repeatable PT pathways within the primary loop, depending on initial conditions. This behaviour, although often not investigated (primary loops only being reported), has been studied in detail for gas adsorption/desorption (Mason 1988). However, as far as we are aware, there is little (if any) comparable data for solid–liquid equilibria available.

Origins of hysteresis

We have shown previously (Anderson et al. 2003b) that, in agreement with the Gibbs–Thomson equation (1), mean pore diameter CH_4, CO_2 and CH_4–CO_2 clathrate hydrate dissociation (and ice melting data) for CPG samples shows a linear correlation between $\Delta T_p/T_b$ and $1/r$, giving a consistent and thermodynamically predictable relationship between hydrate dissociation conditions and pore size (Llamedo et al. 2004). However, data presented here show that hydrate growth conditions are depressed to significantly lower temperatures compared with dissociation, resulting in a distinct PT hysteresis between opposing transitions. To predict this behaviour, and assess its potential implications for hydrates in the seafloor environment, then it is first necessary to establish its origins.

The causes of hysteresis in porous media are still poorly understood (Everett 1954; Mason 1982, 1988; Christensen 2001; Ravikovitch & Neimark 2002). A significant part of the problem lies in the complexity of pore structures, which may comprise various heterogeneous (at the pore scale) pore geometries, a wide range of pore diameters and varying degrees of interconnectivity. To precisely predict hysteresis behaviour for a particular medium, we can imagine that it might be necessary to have an intimate knowledge of the pore space in terms of all these factors. A detailed analysis of CPG pore structures is beyond the scope of this paper; however we can speculate as to the origins of the observed hysteresis patterns based on accepted capillary theory.

The most basic pore model assumes single, simple pore shapes, such as spheres or cylinders. To introduce pore interconnectivity as a factor, it

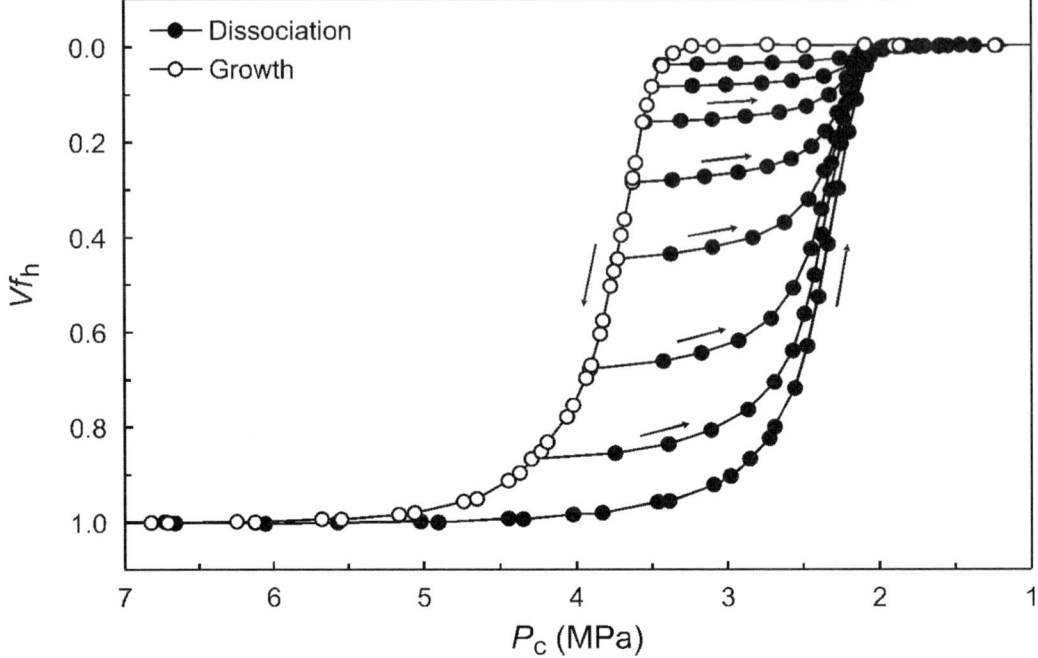

Fig. 6. Volume fraction pore hydrate versus calculated capillary pressure for secondary scanning dissociation curves originating from the primary growth curve (30.6 nm mean pore diameter CPG sample).

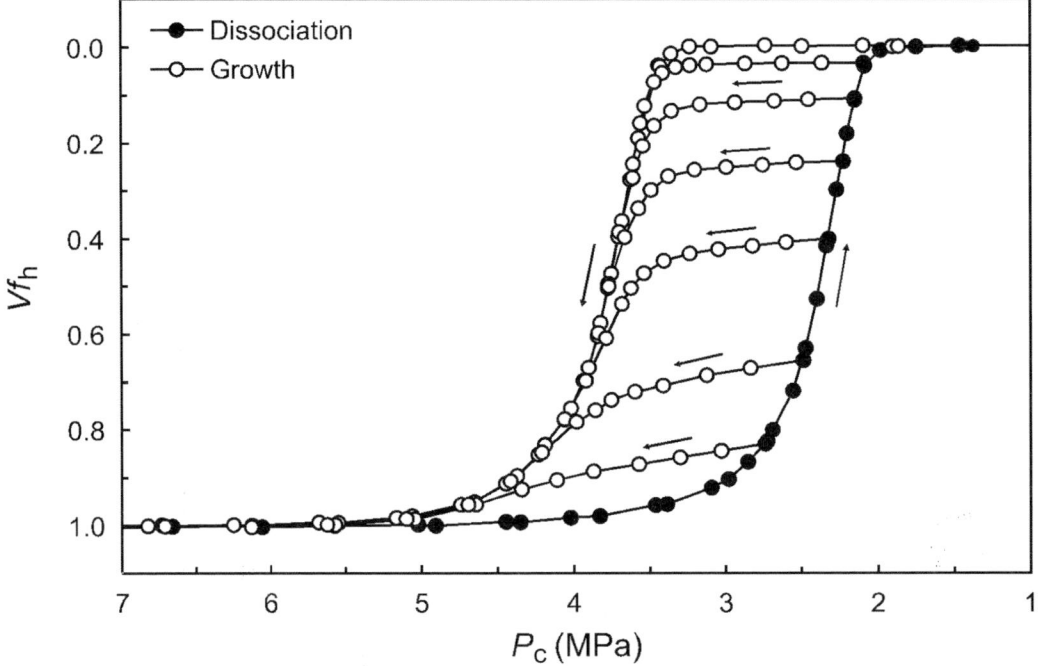

Fig. 7. Volume fraction hydrate versus capillary pressure for secondary scanning growth curves originating from the primary dissociation curve (30.6 nm mean pore diameter CPG sample).

is common to consider a matrix of spherical-like nodes connected by cylindrical-like bonds (Mason 1988; Vidales et al. 1995). Based on scanning electron microscopy (SEM) images and molecular dynamics simulations (Gelb & Gubbins 1998), this type of model might give a reasonable representation of controlled pore glasses. In such a media, we can consider two particular factors which may contribute to hysteresis: (1) pore geometry, and (2) pore blocking.

Influence of pore geometry

The geometry of a pore will have a major influence on the interface curvatures of confined phases, thus capillary pressures. For a media containing a notable component of cylindrical-like capillaries, hysteresis could potentially arise due to differences in solid–liquid interface curvatures during crystallization and melting (Brun et al. 1977; Jallut et al. 1992; Faivre et al. 1999). The interface shape factor F in equations (1) and (2) is defined by the solid–liquid interfacial curvature, κ, in terms of the pore radius by:

$$F = \kappa r \qquad (4)$$

with κ being defined by:

$$\kappa = \left(\frac{1}{r_1} + \frac{1}{r_2}\right) \qquad (5)$$

where r_1 and r_2 are the two orthogonal radii that describe the interface at any point. For solid–liquid transitions in a spherical pore, r_1 and r_2 are equal during both solid growth and melting, thus mean curvature is $2/r$ for both cases. In contrast, as shown in Figure 6, for solid growth in cylindrical pores, if the solid–liquid interface is considered a hemispherical cap, then r_1 and r_2 are equal, giving a mean curvature of $2/r$. However, for melting, although r_1 remains constant, r_2 is infinite ($1/r_2 \to 0$), thus total curvature is $1/r$. A curvature of $1/r$ implies that the solid–liquid interface should not retreat through a pore upon melting, but rather the solid cylinder should instantaneously melt along its length when stability conditions for the appropriate pore radius are surpassed. This concept is analogous to gas/oil phase 'snap-off' in (water-wet) cylindrical pores of reservoir rocks as hydrocarbon saturation is reduced (Blunt 1997; Hui & Blunt 2000). We can account for this geometrical control in equations (1) and (2) by modifying appropriately; $F = 2$ for growth and 1 for dissociation.

Based on the above, we can envisage that a media containing a notable proportion of cylindrical-like capillaries should display a temperature (or pressure) hysteresis between solid-phase crystallization and decomposition. Applying this to the results for CPG detailed here, then, if cylindrical pores are the cause of the observed hysteresis, we might expect to observe that for an appropriate volume of fraction of pore hydrate, Vf_h, capillary pressure during hydrate growth, $P_{c,g}$, should be around double that for dissociation, $P_{c,d}$. From examination of the data presented in Figures 5–7, this is clearly not the case: capillary pressures during hydrate growth are considerably less than double that for dissociation, with $P_{c,g}$ to $P_{c,d}$ ratios decreasing with decreasing sample mean pore diameter.

Figure 8 shows a plot of $P_{c,g}$ v. $P_{c,d}$ for equal pore volume fractions of clathrate present. As can be seen, rather than $P_{c,g}$ being a multiple of $P_{c,d}$ (e.g. $P_{c,g} = 2P_{c,d}$) the relationship appears to be additive, that is, $P_{c,g} = P_{c,d} + x$, where x is relatively constant for a specific CPG sample, but variable between samples, and decreases with mean pore diameter. Furthermore, if CPG is composed primarily of cylindrical-like capillaries, then we might expect that, upon heating from the primary growth curve, dissociation would begin only when the primary dissociation curve was reached, that is, hydrate which had grown into progressively smaller cylindrical pores to radius r at $P_{c,g}(r)$ would only melt on heating when $P_{c,d}(r)$ was reached, with $P_{c,g} = XP_{c,d}$ (X being 2 for an ideal cylinder as detailed). However, Figure 6 shows that hydrate dissociation begins in earnest almost immediately on heating from the primary growth curve, suggesting the presence of a significant proportion of pores with interface curvatures which are approximately equal on growth and dissociation, that is, spherical-like rather than cylindrical.

Data thus suggest that differences in interface curvature for growth and dissociation in cylindrical capillaries is not the sole mechanism responsible for the observed hysteresis. However, results do not preclude this as being at least partly responsible for the phenomena. For any given point on the primary growth curve, the capillary pressure is at least double that for the associated point of complete pore hydrate dissociation achieved on heating, as can be seen in Figures 6 and 9. As shown in Figure 9, data suggest that $P_{c,g}$ is close to $2P_{c,d}$ at the points of initial hydrate growth (on the primary growth curve) and final hydrate dissociation, respectively (although determining these conditions exactly is problematic as the amount of hydrate present in the pores becomes infinitesimally small and within the error in measured pressure change).

As noted, the fact that dissociation begins almost immediately on heating from the primary growth curve suggests hydrate in pores with mean interface curvatures (thus capillary pressures) which are similar on growth and dissociation,

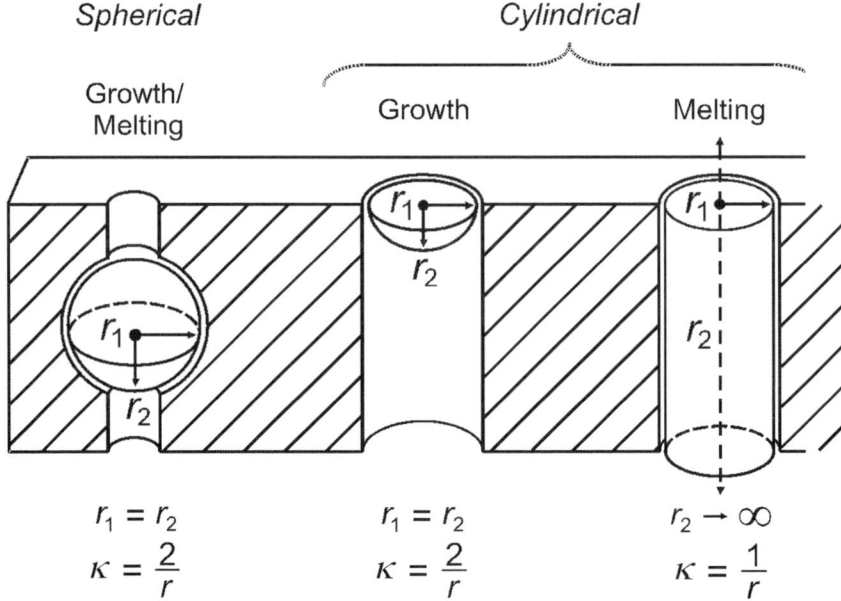

Fig. 8. Illustration of the difference in interface curvatures for hydrate growth and dissociation in ideal spherical and cylindrical pores.

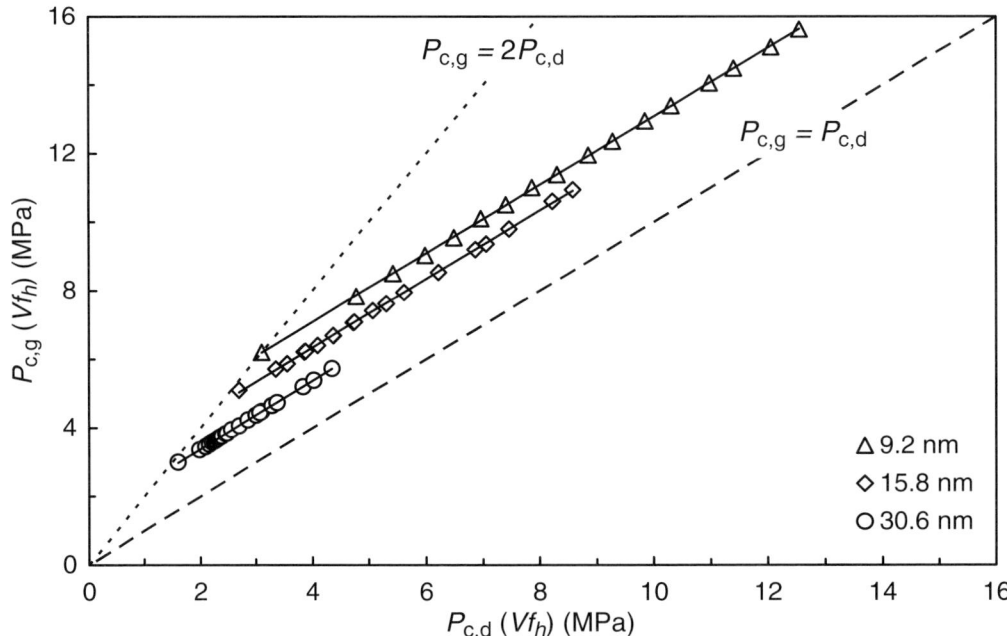

Fig. 9. Plot of capillary pressure during hydrate ($P_{c,g}$) growth v. that for dissociation ($P_{c,d}$) for equal volume fractions (Vf_h) of pore hydrate present. At the points corresponding to initial growth/final dissociation ($P_{c,g}$ and $P_{c,d}$ minima), $P_{c,g}$ approaches $2P_{c,d}$.

that is, spherical-like pores. Hydrate in these pores should theoretically grow and melt at the same capillary pressure condition (or ΔT_p). However, data in Figure 6 shows that this is not the case as P_c at conditions for growth is much higher than that for dissociation for the same volume of hydrate present. In light of this, it is necessary to consider mechanisms which could cause the observed hysteresis that are not primarily related to the interface curvature/geometry of individual pores. A potential candidate for this is 'pore blocking'.

Pore blocking effects

Pore blocking has been proposed by a number of authors as a primary cause of the hysteresis commonly observed for liquid–vapour phase transitions in porous media (Mason 1988; Ravikovitch & Neimark 2002). The classic example of pore blocking is that for 'ink-bottle' pores (large pores with narrow necks) which cannot drain (desorb) until the capillary pressure reaches that needed for vapour phase entry into the narrow pore neck. We can apply this same theory to solid–liquid transitions if we consider the solid hydrate phase penetrating liquid-filled pores as analogous to vapour phase penetration during desorption.

Figure 10 shows a simple illustration of the mechanisms by which pore blocking could be envisaged to occur in a hydrate–liquid system.

If we consider a large pore of radius r_a accessible to the bulk only via smaller pores of $r < r_a$, then, in the absence of heterogeneous nucleation within the pore space, hydrate growth conditions for the large pore will be determined by the capillary entry pressure required for clathrate penetration of the smallest access pore throat, in this case of radius r_d. This means that hydrate growth in the large pore will take place at a temperature much lower than its 'unblocked' equilibrium freezing/melting temperature, as predicted by equation (1). On heating, however, equilibrium dissociation conditions of hydrate in the large pore will depend on its own radius, r_a. If we consider this blocking mechanism acting in a porous medium with a wide distribution of interconnected pores of different radii, it can be envisaged that many pores of large radius may only be accessible to growth fronts propagating from the bulk by means of narrower pore throats. In this case, it would be expected that a significant hysteresis would develop between solid growth and melting temperatures.

In Figure 11, data for selected individual secondary dissociation (scanning) curves for the 30.6 nm

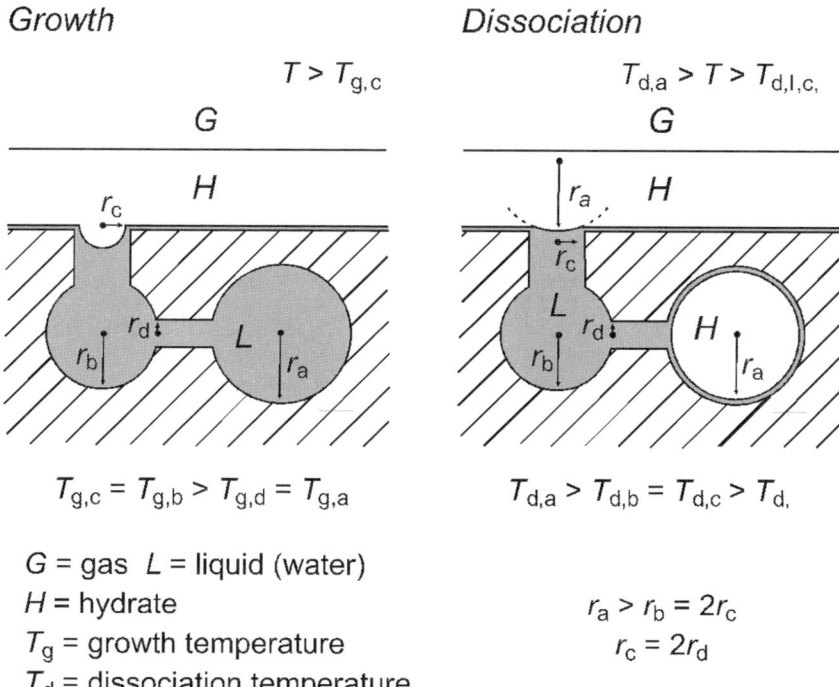

Fig. 10. Illustration of pore blocking effects in interconnected pores of different radii and geometry (cylindrical or spherical). See text for discussion.

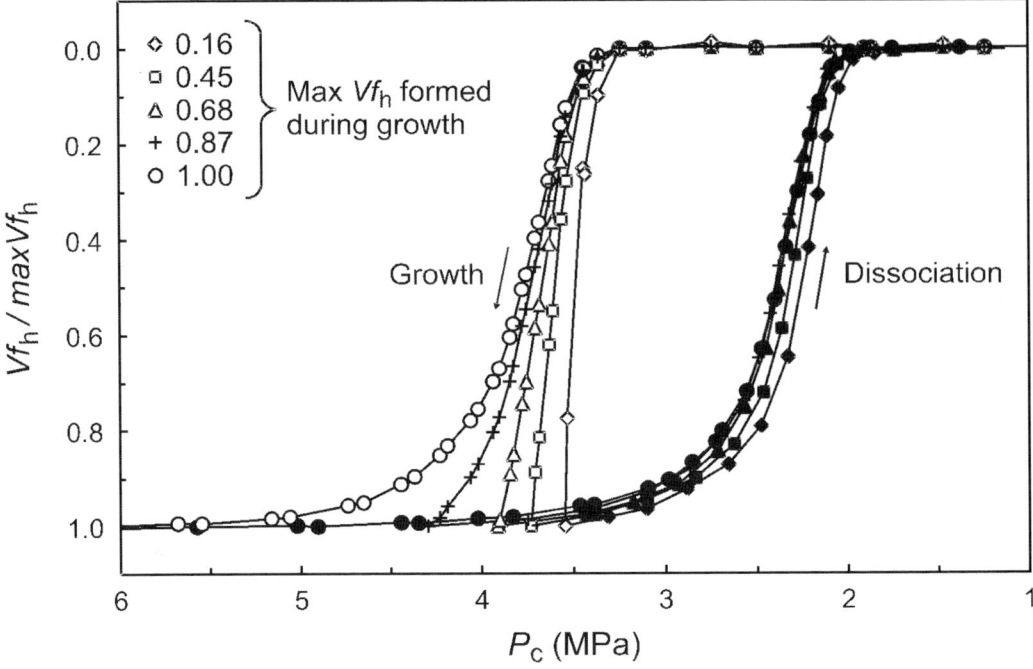

Fig. 11. Normalized volume fractions of pore hydrate versus capillary pressure for selected secondary scanning dissociation curves originating on the primary growth curve. Plotted Vf_h data are normalized individually based on maximum pore volume fraction of hydrate (max Vf_h) formed during growth (open symbols) on the primary growth curve before dissociation was initiated along a scanning curve (solid symbols).

sample have been normalized with respect to appropriate maximum volume fraction of hydrate formed on the primary growth curve in each case. Also shown are associated primary growth curve data for each. Heating curves essentially represent dissociation across the cumulative pore size/volume distributions (PSD) for pores in which hydrates have formed during cooling along the primary growth curve to the starting P_c condition. We can see from Figure 11 that, irrespective of initial starting conditions on the primary growth curve, dissociation curves are strikingly similar, suggesting that, in each case, hydrate decomposition takes place across a PSD closely representative of the media as a whole. As the P_c (thus ΔT_p) reached during primary growth is increased, so a larger volume of hydrate formed in pores of smaller radii (increased capillary pressure) is added to the total hydrate volume, as evidenced by associated secondary scanning dissociation curves shifting to higher capillary pressures. This pattern strongly supports pore blocking as a cause of the observed hysteresis.

Based on the above, we can envisage that the primary growth curve represents hydrate penetration into the media as a function of the pore throat entry radius distribution and associated accessible, 'freezable' volume (i.e. pore volume of water which can be converted to gas hydrate). As the system is cooled, the capillary entry pressure for progressively smaller 'access' pore throats is achieved, allowing the growth front to penetrate further from the bulk into the media, converting an additional fraction of the pore volume to hydrate at each stage, with each volume fraction converted being closely representative of the pore size distribution as a whole.

Secondary growth scanning curves originating on the primary dissociation curve add support to a pore blocking model (Fig. 7). It can be envisaged that, during dissociation, hydrate in some large pores (dissociation temperatures not yet reached) should become isolated (e.g. consider the large pore of radius, r_a in Fig. 10). On cooling, hydrate in these large pores will act as secondary sites for initiation of the hydrate growth front as it starts to penetrate back into the media. For many regions of the pore network, this may mean that the capillary pressure, thus ΔT_p required to initiate hydrate growth, is considerably less than that which would normally be required for conditions where the front penetrates from the bulk alone (i.e. primary growth curve conditions). As such, re-growth on secondary scanning curves should be more pronounced and occur at lower $P_{c,g}$ than for the primary growth curve. This behaviour is observed in Figure 7; for secondary scanning growth curves initiated on the primary dissociation curve, hydrate formation (increasing Vf_h) at lower P_c becomes increasingly pronounced as starting P_c is reduced (i.e. less hydrate dissociated before regrowth initiated), with the 'knees' which represent breakthrough pressures becoming increasingly flattened.

Significance for seafloor hydrate systems

It is beyond the scope of this paper to investigate in detail the potential effects of the observed hysteresis on gas hydrate growth dissociation conditions in the natural sedimentary environment. However, some preliminary comments can be made based on the results presented here.

Fine-grained silts, muds and clays which commonly host gas hydrates can have quite narrow mean pore diameters (0.1 μm) (Griffiths & Joshi 1989; Clennell et al. 1999). For curvatures of $2/r$ (spherical) and $1/r$ (cylindrical), our results suggest that pore diameters of 0.1 μm could reduce hydrate stability (dissociation) by 15 m (c. 0.5 °C) and 30 m (c. 0.9 °C) respectively in areas of moderate geothermal gradient (30 °C/km). This is a significant potential displacement. Depending on the extent to which pore blocking (and the locus of hydrate formation within pore space) plays a role, then temperature restrictions for hydrate growth could be notably greater.

Results strongly suggest that hydrate formation in narrow pores is characterized by progressive solid growth front penetration from the bulk or larger voids into the media. Front progression (as temperature is decreased or pressure increased) will be dependent upon the distribution of narrow pore throats relative to associated accessible voids. One would imagine these factors to be quite media-specific. This, and the fact that pore hydrate dissociation conditions are independent of interconnectivity, suggests the pore space of natural sediments must be characterized in terms of both pore throat entry radius distribution and specific pore radius/volume distribution if we are to accurately predict both hydrate growth and dissociation conditions for a particular media.

Regarding the proposed pore blocking phenomena, it should be noted that the process requires nucleation to be restricted within pores. For the mesoporous materials examined here (maximum pore diameters of 0.05 μm or 50 nm), it seems that growth front propagation is favoured over nucleation, although it might be expected that, in much larger pores/voids, heterogeneous nucleation may be the preferred mechanism for hydrate crystallization.

Conclusions

We have reported the results of a detailed experimental investigation of methane hydrate growth and dissociation conditions in synthetic mesoporous silica glasses. Data demonstrates that hydrate formation and decomposition in narrow pore networks are characterized by a distinct hysteresis between opposing transitions – hydrate growth taking place at temperatures considerably lower (or pressures higher) than those of dissociation. The hysteresis is an equilibrium phenomenon, and takes the form of irreversible, repeatable closed primary bounding growth/dissociation PT loops within which various characteristic secondary growth and dissociation specific 'scanning' PT pathways may be followed, depending on initial conditions. Similar hysteretic phenomena have been reported for ice growth and melting in the pores of cement pastes, and the behaviour appears to be closely analogous to that commonly observed for liquid–vapour transitions (gas adsorption–desorption) in mesoporous materials.

A detailed experimental analysis suggests that hysteresis arises primarily as a result of pore blocking during hydrate growth, although differences in interface curvatures during solid growth and decomposition resulting from pore geometry constraints (e.g. cylindrical pores) are likely to also contribute.

Results show that hydrate growth is characterized by capillary pressure-controlled progressive solid growth penetration from the bulk (or larger pores/voids) into the pore network as a function of decreasing temperature (increasing capillary pressure) with heterogeneous nucleation in pores not being favoured. In contrast, pore hydrate dissociation conditions appear to be principally controlled by interface curvatures as determined by individual pore geometry. As this behaviour has been observed for both synthetic and (more) natural (i.e. cement pastes composed of a variety of natural minerals), for both hydrates and ice, it is very likely that similar phenomena will occur during hydrate growth and dissociation in fine-grained natural sediments.

This work was funded by the UK Engineering and Physical Sciences Research Council (EPSRC grant no. EP/D052556), whose support is gratefully acknowledged. The authors would like to thank Jim Pantling and Colin Flockhart for manufacture and maintenance of experimental equipment.

References

ALADKO, E. Y., DYADIN, Y. A. ET AL. 2004. Dissociation conditions of methane hydrate in mesoporous silica gels in wide ranges of pressure and water content. *Journal of Physical Chemistry B*, **108**, 16540–16547.

ANDERSON, R., LLAMEDO, M., TOHIDI, B. & BURGASS, R. W. 2003a. Experimental measurement of methane and carbon dioxide clathrate hydrate equilibria in mesoporous silica. *Journal of Physical Chemistry B*, **107**, 3507–3514.

ANDERSON, R., LLAMEDO, M., TOHIDI, B. & BURGASS, R. W. 2003b. Characteristics of clathrate hydrate equilibria in mesopores and interpretation of experimental data. *Journal of Physical Chemistry B*, **107**, 3506–3509.

BOOTH, J. S., ROWE, M. M. & FISHER, K. M. 1996. Offshore gas hydrate sample database with an overview and preliminary analysis. *US Geological Survey Open File Report*, **96–272**.

BLUNT, M. J. 1997. Pore level modeling of the effects of wettability. *SPE Journal*, **2**, 494–510.

BREWER, P. G., FRIEDERICH, C. & PELTZER, E. T. 1999. Direct experiments on the ocean disposal of fossil fuel CO_2. *Science*, **284**, 943–945.

BRUN, M., LALLEMAND, A., QUINSON, J.-F. & EYRAUD, C. 1977. A new method for the simultaneous determination of the size and shape of pores: The thermoporometry. *Thermochimica Acta*, **21**, 59–88.

CHRISTENSEN, H. K. 2001. Confinement effects on freezing and melting. *Journal of Physics: Condensed Matter*, **13**, R95–133.

CISCONE, S., KIRBY, S. H. & STERN, L. A. 2005. Direct measurements of methane hydrate composition along the hydrate equilibrium boundary. *Journal of Physical Chemistry B*, **109**, 9468–9475.

CLENNELL, M. B., HOVLAND, M., BOOTH, J. S., HENRY, P. & WINTERS, W. J. 1999. Formation of natural gas hydrates in marine sediments. Part 1: Conceptual model of gas hydrate growth conditioned by host sediment properties. *Journal of Geophysical Research B*, **104**, 22985–23003.

DEATON, W. M. & FROST, E. M. 1946. Gas hydrates and their relation to the operation of natural gas pipelines. *US Bureau of Mines Monograph*, **8**, 101.

DICHARRY, C., GAYET, P., MARION, G., GRACIAA, A. & NESTEROV, A. N. 2005. Modeling heating curve for gas hydrate dissociation in porous media. *Journal of Physical Chemistry B*, **109**, 17205–17211.

DICKENS, G. R. 2003. Methane hydrates in quaternary climate change – the clathrate gun hypothesis. *Science*, **299**, 1017.

DORE, J. C., WEBBER, J. B. W. & STRANGE, J. H. 2004. Characterisation of porous solids using small-angle scattering and NMR cryoporometry. *Colloids and Surfaces A*, **241**, 191–200.

ENÜSTÜN, B. V., ŞENTÜRK, H. S. & YURDAKUL, O. 1978. Capillary freezing and melting. *Journal of Colloid and Interface Science*, **65**, 509–516.

EVERETT, D. H. 1954. A general approach to hysteresis. Part 3: A formal treatment of the independent domain model of hysteresis. *Transactions of the Faraday Society*, **50**, 1077–1096.

FAIVRE, C., BELLET, D. & DOLINO, G. 1999. Phase transitions of fluids confined to porous silicon: A differential calorimetry investigation. *European Physical Journal B*, **7**, 19–36.

GELB, L. D. & GUBBINS, K. E. 1998. Characterization of porous glasses: Simulation models, adsorption isotherms, and the BET analysis method. *Langmuir*, **14**, 2097–2111.

GRIFFITHS, F. J. & JOSHI, R. C. 1989. Change in pore-size distribution due to consolidation of clays. *Geotechnique*, **39**, 159–167.

HANDA, Y. P. 1986. Compositions, enthalpies of dissociation, and heat capacities in the range 85 to 270 K for clathrate hydrates of methane, ethane, and propane, and enthalpy of dissociation of isobutane hydrate, as determined by a heat-flow calorimeter. *Journal of Chemical Thermodynamics*, **18**, 915–921.

HANDA, Y. P. & STUPIN, D. 1992. Thermodynamic properties and dissociation characteristics of methane and propane hydrates in 70-Å-radius silica gel pores. *Journal of Physical Chemistry*, **96**, 8599–8603.

HENRY, P., THOMAS, M. & CLENNELL, M. B. 1999. Formation of natural gas hydrates in marine sediments. Part 2: Thermodynamic calculations of stability conditions in porous sediments. *Journal of Geophysical Research B*, **104**, 23005–23022.

HUI, M.-H. & BLUNT, M. J. 2000. Pore-scale modeling of three-phase flow and the effects of wettability. *SPE*, **59309**.

HUNTER, K. A. 1999. Direct disposal of liquefied fossil fuel carbon dioxide in the ocean. *Marine and Freshwater Research*, **50**, 755–760.

JACKSON, C. L. & MCKENNA, G. B. 1990. The melting behavior of organic materials confined in porous solids. *Journal of Chemical Physics*, **93**, 9002–9011.

JACKSON, C. L. & MCKENNA, G. B. 1996. Vitrification and crystallization of organic liquids confined to nanoscale pores. *Chemistry of Materials*, **8**, 2128–2137.

JALLUT, C., LENOIR, J., BARDOT, C. & EYRAUD, C. 1992. Thermoporometry: Modeling and simulation of a mesoporous solid. *Journal of Membrane Science*, **68**, 271–282.

KVENVOLDEN, K. A. 1999. Potential effects of gas hydrate on human welfare. *Proceedings of the National Academy of Sciences of the USA*, **96**, 3420–3426.

LEE, S.-Y. & HOLDER, G. D. 2001. Methane hydrates potential as a future energy source. *Fuel Processing Technology*, **71**, 181–186.

LIEVOIS, J. S., PERKINS, R., MARTIN, R. J. & KOBAYASHI, R. 1990. Development of an automated, high pressure heat flux calorimeter and its application to measure the heat of dissociation and hydrate numbers of methane hydrate. *Fluid Phase Equilibria*, **59**, 73–97.

LLAMEDO, M., ANDERSON, R. & TOHIDI, B. 2004. Thermodynamic prediction of clathrate hydrate dissociation conditions in mesoporous media. *American Mineralogist*, **89**, 1264–1270.

MASON, G. 1982. The effect of pore space connectivity on the hysteresis of capillary condensation in adsorption-desorption isotherms. *Journal of Colloid and Interface Science*, **88**, 36–46.

MASON, G. 1988. Determination of the pore-size distributions and pore-space interconnectivity of Vycor porous glass from adsorption–desorption hysteresis capillary condensation isotherms. *Proceedings of the Royal Society of London A*, **415**, 453–486.

MAX, M. D. (ed.). 2000. *Natural Gas Hydrate in Oceanic and Permafrost Regions*. Kluwer Academic, Dordrecht.

MCLEOD, H. O. & CAMPBELL, J. M. 1961. Natural gas hydrates at pressures to 10,000 psia. *Journal of Petroleum Technology*, **222**, 590–594.

MILKOV, A. V., SASSEN, R., NOVIKOVA, I. & MIKHAILOV, E. 2000. Gas hydrates at minimum stability depths in the Gulf of Mexico: Significance to geohazard assessment. *Gulf Coast Association of Geological Societies Transactions*, **50**, 217–224.

MORISHIGE, K. & KAWANO, K. 1999. Freezing and melting of water in a single cylindrical pore: The pore-size dependence of freezing and melting behavior. *Journal of Chemical Physics*, **110**, 4867–4872.

ØSTERGAARD, K. K., ANDERSON, R., LLAMEDO, M. & TOHIDI, B. 2002. Hydrate phase equilibria in porous media: Effect of pore size and salinity. *Terra Nova*, **14**, 307–312.

PAULL, C. K., MATSUMOTO, R., WALLACE, P. J. & DILLON, W. P. (eds). 2000. *Proceedings of the Ocean Drilling Program: Scientific Results*, **164**. Ocean Drilling Program, College Station, TX.

RAVIKOVITCH, P. & NEIMARK, A. V. 2002. Experimental determination of different mechanisms of evaporation from ink-bottle type pores: Equilibrium, pore blocking, and cavitation. *Langmuir*, **18**, 9830–9837.

RENNIE, G. K. & CLIFFORD, J. J. 1977. Melting of ice in porous solids. *Journal of the Chemical Society, Faraday Transactions 1*, **73**, 680–689.

RUPPEL, C. 1997. Anomalously cold temperatures observed at the base of the gas hydrate stability zone on the U.S. passive Atlantic margin. *Geology*, **25**, 699–702.

SCHULSON, E. M., SWAINSON, I. P., HOLDEN, T. M. & KORHONEN, C. J. 2000. Hexagonal ice in a hardened cement. *Cement and Concrete Research*, **30**, 191–196.

SEO, Y. & LEE, H. 2003. Hydrate phase equilibria of the ternary $CH_4 + NaCl +$ water, $CO_2 + NaCl +$ water and $CH_4 + CO_2 +$ water mixtures in silica gel pores. *Journal of Physical Chemistry B*, **107**, 889–894.

SEO, Y., LEE, H. & UCHIDA, T. 2002. Methane and carbon dioxide hydrate phase behaviour in small porous silica gels: Three phase equilibrium determination and thermodynamic modelling. *Langmuir*, **18**, 9164–9170.

SESHADRI, K., WILDER, J. W. & SMITH, D. H. 2001. Measurements of equilibrium pressures and temperatures for propane hydrate in silica gels with different pore size distributions. *Journal of Physical Chemistry B*, **105**, 2627–2631.

SMITH, D. H., WILDER, J. W. & SESHADRI, K. 2002a. Methane hydrate equilibria in silica gels with broad pore size distributions. *AIChE Journal*, **48**, 393–400.

SMITH, D. H., WILDER, J. W. & SESHADRI, K. 2002b. Thermodynamics of carbon dioxide hydrate formation in media with broad pore size distributions. *Environmental Science and Technology*, **36**, 5192–5198.

SMITH, D. H., SESHADRI, K., UCHIDA, T. & WILDER, J. W. 2004. Thermodynamics of methane, propane and carbon dioxide hydrates in porous glass. *AIChE Journal*, **50**, 1589–1598.

SWAINSON, I. P. & SCHULSON, E. M. 2001. A neutron diffraction study of ice and water within a hardened cement paste during freeze–thaw. *Cement and Concrete Research*, **31**, 1821–1830

TOHIDI, B., BURGASS, R. W., DANESH, A., TODD, A. C. & ØSTERGAARD, K. K. 2000. Improving the accuracy of gas hydrate dissociation point measurements. *In*: HOLDER, G. D. & BISHNOI, P. R. (eds) *Gas Hydrate Challenges for the Future*. Annals of the New York Academy of Sciences, **912**, 924–931.

TRÉHU, A. M. & SHIPBOARD SCIENTIFIC PARTY. 2003. *Proceedings of the Ocean Drilling Program: Initial Reports*, **204**. Ocean Drilling Program, College Station, TX.

UCHIDA, T., EBINUMA, T. & ISHIZAKI, T. 1999. Dissociation conditions of methane hydrate in confined small pores. *Journal of Physical Chemistry B*, **103**, 3659–3662.

UCHIDA, T., EBINUMA, T., NAGAO, J. & NARITA, H. 2002. Effect of pore sizes on dissociation temperatures and pressures of methane, carbon dioxide and propane hydrates in porous media. *Journal of Physical Chemistry B*, **106**, 820–826.

UCHIDA, T., TAKEYA, S. *ET AL*. 2004. Decomposition of methane hydrates in sand, sandstone, clays, and glass beads. *Journal of Geophysical Research B*, **109**, B05206.

VIDALES, A. M., FACCIO, R. J. & ZGRABLICH, G. 1995. Capillary hysteresis in porous media. *Journal of Physics: Condensed Matter*, **7**, 3835–3843.

WILDER, J. W. & SMITH, D. H. 2002. Dependencies of clathrate hydrate dissociation fugacities on the inverse temperature and inverse pore radius. *Industrial and Engineering Chemistry Research*, **41**, 2819–2825.

WILDER, J. W., SESHADRI, K. & SMITH, D. H. 2001a. Modelling hydrate formation in media with broad pore size distributions. *Langmuir*, **17**, 6729–6735.

WILDER, J. W., SESHADRI, K. & SMITH, D. H. 2001b. Resolving apparent contradictions in equilibrium measurements for clathrate hydrates in porous media. *Journal of Physical Chemistry B*, **105**, 9970–9972.

ZHANG, W., WILDER, J. W. & SMITH, D. H. 2002. Interpretation of ethane hydrate equilibrium data for porous media involving hydrate–ice equilibria. *AIChE Journal*, **48**, 2324–2331.

ZHANG, W., WILDER, J. W. & SMITH, D. H. 2003. Methane hydrate–ice equilibria in porous media. *Journal of Physical Chemistry B*, **107**, 13084–13089.

Gas hydrate crystallite size investigations with high-energy synchrotron radiation

S. A. KLAPP[1,2], H. KLEIN[1] & W. F. KUHS[1]*

[1]*Abteilung Kristallographie, Geowissenschaftliches Zentrum der Universität Göttingen Goldschmidtstrasse 1; D-37077 Göttingen, Germany*

[2]*Present address: MARUM, Center for Marine Environmental Sciences, University of Bremen, D-28359 Bremen, Germany*

*Corresponding author (e-mail: wkuhs1@gwdg.de)

Abstract: The grain sizes of gas hydrate crystallites are largely unknown in natural samples. Single grains are hardly detectable with electron or optical microscopy. For the first time, we have used high-energy synchrotron diffraction to determine grain sizes of six natural gas hydrates retrieved from the Bush Hill region in the Gulf of Mexico and from ODP Leg 204 at the Hydrate Ridge offshore Oregon from varying depth between 1 and 101 metres below seafloor. High-energy synchrotron radiation provides high photon fluxes as well as high penetration depth and thus allows for investigation of bulk sediment samples. Gas hydrate grain sizes were measured at the Beam Line BW 5 at the HASYLAB/Hamburg. A 'moving area detector method', originally developed for material science applications, was used to obtain both spatial and orientation information about gas hydrate grains within the sample. The gas hydrate crystal sizes appeared to be (log-)normally distributed in the natural samples. All mean grain sizes lay in the range from 300 to 600 μm with a tendency for bigger grains to occur in greater depth. Laboratory-produced methane hydrate, aged for 3 weeks, showed half a log-normal curve with a mean grain size value of *c.* 40 μm. The grains appeared to be globular shaped.

The grain sizes of gas hydrate crystallites are largely unknown in natural samples despite their possible importance for our understanding of gas hydrate formation and the physical properties of gas hydrate aggregates. Grain sizes and shapes of gas hydrates yield insights for geosciences, glaciology and chemistry. The understanding of gas hydrate crystal growth could be significantly enhanced by knowing the typical sizes of gas hydrate crystals. This information may also help in understanding possible time-dependent continued growth processes; gas hydrate crystal growth might resemble a ripening process similar to Ostwald ripening (Lifshitz & Slyozov 1961; Wagner 1961). In such a process, large grains grow at the expense of smaller grains in order to minimize the free energy within a system. This happens both because of the higher solubility of smaller particles compared with larger particles (Wagner 1961) and because the grain boundary energy of larger particles is relatively less than that of smaller particles. The ripening of crystals will be a function of time, thus larger grains might be older than smaller grains assuming that the initial grain size was similar (i.e. the formation conditions are similar). Therefore, grain size information may well contribute to evaluating the formation ages of gas hydrates once the formation processes and conditions have been established. Knowing the size of gas hydrate crystals also helps in understanding the processes taking place on grain boundaries, e.g. mass transport and diffusion. Many kinetic and thermodynamic considerations involve surface processes and depend on the available surface area (Lasaga 1998), e.g. gas exchange reactions that result in readjustment of the composition and cage filling of gas hydrates in response to changing environmental and physical conditions. The application of these considerations to gas hydrates will require knowledge of the grain size. This information is scarce for gas hydrates. In particular, statistically representative quantitative grain size information is unavailable.

In this study, grains are understood to be represented as single crystals, separated by grain boundaries. To explore the nature of these single grains, a sample of artificial gas hydrate, sieved to provide a particle size range between 200 and 400 μm and comprising single grains or agglomerates, was investigated.

Some grain size information on synthetic methane hydrates was reported in an earlier study by Klapproth (2002), who estimated the typical crystal size to be 15 μm using X-ray diffractometric techniques. The samples were produced from ice and methane gas by the methods described by Kuhs et al. (1992) & Stern et al. (1996). Although difficult, Staykova et al. (2003) investigated methane hydrates (synthesized in a similar way, using a well-defined particle size and shape for the initial ice particles) using scanning electron microscopy (SEM), which is possible in the case of synthetic hydrates. Boundaries between hydrate crystallites are difficult to identify; only some freely grown euhedral hydrate crystals could be measured to have a size of 30–40 μm. Since these visual measurements were too few to have statistical significance, we have to assume that synthetic methane hydrate crystals, fabricated as described by Kuhs et al. (1992) & Stern et al. (1996), are typically 15 μm in size with individual crystals going up to at least 40 μm.

One objective of this study is to use a sample with a known particle size constrained between 200 to 400 μm to test the application of synchrotron techniques for determining the particle size of gas hydrates. Results within the order of magnitude as reported by Klapproth (2002) & Staykova et al. (2003) are expected for the measurement of the synthetic sample. The results within the anticipated range should underline the reliability and precision of the technique and serve also as a calibration.

The six natural samples investigated were from two different locations. One sample originates from the Bush Hill region in the Gulf of Mexico and was retrieved from the seafloor during the RV Sonne Cruise SO 174 in 2003 (Bohrmann & Schenck 2004). The other five natural samples are from Hydrate Ridge at the Cascadia Margin offshore Oregon. These samples were obtained from four different drilling sites of the ODP Leg 204 and originate from varying depths between 1 and 101 m below the seafloor (mbsf) (Shipboard Scientific Party 2002).

To enable good evaluation of the measurements, natural samples with considerable amounts of gas hydrate were chosen. A pre-selection was done based on 'on-the-catwalk' observations after sample recovery and computer-tomography (Abegg et al. 2006). For the purpose of this work, samples containing sufficiently large volumes of dense gas hydrate were needed, that is samples containing discrete nodules or lenses as described by Abegg et al. (2007). Bohrmann et al. (2007) also undertook quantitative phase analysis and scanning electron microscopy on some of the samples investigated in this study. Based on their results, samples with high proportions of gas hydrate were finally selected and used in this work.

Experimental methods

The grain sizes of one synthetic sample and six natural samples were investigated. The synthetic sample was prepared as in previous laboratory investigations on synthetic hydrates (Kuhs et al. 1992 & Stern et al. 1996) using ice grains (≤ 400 μm) and methane gas at a pressure of 60 bar for 3.5 weeks. During this time the temperature was raised in steps from -5 to $+2$ °C each step being taken after the gas consumption at the previous temperature step had almost ceased.

Because the objective for investigating the natural samples was to evaluate the sizes of individual gas hydrate crystals, it was therefore important not to change the state of the samples (e.g. by crushing the sample). Consequently, small pieces of intact material about about 1 cm^2 in size were measured.

The grain sizes of gas hydrates were measured by imaging the length of the crystals with high-energy synchrotron radiation. Initially, however, the specimens were investigated with a cryo-scanning electron microscope. We used a field-emission scanning electron microscope (FE-SEM; *Zeiss Leo 1530 Gemini*). The FE-SEM is designed for work at low acceleration voltages of less than 2 kV, which is important for ice or gas hydrates in order to minimize sample alteration due to beam damage (Kuhs et al. 2004; Stern et al. 2004). During the measurements, the samples were placed on a liquid-nitrogen cooled sample stage inside a vacuum chamber at temperatures of about 90 K (-188 °C) and a pressure of about 1×10^{-6} bar. The uncoated pieces of gas hydrate bearing samples with dimensions of about $0.5 \times 0.5 \times 0.5$ cm were examined in the cryo-SEM. The SEM instrument was equipped with a thin-window energy-dispersive X-ray microanalysis (EDX) detector allowing for microchemical analyses of low-Z elements.

The goal of SEM-imaging was to define the grain boundaries between gas hydrates on the surface of the specimen. Gas hydrate could be distinguished from other phases like ice and sediments by the presence of typical homogeneously distributed micropores within the gas hydrate phase. Submicrometre-sized micropores were described by Kuhs et al. (2000, 2004) & Techmer et al. (2005) within the gas hydrate phase as a typical feature and can be regarded as a potential tool for identifying gas hydrates. The pores are not fully connected and have diameters ranging from 200 to 400 nm in methane hydrate (Kuhs et al. 2000).

For the synchrotron measurements, pieces of pure hydrate/ice parts were broken off each of the six natural samples and then fixed into aluminium cans of 7 mm inner diameter and 40 mm length in an arbitrary orientation. In order to measure the natural sizes of the grains, only large pieces coming close to the inner diameter of the cans were selected. The gas hydrate grain sizes were measured at the Beam Line BW 5 at the Hamburg Synchrotron Laboratory (HASYLAB) of the Deutsche Elektronen Synchrotron (DESY). High-energy synchrotron radiation provides a short X-ray wavelength and thus allows investigating bulk sediments up to some centimetres thickness (for a wavelength of c. 0.12 Å) in X-ray diffraction experiments. This depth of penetration is essential because the samples are 7 mm thick and are also located in a cryostat with Al shields (Fig. 1). Since the crystal sizes of natural samples were unknown, but expected to be quite small, a particularly good angular and spatial resolution was needed. This is provided by high collimation, i.e. parallel radiation of the high-energy synchrotron radiation and a high photon flux allowing small beam diameters.

During measurements, the cryostat stage was cooled by a closed cycle helium system to temperatures of 70 K (-208 °C). A Lake Shore (Cryotronics Inc.) controller was used to regulate the temperature of the sample. A vacuum of about 1×10^{-5} bar between the sample can and the shields of the cryostat insulates the sample from external heat.

The grain size measurements were performed with the synchrotron radiation using the moving area detector method, which was developed for material science purposes such as metals or ceramics (Bunge et al. 2002; Klein et al. 2004). It enables the simultaneous measurement of the grain sizes of many crystals in a single sample in relatively short time, and the measurement of grain sizes of many crystallites in one run gives statistically significant results. The diffracted Debye–Scherrer cones of individual crystal planes of sample crystallites are imaged as rings on a two-dimensional image plate detector. For this study, a *mar345* image plate detector was used. A diffraction image of a gas hydrate sample is shown in Figure 2.

The Debye–Scherrer ring corresponding to the gas hydrate structure I (321)-reflex is used for grain size measurements in the moving area detector method, because it is strong and does not overlap with other phases. In order to exclude all other reflections except the gas hydrate (321)-reflex,

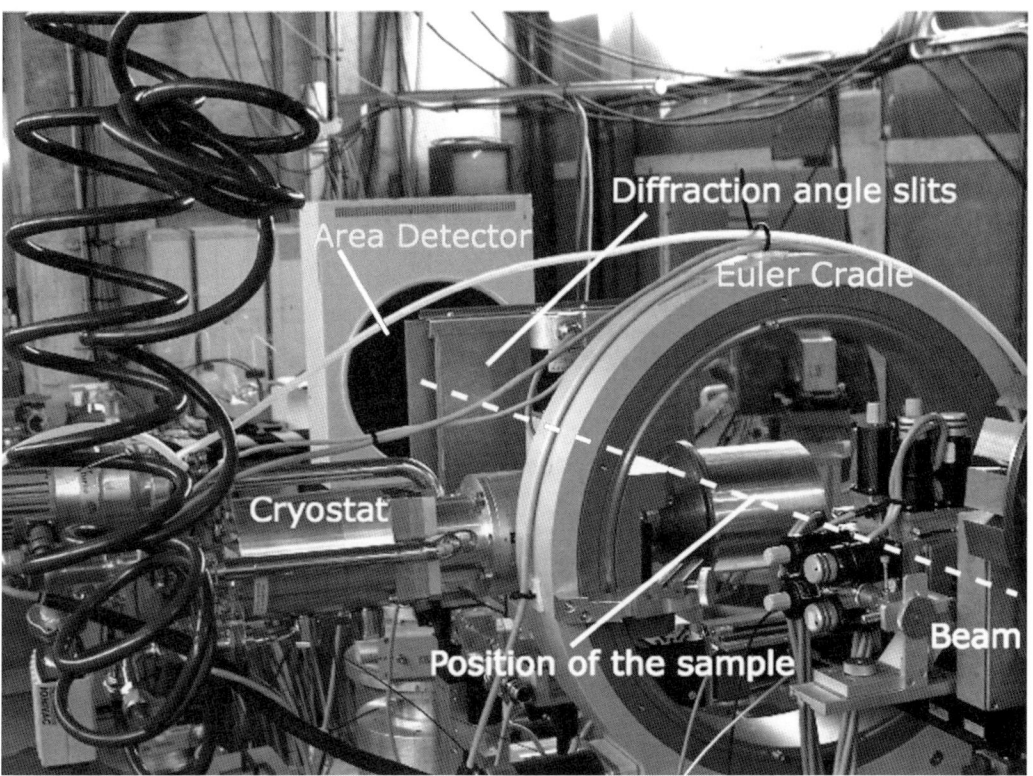

Fig. 1. Set up at BW5, the photograph is taken parallel to the beam direction. The gas hydrate sample is stored during measurements in a cylindrical (7.7 mm diameter) size aluminium can which is located within the cryostat.

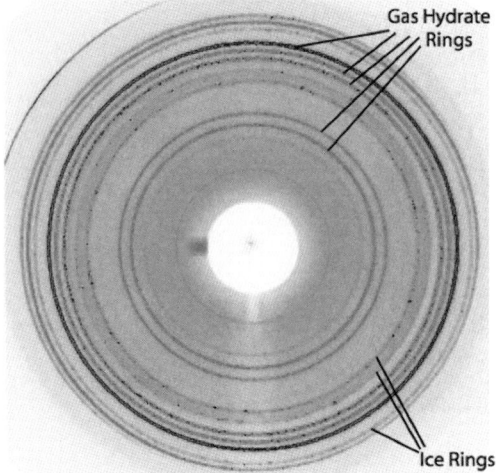

Fig. 2. Debye–Scherrer cones are imaged as rings on the area detector; each diffraction ring represents an (hkl) plane. One such ring is then selected for further grain size measurements (sample: ODP Leg 204, 74.5 mbsf). The sample was exposed to the synchrotron radiation for 90 s, beam energy was 100 keV.

Bragg slits are positioned between sample and detector (Figs 1 & 3). A 'location scan method' determines grain sizes in the scanning direction (Bragg-angle slit system, Fig. 3). Both the sample and the detector are moved perpendicular to the beam; consequently, all crystal planes are imaged as streaks as long as they stay within the beam line. The streaks correspond to individual grains, their lengths represent the grain diameters in the scanning direction. Figure 3 shows this method as a scheme; Figure 4 depicts an orientation–location scan from a natural methane gas hydrate from ODP Leg 204 (1.5 mbsf). The samples were moved 10 mm along the cylinder axis of the cans. Typically, the central part of the 40 mm can was measured. In order to inspect whether crystals might be elongated in one direction, a second set of measurements was carried out for one sample with the sample rotated 90° so that the y-axis and z-axis were changed (see Fig. 3). For these latter measurements, several sections through the diameter of the can were measured, each section 3.5 mm long, making half of the inner diameter of the cylindrical can. In order to obtain reflections from differently oriented crystallites, three to six scans at different ω-positions (Fig. 3) were done for one sample. Each time, the sample was rotated 0.2° along the length axis of the can.

Data processing

Detector images were saved as digital images and were processed by the *marView* software (Klein 2004). Measuring and counting of the streaks was done using the *Image-Pro* software (Media Cybernetics 2002), allowing an automatic detection and measurement of certain features within an image.

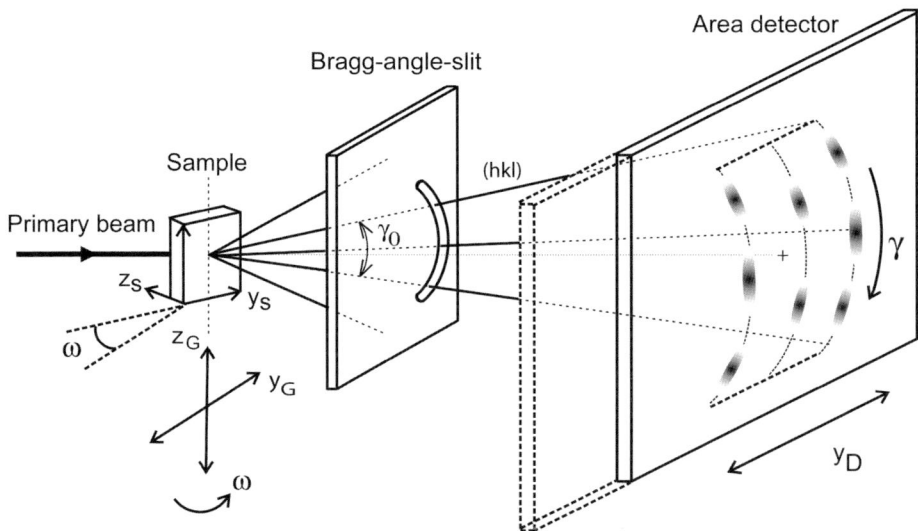

Fig. 3. Schematic drawing of the moving area detector method. Both sample and area detector are moved while the beam remains constant. Subsequently, crystals are moved in the y-direction through the beam and their reflections reach the detector; there, a continuous image of the scanned volume is recorded (Fig. 7). In succeeding measurements the sample is rotated around the angle ω. The Bragg-angle-slit lets pass only the chosen Debye–Scherrer ring.

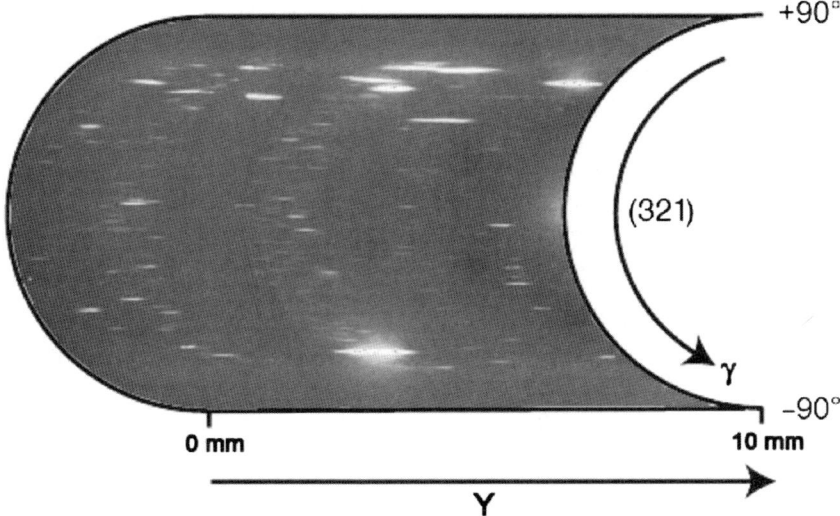

Fig. 4. Detector image of a sample (204 1249C-1H-CC, 0–10 cm). The streaks correspond to individual crystals; only reflections from the crystal plane (321) are recorded. The sample was scanned 10 mm in the y-direction. The total exposure time for this image was 85 min; the beam energy was 100 keV.

However, gas hydrates are weak scatterers unlike technical materials such as metals, for which the Moving Area Detector Method was developed (Bunge et al. 2003). The detector image processing and the measurement of the streaks are the most time-consuming steps in the data treatment process, because two challenges need to be met. First, the weak scattering of the synchrotron radiation by the gas hydrate crystal planes results in low intensities. Consequently, the data processing needs to enhance the contrast between background and reflections. Second, by enhancing the displayed intensity, artefacts on the image plate detector become more prominent and start to resemble features from true reflections. Artefacts are pixels with higher than background intensities, which do not belong to gas hydrate reflections. Unlike true reflections, they are not extended into the scanning direction or have a gradual increase in intensity towards the centre of the reflection.

The problems were solved in three succeeding steps for each individual sample. First, the background noise of the raw data from the synchrotron beamline was reduced by the image processing software. In the next step, objects with high intensities were counted and measured on the whole detector image of the sample. By doing that, intense gas hydrate reflections were measured, but not artefacts or weak gas hydrate reflections. To include streaks of gas hydrates with low intensities, small-scale areas of interest (AOIs) were drawn into the image encircling gas hydrate streaks. For these AOIs, the intensity threshold was decreased, which allowed the objects to be added to the results. In a third step, the results were double-checked to determine if a measured object was a gas hydrate reflection or an artefact. This was done by switching between a results list and the corresponding objects in the images.

A reflected crystallite on the detector of just one pixel is equivalent to about 6 μm of crystal size. Theoretically, a crystal of that size could be imaged on the screen; practically, streak lengths starting at four pixels (which is equivalent to 24 μm) are included in the data. This was done to ensure that all artefacts are kept out of the data.

Results

Gas hydrates were found in all samples by cryo-SEM. Areas of gas hydrate content could be identified both by carbon detected in the EDX analysis and as patches of sub-micrometre pore sizes (Fig. 5a, b); these pores are typically seen in gas hydrates and are well described features of gas hydrate occurrences (Kuhs et al. 2000; Techmer et al. 2005). These pores are not fully connected and have similar pore diameters (Kuhs et al. 2004). Boundaries between the gas hydrates and surrounding phases are indicated by a sharp change in the pore size. Dense material surrounding patches of sub-micrometre sized gas hydrate have fewer pores; also, these pores are of the order of tens of micrometres in diameter. EDX analysis detected no carbon signal in the dense parts of the

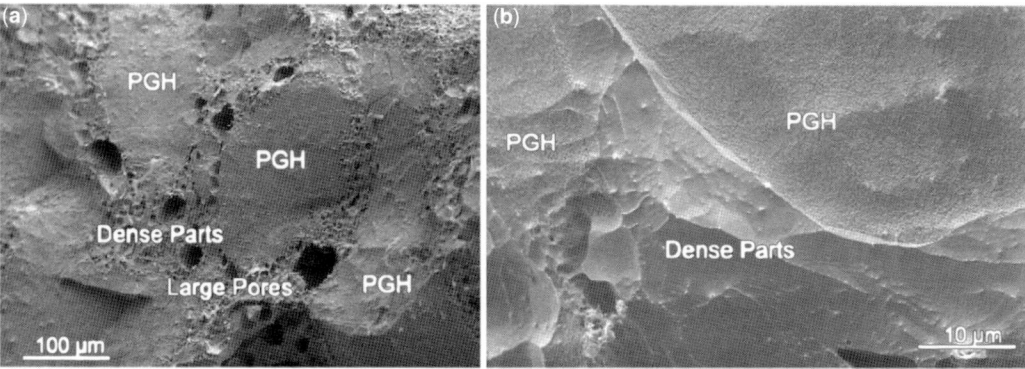

Fig. 5. Images of gas hydrate samples. (**a**) Overview of patches of gas hydrate (PGH), characterized by sub-micrometre-sized pores. The patches are surrounded by dense ice parts containing larger pores of tens of micrometre size in diameter. No information can be obtained about the actual size of the hydrate crystallites within the patches, nor is it certain that larger pores within the dense matrix are forming grain boundaries between gas hydrate grains. Sample ODP 204 1247B-12H-2, 41–51 cm. (**b**) Boundary between dense ice parts and sub-micrometre sized porous gas hydrate in the synthetic methane hydrate sample. Such clear boundaries are generally scarce and can hardly be found in natural samples.

samples (Fig. 5b). Bohrmann et al. (2007) describe these parts as ice, frozen from porewater during quenching in liquid nitrogen upon sample recovery. Although gas hydrates are detectable in electron microscopy, grain boundaries between single crystals are usually not clearly imaged. Patches of gas hydrate can be partially covered by crusts of salt or sediments. Most important, however, is that no boundaries are discernable within the sub-micrometre-sized patches of gas hydrate. This is why high-energy synchrotron radiation was applied as an alternative method.

The grain sizes measured by synchrotron radiation were collected and plotted as histograms. Different data sets from one sample, coming from measurements with different ω-positions as explained above, are summarized as follows. For the natural samples, the total number of counts for each sample varies and lies in the range of a few hundred to about 600. This does not include the sample that was tilted 90°, since due to the shorter distances fewer crystals were scanned. This also accounts for the increased standard deviation of the corresponding results (Table 1). Although the distributions of the crystal sizes from natural samples look normally distributed (Fig. 6a, b), they cannot be mathematically fitted due to the number of streaks measured, which would allow both normal and log-normal fitting. The mean value was calculated by dividing the sum of all crystal sizes of a sample through the frequency of occurrence. For all natural samples, the crystals are about 300–600 µm in diameter as measured by the location scan method. Table 1 gives the crystal sizes for the gas hydrates of the seven different samples. Figure 7 displays the crystal size v. depth.

The synthetic gas hydrate sample was prepared in a way that most of the ice reacted with the methane gas to form methane hydrate. The sample was fully

Table 1. *Crystal sizes of measured samples*

Sample-ID	Depth (mbsf)	Location-scan-method	
		Size (µm)	Standard deviation (µm)
204 1248C-8H-6, 68–87 cm	74.46	361	79
204 1247B-12H-2, 41–51 cm	93.01	592	374
204 1249C-1H-CC, 0–10 cm	1.5	373	134
204 1248C 11H-5 (no tilt)	100.89	517	176
204 1248C 11H-5 (90° tilted)	100.89	569	342
204 1250C 2H-CC 0-1 cm	5.06	451	132
Bush Hill, TVG 10 GH 1	Seafloor	301	114
CH_4-Hydrate	Synthetic	43	24

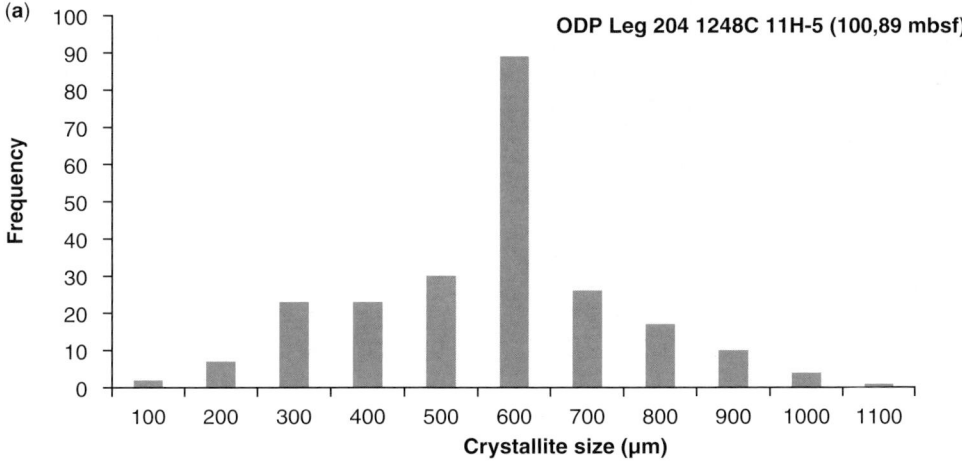

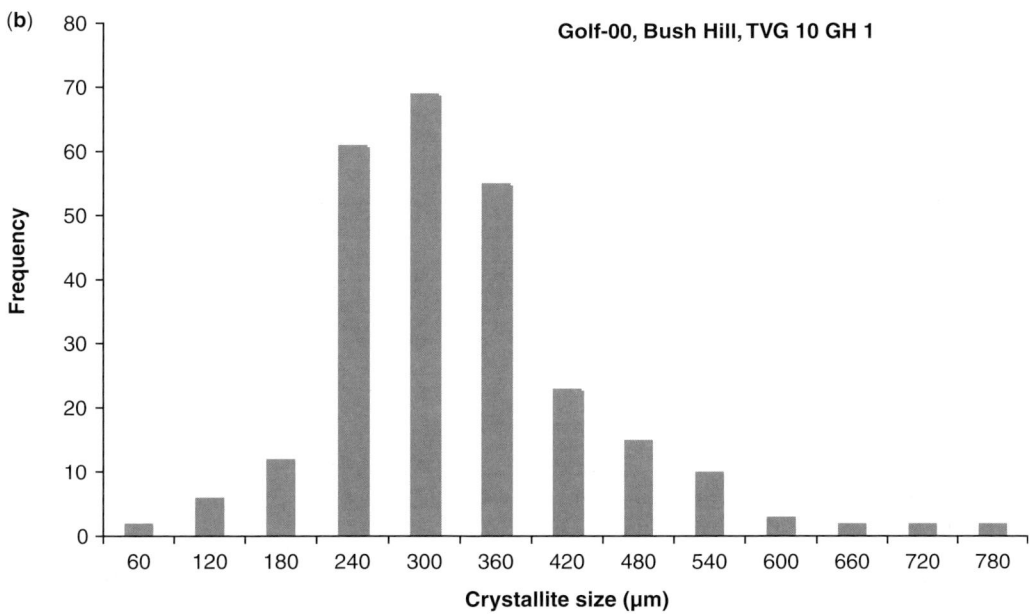

Fig. 6. Examples of grain size distributions. (**a**) Sample ODP 204 1248C 11H-5 (100.89 mbsf). The crystallite sizes appear to be normally distributed with a mean size of 517 μm. (**b**) Sample Gulf of Mexico, Bush Hill, TVG 10 GH 1 (seafloor). The crystallite sizes appear to be normally distributed with a mean size of 301 μm. (**c**) Synthetic methane hydrate sample. The size distribution is shaped like half a normal distribution; the mean grain size is 43 μm assuming a log-normal distribution.

reacted and thus essentially free of ice. A large number of crystallites were measured in one run as a consequence of the distinctly smaller grain sizes as compared with the natural samples (Fig. 6c). However, although the sizes of the individual grains, which were measured with synchrotron radiation, is a sieved fraction of 200–400 μm, the actual crystal sizes came out as well below these sizes with a mean value of c. 40 μm. This shows that the crystal sizes lay in the range of 15–50 μm, as described by Klapproth (2002) & Staykova et al. (2003).

The crystal size distribution of the synthetic sample, unlike the natural samples, is shaped as half a log-normal distribution (Fig. 6c). Since the crystals were freshly prepared in the laboratory, the initial grain size is quite small with a mean

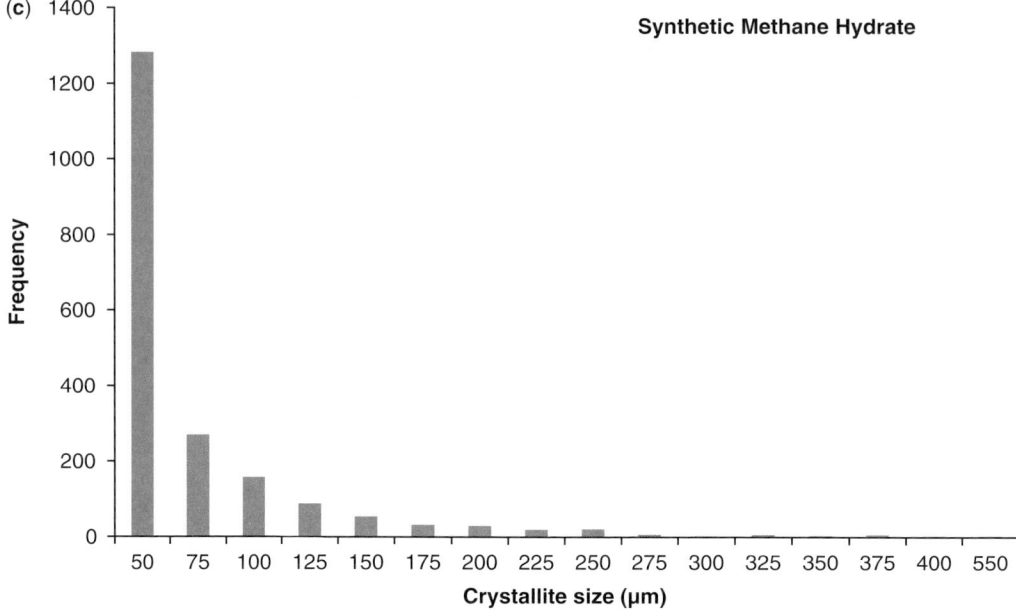

Fig. 6. (*Continued*).

size of about 40 μm, which is close to the detection limit of the measurements, as explained above.

Discussion

Exploring the nature of gas hydrate crystal growth and of processes involving surfaces of gas hydrate crystals requires information about the actual size of crystallites. Scanning electron microscopy is an effective tool for investigating gas hydrates when it comes to observation of the hydrate surface, like studying the microstructure or phenomena related to the decomposition of gas hydrates. Gas hydrate patches could be identified by sub-micrometre-sized pores and EDX analysis. The patches are surrounded by a dense matrix, which could lead to the impression that a single patch is a single grain. Actually, single grains, which are connected to crystal agglomerates, will have the same appearance in scanning electron microscopic images; also, the patches are likely to be shaped by partial decomposition of gas hydrates during retrieval. Accordingly, single patches of hydrate cannot necessarily be addressed as single gas hydrate crystals. However, crystals, which are attached to each other, can be distinguished by different orientation of the crystals. This is used by synchrotron measurements.

The moving area detector method is already regularly applied to materials such as metals or ceramics (e.g. Bunge *et al.* 2003; Klein *et al.* 2004) and, as shown here, it can successfully be used also for gas hydrates. Previous studies on laboratory-made hydrates that revealed sizes of 15–40 μm (Klapproth 2002; Staykova *et al.* 2003) were confirmed by the moving area detector method with high-energy synchrotron radiation (Table 1).

Repeated measurement series of a sample turned by 90° indicates that that the crystallites are not elongated into a preferred direction or orientation. This gives confidence that the samples are of globular shape, meaning that they have about the same extensions into the two orthogonal, measured directions. The synchrotron radiation measurements on hundreds of crystals indicate that the sizes of six different natural samples, ranging between 300 and 600 μm, are actually much larger than that of the synthetic sample.

Conclusions

The moving area detector method, applied to high-energy synchrotron radiation, is suggested as a new tool for the investigation of both synthetic and natural gas hydrates.

The grain sizes of natural and freshly laboratory-prepared samples differ by about an order of magnitude. Whether this is predominantly a consequence of some ripening process within the natural gas hydrates or due to different initial growth processes must remain an open question at this stage. In any case, some caveats need to be formulated concerning the relevance of physical properties of gas

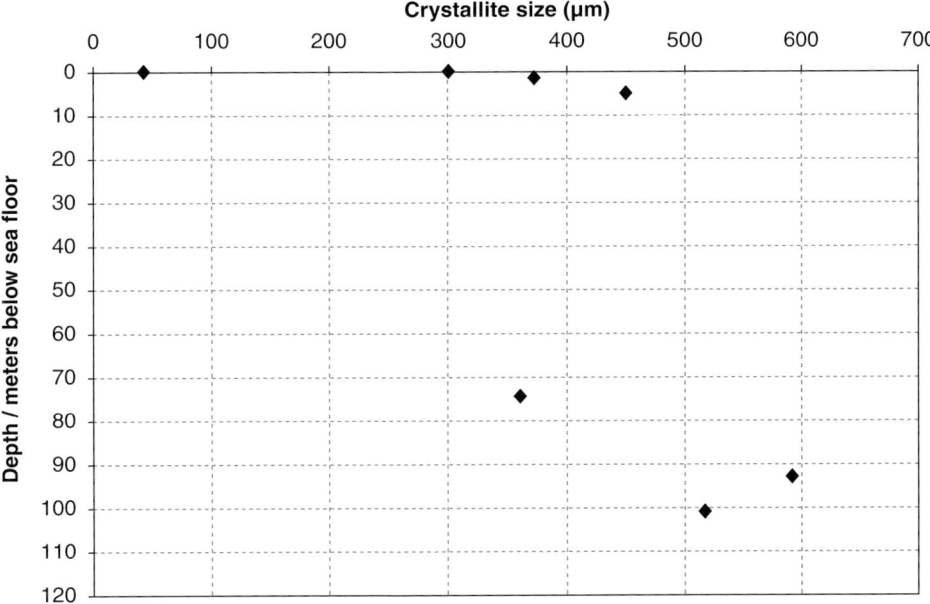

Fig. 7. Mean crystallite size v. depth: no clear correlation of size with depth is present, although larger grains tend to occur in greater depth.

hydrates measured on synthetic samples in applications on natural samples. Judging from the marine samples investigated, surface processes like diffusion or other kinds of mass transport, including mechanical properties, may take place differently in nature than observed in laboratory experiments.

Moreover, the results of this study suggest that growth and ripening of gas hydrate crystals could well take place in oceanic environments. Exploring the time involved for the ripening could lead to the approximate determination of the age of the gas hydrates. Such ripening processes have been observed in air hydrates enclosed in deep ice cores of Arctic and Antarctic ice sheets (Kipfstuhl et al. 2001). The primary crystallites are small with a typical crystallite size of a few tens of micrometres, which then transform into one single crystal of typically a few hundred micrometres; as the ice cores are dated, the time-scales of re-growth processes may be obtained by further analyses.

We thank Gerhard Bohrmann from MARUM, Center for Marine Environmental Sciences, in Bremen for providing the samples. Credit is also given to Andrea Preusser and Lars Raue for support at the HASYLAB and Kirsten Techmer for help at the FE-SEM (all GZG Göttingen). We thank the Deutsche Forschungsgemeinschaft for funding and HASYLAB at DESY for providing beam time and support. This work was funded by the Deutsche Forschungsgemeinschaft and the German Ministry of Education and Research in the programme Geotechnologien, of which this is publication no. GEOTECH-322.

References

ABEGG, F., BOHRMANN, G., FREITAG, J. & KUHS, W. F. 2007. Fabric of gas hydrate in sediments from the Hydrate Ridge – results from ODP 204 samples. *Geo-Marine Letters*, **27**, 269–277, doi: 10.1007/s00367-007-0080-4.

ABEGG, F., BOHRMANN, G. & KUHS, W. F. 2006. Data report: Shapes and structures of gas hydrates imaged by computed tomographic-analyses, ODP Leg 204, Hydrate Ridge. *In*: TRÉHU, A., BOHRMANN, G., COLWELL, F. S. & TORRES, M. (eds) *Proceedings of ODP, Scientific Results*, **204**, 1–8. World Wide Web Address: http://www-odp.tamu.edu/publications/204_SR/122/122.htm.

BOHRMANN, G. & SCHENCK, S. 2004. *GEOMAR Cruise Report SO 174, OTEGA II, RV 'SONNE'*. GEOMAR Report, Kiel.

BOHRMANN, G., KUHS, W. F. ET AL. 2007. Appearance and preservation of natural gas hydrate from Hydrate Ridge sampled during ODP Leg 204 drilling. *Marine Geology*, **244**, 1–14.

BUNGE, H. J., WCISLAK, L., KLEIN, H., GARBE, U., NOWAK, R. & SCHNEIDER, J. R. 2002. Orientation imaging of crystals in polycrystalline materials. *Advanced Engineering Materials*, **4**, 300–305.

BUNGE, H. J., WCISLAK, L., KLEIN, H., GARBE, U. & SCHNEIDER, J. R. 2003. Texture and microstructure imaging in six dimensions with high-energy synchrotron radiation. *Journal of Applied Crystallography*, **36**, 1240–1255.

KIPFSTUHL, S., PAUER, F. & KUHS, W. F. 2001. Air bubbles and clathrates in the transition zone of the NGRIP deep ice core. *Geophysical Research Letters*, **28**, 591–594.

KLAPPROTH, A. 2002. *Strukturuntersuchungen an Methan- und Kohlenstoffdioxid-Clathrat-Hydraten*. Ph.D. dissertation, Georg-August-Universität, Göttingen.

KLEIN, C. 2004. mar345dtb. User's Guide. Version 4.3. World Wide Web Address: http://www.marresearch.com/mar345dtb/mar345dtb.htm [online in this version since 28 October 2004, downloaded March 2006].

KLEIN, H., PREUSSER, A., BUNGE, H. J. & RAUE, L. 2004. Recrystallisation texture and microstructure in Ni and AlMg1Mn1 determined with high-energy synchrotron radiation. In: *Proceedings of the 2nd International Conference on Recrystallization and Grain Growth*. Material Science Forum, **467–470**, 1379–1384.

KUHS, W. F., DORWARTH, R., LONDONO, D. & FINNEY, J. L. 1992. *In-situ* study on composition and structure of Ar-clathrate. In: MAENO, N. & HONDOH, T. (eds) *Physics and Chemistry of Ice*. Hokkaido University Press, Sapporo, 126–130.

KUHS, W. F., GENOV, G. Y., GORESHNIK, E., ZELLER, A., TECHMER, K. & BOHRMANN, G. 2004. The impact of porous microstructures of gas hydrates on their macroscopic properties. *International Journal of Offshore and Polar Engineering*, **14**, 305–309.

KUHS, W. F., KLAPPROTH, A., GOTTHARDT, F., TECHMER, K. & HEINRICHS, T. 2000. The formation of meso- and macroporous gas hydrates. *Geophysical Research Letters*, **27**, 2929–2932.

LASAGA, A. C. 1998. *Kinetic Theory in the Earth Sciences*. Princeton Series in Geochemistry. Princeton University Press, Princeton, NJ, 811.

LIFSHITZ, I. M. & SLYOZOV, V. V. 1961. The kinetics of precipitation from supersaturated solid solution. *Journal of Physics and Chemistry of Solids*, **19**, 35–50.

MEDIA CYBENETICS, INC. 2002. *Image-Pro Plus. Version for Windows. Start-Up Guide*.

SHIPBOARD SCIENTIFIC PARTY. 2002. Preliminary report. ODP Prelim. Report **204**. World Wide Web Address: http://www-odp.tamu.edu/publications/prelim/204_prel/204toc.html (accessed 15 June 2006).

STAYKOVA, D. K., KUHS, W. F., SALAMATIN, A. N. & HANSEN, T. 2003. Formation of porous gas hydrates from ice powders: Diffraction experiments and multistage model. *Journal of Physical Chemistry B*, **107**, 10299–10311.

STERN, L., KIRBY, S. H. & DURHAM, W. B. 1996. Peculiarities of methane clathrate hydrate formation and solid-state deformation, including a possible superheating of water ice. *Science*, **273**, 1843–1848.

STERN, L. A., KIRBY, S. H., CIRCONE, S. & DURHAM, W. B. 2004. Scanning electron microscopy investigations of laboratory-grown gas clathrate hydrates formed from melting ice, and comparison to natural hydrates. *American Mineralogist*, **89**, 1162–1175.

TECHMER, K., HEINRICHS, T. & KUHS, W. F. 2005. Cryo-electron microscopic studies on the structures and composition of Mallik gas-hydrate-bearing samples. In: DALLIMORE, S. R. & COLLETT, T. S. (eds) *Scientific Results from JAPEX/JNOC/GSC et al. Mallik as Hydrate Production Research Well Program, Mackenzie Delta, Northwest Territories, Canada*. Geological Survey of Canada Bulletins, **585**.

WAGNER, C. 1961. Theorie der Alterung von Niederschlägen durch Umlösen (Ostwald-Reifung). *Zeitschrift für Elektrochemie*, **65**, 581–591.

Can CO$_2$ hydrate assist in the underground storage of carbon dioxide?

C. A. ROCHELLE[1]*, A. P. CAMPS[1,2], D. LONG[3], A. MILODOWSKI[1], K. BATEMAN[1], D. GUNN[1], P. JACKSON[1], M. A. LOVELL[2] & J. REES[1]

[1]*British Geological Survey, Keyworth, Nottingham NG12 5GG, UK*
[2]*Department of Geology, University of Leicester, Leicester LE1 7RH, UK*
[3]*British Geological Survey, Edinburgh EH9 3LA, UK*
*Corresponding author (e-mail: caro@bgs.ac.uk)

Abstract: The sequestration of CO$_2$ in the deep geosphere is one potential method for reducing anthropogenic emissions to the atmosphere without necessarily incurring a significant change in our energy-producing technologies. Containment of CO$_2$ as a liquid and an associated hydrate phase, under cool conditions, offers an alternative underground storage approach compared with conventional supercritical CO$_2$ storage at higher temperatures. We briefly describe conventional approaches to underground storage, review possible approaches for using CO$_2$ hydrate in CO$_2$ storage generally, and comment on the important role CO$_2$ hydrate could play in underground storage. Cool underground storage appears to offer certain advantages in terms of physical, chemical and mineralogical processes, which may usefully enhance trapping of the stored CO$_2$. This approach also appears to be potentially applicable to large areas of sub-seabed sediments offshore Western Europe.

It is now widely accepted the rising levels of carbon dioxide (CO$_2$) in the Earth's atmosphere are causing global climate change, and this is a subject of international concern (e.g. IPCC 1990, 2007). Furthermore, if something is not done to reduce emissions of greenhouse gases to the atmosphere, predictions suggest an unprecedented rate of future temperature increase, with unknown, but possibly rapid, consequences for the global climate. Measurements show that global temperatures rose by 0.3–0.6 °C in the twentieth century. If the trends in current emissions continue there are suggestions (Karl *et al.* 2000; RCEP 2000) that the global mean temperature is likely to be about 3 °C higher than at present by the end of the twenty-first century. The main difficulty in attempting to combat climate change is the world population's high dependence on fossil fuels as an energy source. Alternatives such as solar energy and other renewables are making a useful contribution, and some countries presently rely heavily on nuclear power, nonetheless, the culture and lifestyle of many countries appear to be strongly linked to fossil fuel usage for many years to come.

Assuming that we continue to burn fossil fuels, yet wish to mitigate CO$_2$ emissions to the atmosphere, we are faced with a limited number of alternatives:

(1) Reduce our CO$_2$ emissions by using lower carbon fuels (e.g. gas instead of coal).
(2) Utilize the produced CO$_2$.
(3) Dispose of the CO$_2$ in another domain of the planet, such as the geosphere, the terrestrial biosphere or the oceans.

In order to stabilize atmospheric CO$_2$ concentrations at current values, it may be necessary to reduce CO$_2$ emissions by 60% or more over the next 50 years (RCEP 2000). Although many countries are making strenuous efforts to reduce their CO$_2$ emissions, this is proving extremely difficult because all countries, and not just the developing ones, continue to strive for economic growth, which requires energy. Even those countries that have managed to make significant reductions in their energy intensity have struggled to reduce overall emissions. Almost the only exceptions are countries that have greatly reduced their use of coal or have developed a substantial nuclear energy base (e.g. Sweden). Therefore, it seems likely there will be no reduction in the production of CO$_2$, at least in the short term.

Although large-scale utilization of waste CO$_2$ is initially attractive, it has major problems, as converting it into useful substances requires large energy inputs. In many cases, if these energy inputs are in the form of fossil fuels, a net saving of CO$_2$ emissions becomes impossible. Alternatively, we could prevent the CO$_2$ entering the atmosphere in the first place. At present we are disturbing the balance of the natural 'carbon cycle' where carbon is slowly cycled between atmosphere, biosphere, hydrosphere and geosphere, by rapidly transferring large amounts of carbon from the geosphere into the atmosphere. By storing (or sequestering) vast volumes of CO_2 securely without any land use or verification problems, we may be able to redress some of this imbalance. One possible location for such a store is within porous rocks underground (the geosphere). In essence, geological storage aims to put the carbon directly back into the place from which it originally came (in the form of fossil fuels), thereby avoiding the atmospheric part of the carbon cycle.

The underground storage of CO_2

Underground storage is a feasible means of sequestering very large quantities of CO_2 produced by point sources such as fossil fuel fired power plants (e.g. Freund & Ormerod 1996; Haugen & Eide 1996; Holloway 1996a, b; Baines & Worden 2004a and references therein). Currently this is already being demonstrated, with 1–2 Mtonnes of CO_2 injected annually at both the Sleipner gas field, North Sea (Baklid et al. 1996; IEA GHG 1998) and the Weyburn oil field, southern Saskatchewan, Canada (Malik & Islam 2000; Moberg 2001; Wilson & Monea 2004).

The concept of underground sequestration (in its conventional form) involves first capturing the CO_2 at source to produce a pure CO_2 stream (e.g. via amine scrubbing of power plant flue gases). This is followed by compression to liquefy the CO_2 prior to transportation by pipeline to the injection site. Once at the injection site, the CO_2 can be injected via wells into deep reservoir rocks capped by very low permeability seals such as shales or clays. Injection could be into traps directly analogous to oil or natural gas fields (e.g. Bergman et al. 1996), or as at Sleipner into large aquifers.

If the CO_2 is injected at a depth of about 800 m or more, and assuming average geological conditions in the world's sedimentary basins, pressure and temperature will increase beyond the point where CO_2 becomes *supercritical* (approximately 31 °C, 74 bars) (Fig. 1). The density of supercritical CO_2 varies depending on pressure and temperature. However, for many currently envisaged storage conditions it is likely to have a density in the order of 700 kg m^{-3} – far denser than gaseous CO_2 (approximately 2 kg m^{-3} at Earth surface conditions), but less dense than formation porewater (approximately just greater than 1000 kg m^{-3} depending on salinity). As a consequence, stored CO_2 will occupy much less volume than gaseous CO_2, effectively greatly increasing the storage potential of sedimentary basins.

After injection, the CO_2 will initially be stored in a free state as a buoyant 'pure' phase below an impermeable caprock (described as 'physical trapping' by Bachu et al. 1994), in much the same way as methane in natural gas fields. However, over time it will dissolve in the formation water of the reservoir ('solubility trapping'). Once dissolved, it will no longer be buoyant, and hence its migration will only be driven by very slow regional-scale groundwater flow. The dissolved CO_2 will lower the pH of the formation water, and over even longer timescales (measurable in hundreds or thousands of years) this dissolved CO_2 will react with minerals within the rocks to precipitate calcite or other carbonate minerals (described as 'mineral trapping' by Bachu et al. 1994). This will result in

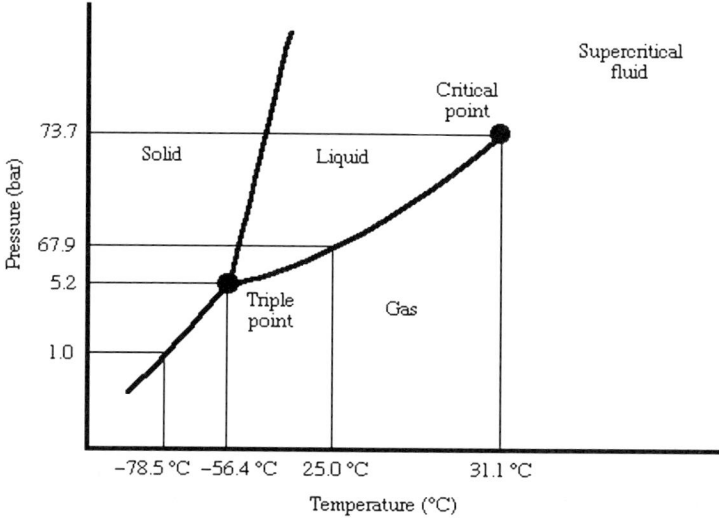

Fig. 1. CO_2 phase diagram (modified after Atkins 1982).

the immobilization of at least a proportion of the carbon for geologically significant timescales (e.g. Gunter et al. 1993, 1997; Baker et al. 1995; Czernichowski-Lauriol et al. 1996a, b; Rochelle et al. 1999, 2004). The extent of such reactions will depend upon various factors such as the composition of the porewater, the composition of the rocks and minerals it encounters, as well as the *in-situ* pressure and temperature.

If the underground storage of CO_2 is to be a practicable large-scale disposal method, there is a need to ensure it will remain safely underground, and not return to the atmosphere within relatively short geological timescales (i.e. thousands of years). This would allow natural buffering processes (e.g. oceanic and forestry sinks) to have sufficient time to reduce global atmospheric CO_2 levels to environmentally acceptable levels. Indeed, acceptable performance will need to be demonstrated in order to satisfy operational, regulatory and public acceptance criteria. The generally good track record of CO_2-assisted enhanced oil recovery operations and purpose-designed underground storage of natural gas shows underground storage can be practicable and leakage minimised, at least over anthropogenic or 'industrial' decadal timescales.

With underground sequestration, it is possible that the CO_2 could be retained for timescales of tens of thousands to millions of years (e.g. Pearce et al. 1996; Holloway 1997). Indeed, many *natural* CO_2 fields have been discovered that are far older than this (e.g. Pearce et al. 1996, 2004; Czernichowski-Lauriol et al. 1996a; Zheng et al. 2001; Baines & Worden 2004b). For example, the CO_2 in the natural carbon dioxide field at Pisgah Anticline in Central Mississippi, USA is thought to have originated from thermal metamorphism of Jurassic carbonates by the Jackson Dome igneous intrusion during late Cretaceous times (Studlick et al. 1990), which ended some 65 million years ago. Thus, given appropriate geological structures, the underground storage of CO_2 appears to be a safe and practicable way of reducing anthropogenic emissions of CO_2 to the atmosphere.

How can hydrates help with CO_2 sequestration?

Until recently, the majority of hydrate studies within the natural environment have concentrated on methane hydrate (CH_4) an ice-like compound naturally stable in certain seafloor sediments. Many of these studies have focused on the release of methane through natural processes, possibly linked to long-term changes in climate or through human activity (i.e. global warming or production of CH_4 as an energy source). Over the past few years however, there has been growing interest in hydrates as a store for anthropogenic CO_2, locking up CO_2 in an easily formed solid phase in domains of the planet where it will not be released to the atmosphere over relatively short timescales.

In addition to the current approaches to deep CO_2 storage described previously, there is another approach involving storage at cooler temperatures, but still at high-pressure conditions beneath permafrost regions or in sediments below the floor of deep oceans. This 'cool storage' approach has received relatively little attention even though it may offer certain advantages in terms of long-term containment of CO_2. In particular, under appropriate conditions (typically <10 °C and with hydrostatic heads >400 m) CO_2 hydrate becomes stable, and this could help immobilize CO_2 for geologically important timescales (e.g. Koide et al. 1997).

Trapping CO_2 as a hydrate phase on the ocean floor

Prior to discussing the benefits of immobilizing CO_2 as a hydrate *within sediments*, it is first useful to summarize other proposed methodologies for storing CO_2 as a hydrate phase. Much research has focused on releasing liquid CO_2 into the deep oceans, either as droplets within the water column, or as pools on the ocean floor (e.g. Austvik & Løken 1992; Hirai et al. 1997; Brewer et al. 1999; Warzinski et al. 2000; IPCC 2005). Interaction of the CO_2 with seawater under the *in-situ* pressure and temperature conditions would favour CO_2 hydrate formation, either as 'skins' around liquid CO_2 or as more solid masses over longer times. Although the CO_2 hydrate would eventually dissipate through equilibration with the seawater, the slow rate of reaction and slow turnover of the deep oceans may allow the CO_2 to be locked up in hydrate form for timescales measurable in at least hundreds of years (e.g. Wilson 1992; Herzog et al. 1996). There are two main limitations in applying this approach. Firstly, there has been much concern about the impact that large quantities of CO_2 would have on marine, especially benthic, organisms. Secondly, the emplacement of large quantities of waste CO_2 into the deep oceans is currently prohibited under the terms of international agreements, such as the 'London Dumping Convention' (IMO 1997) and 'OSPAR Convention' (OSPAR 1992).

Other studies have also considered confining CO_2 hydrate directly to shallow sediments on the deep seafloor (e.g. IEA GHG 2004). This approach involves trapping CO_2 in pure hydrate form, transporting the CO_2 hydrate as large blocks on board ships, and then releasing them to fall to the deep seafloor, and possibly even into soft sediments. Again, this approach would not be permissible under the above international conventions.

These approaches involve storage of CO_2 hydrate on the ocean floor or within the top few metres of sediment. One consequence of this is a high potential that dissolution of CO_2 into the bottom waters will reduce seawater pH and adversely impact the marine ecosystem. Such impacts could be avoided if the CO_2 were stored in a stable form with minimum risk of release to the ocean floor. One possible approach would be to create CO_2 hydrate deeper within the sediment, far below the few tens of centimetres of bioturbated sediment, and at a depth where it would not affect marine organisms. Indeed, the presence of significant accumulations of CH_4 hydrate in such sediments testifies to its potential as a long-term store of gas trapped in hydrate form.

Trapping CO_2 as a hydrate phase during methane extraction

One way to help offset the costs of CO_2 storage in sediments would be to combine it with the recovery of hydrocarbons. In the case of hydrates, several studies have investigated the use of injected CO_2 to liberate methane gas (CH_4) from hydrate in sediments, and in the process lock up CO_2 in CO_2 hydrate (e.g. Nakano *et al.* 1998; IEA GHG 2000*a, b*). The methane gas could then be captured and marketed. Although initially attractive, this approach has some potential problems, notably the distributed nature of methane hydrate in marine sediments and the costs of working offshore may make this approach overly expensive. Although sub-permafrost hydrates may be less costly in terms of drilling compared with those below the seabed, the generally remote location of most permafrost areas may mean they are further away from large sources of CO_2, and hence CO_2 transportation or pipeline costs would be higher. Finally, the injected CO_2 and liberated methane would mingle within the sediment, and thus a mixed gas could be produced at the production well. This could necessitate expensive separation equipment to get the methane to saleable quality.

Direct trapping of CO_2 as a hydrate phase within sediments

Applications of CO_2 hydrate storage involve the direct geological disposal of CO_2, and two different scenarios can be envisaged:

(1) As a secondary chemical containment mechanism, resulting from the (unintended) upward migration of CO_2 from a deep, warm storage reservoir (i.e. escape from a deep store of supercritical CO_2);

(2) As a primary containment mechanism, where CO_2 hydrate forms an impermeable 'cap' over a larger quantity of liquid CO_2.

First, secondary chemical containment provides a backup trapping mechanism, should deeper barriers be breached, effectively building 'redundancy' into the storage scheme. As mentioned earlier, the deep storage of CO_2 involves injection into warm rocks that are at least 800 m deep. Although detailed geological characterization of deep storage facilities would be carried out, it is always possible that some CO_2 may migrate upwards at some time. This could occur along unidentified small faults/fractures below the resolution of geophysical imaging, or along poorly sealed boreholes. If the CO_2 store lay below a deep enough and cold enough body of water, or below a region of thick permafrost, then upward-migrating CO_2 could enter a zone of CO_2 hydrate stability within the sediments. The formation of CO_2 hydrate could then enhance any natural low permeability caprock, slowing the ascent of CO_2, or even possibly blocking flow pathways (such as can happen when methane hydrate completely blocks pipelines).

Other studies have suggested that CO_2 hydrate may also be able to form locally, within sediments seemingly too warm to form hydrate. A preliminary study by Pruess (2003) modelled what would happen if liquid CO_2, rising along a flow pathway, started to boil off as it depressurized. The latent heat of vapourization required to boil off the CO_2 would cool the surrounding rocks, possibly to the point at which CO_2 hydrate, or even ice, would form. As a result, flow pathways could be reduced or even blocked. In the Pruess (2003) model, a cool zone several hundreds of metres thick was predicted to form, which could slow the ascent of the CO_2 and cause it to spread out laterally. Pruess (2003) also notes, however, that the preliminary model was somewhat idealized, and needs to be improved with more realistic geological structures.

Second, using CO_2 hydrate as a primary containment mechanism for stored CO_2 has been considered in several studies (e.g. Kiode *et al.* 1997; Sasaki & Akibavashi 2000; IEA GHG 2000*a*; Someya *et al.* 2006). This approach involves injecting (usually liquid) CO_2 into deep-water sediments or sub-permafrost sediments just below the CO_2 hydrate stability zone. As the slightly buoyant liquid CO_2 rises, it would enter cooler rocks lying within the hydrate stability zone. The precipitation of significant amounts of CO_2 hydrate within pore spaces could impede further upward migration of CO_2. On a larger-scale, the liquid CO_2 would spread out forming a 'pool' capped with an impermeable layer of CO_2 hydrate (Fig. 2) (together with any pre-existing natural low permeability caprock).

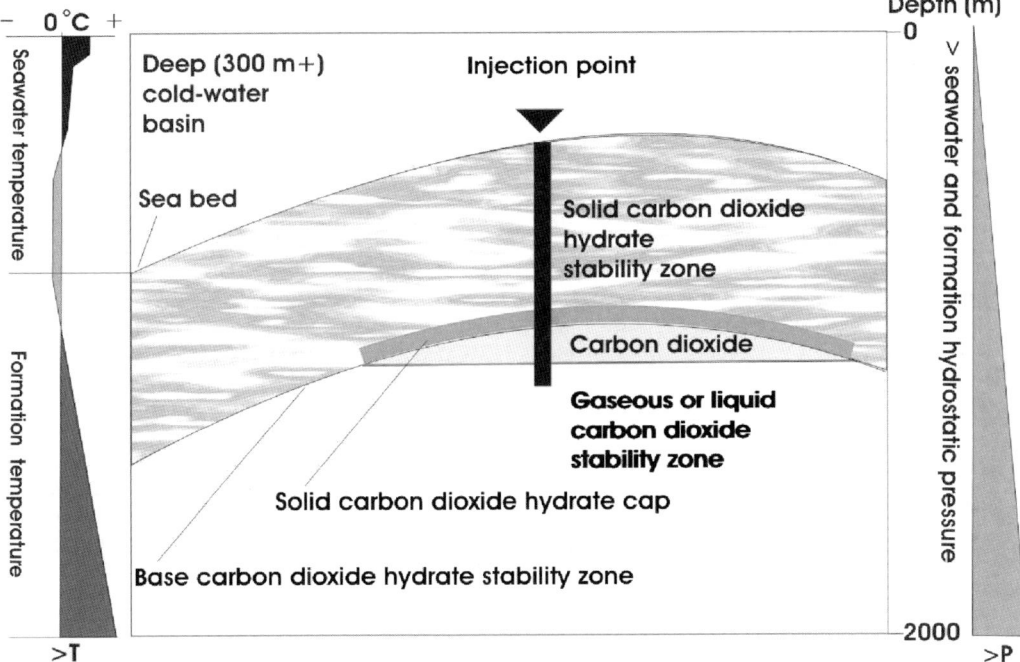

Fig. 2. Schematic diagram showing the relative position of injected liquid CO_2 and associated 'cap' of CO_2 hydrate.

Any liquid CO_2 able to find a way through the cap of CO_2 hydrate or natural caprock would itself react to form CO_2 hydrate as long as it encountered water-rich sediment. With a sufficiently thick hydrate stability zone, therefore, the hydrate cap would have a capacity to self-seal, building redundancy into the storage scheme. For CO_2, this self-sealing would be facilitated by the ability of CO_2 hydrate to form relatively rapidly (probably faster than for methane hydrate) under appropriate conditions (e.g. Sakai *et al.* 1990; Brewer *et al.* 1999; Riestenberg *et al.* 2004; Someya *et al.* 2006). However, this may not be immediately adjacent to the hydrate cap; previous studies have found methane transport along fractures through the lower parts of the hydrate stability zone (Gorman *et al.* 2002), for example. These fractures were hydrate-lined and prevented water reacting with the methane, although they could not be maintained in the shallower, more plastic sediments and gas migration was stopped here due to more extensive hydrate formation.

CO_2 hydrate stability within sediments

Whether CO_2 hydrate acted as a primary or secondary trapping phase (or even its formation during the liberation of methane from CH_4 hydrate), it is necessary to know over what conditions/depths CO_2 hydrate is stable and how CO_2 hydrate forms within sediments. The former is important for large-scale predictions to identify suitable regions with potential for CO_2 storage, assuming the underlying geology is suitable. Indeed, the IEA Greenhouse Gas R&D Programme has already identified the need for such work (IEA GHG 2000*a*). Large-scale predictions will also be necessary to identify regions where methane hydrate may also be stable (e.g. on one hand so as not to 'pollute' an exploitable CH_4 hydrate resource, or on the other hand, to explore areas for possible liberation of methane from CH_4 hydrate during CO_2 hydrate formation).

Mapping hydrate stability zones – the large scale

In an attempt to address the issue of where CO_2 hydrate may be stable on a regional scale, a preliminary theoretical study has been undertaken to estimate CO_2 (and CH_4) hydrate stability zones for sediments offshore western Europe (Rochelle & Camps 2006; Camps 2007). As a basis for the calculations, an empirical relationship between pure methane and pure water was used. This was presented in the JOIDES Pollution Prevention and

Safety Panel report (JOIDES 1992), where the equilibrium is described by the equation:

$$\ln P = A - B/T \qquad (1)$$

where A and B = constants determined by experimental hydrate stability data.

Using this relationship an algorithm was developed to enable calculation of CO_2 and CH_4 hydrate stability zones. An appropriate temperature reduction was included to account for the changes in equilibrium conditions owing to the presence of seawater (salinity reducing hydrate stability). Constants were determined from data constructed using CSMHYD (Sloan 1998), which calculates hydrate equilibrium formation conditions.

For each location (latitude and longitude), the program uses input of water depth (i.e. pressure), bottom-water temperature and geothermal gradient. GEBCO global bathymetry data sets have been used to provide detailed bathymetry for offshore Europe (IOC et al. 2003). To determine bottom-water temperatures CTD cast temperature data have been gathered from various sources, including ICES (the International Council for the Exploration of the Sea), BODC, the Coriolis Data Service and IOS reports (e.g. Saunders & Cooper 1987; Read et al. 1991). Within the model, water depths are relatively well constrained, and based upon high-resolution, detailed datasets. Less information is available on bottom-water temperatures, and the largest area of uncertainty is the resolution of geothermal gradient data because the available information is limited. A value of 30 °C km^{-1} was used in the model, but it is acknowledged that, in reality, certain areas may have higher or lower geothermal gradients, and this would affect predicted hydrate stability, particularly at a local scale (see Camps et al. 2009). For example, a higher geothermal gradient would result in a thinner hydrate stability zone (and vice versa). For the data shown in Figure 3 an increase in temperature of 1 °C results in a decrease in CO_2 hydrate stability zone thickness of about 35–40 m.

The calculated CO_2 (and CH_4) hydrate stability zones were plotted using the contouring package SURFER and output as maps of hydrate thickness (Fig. 3). Calculations predict that CO_2 hydrate will be stable over large regions, with the base of the

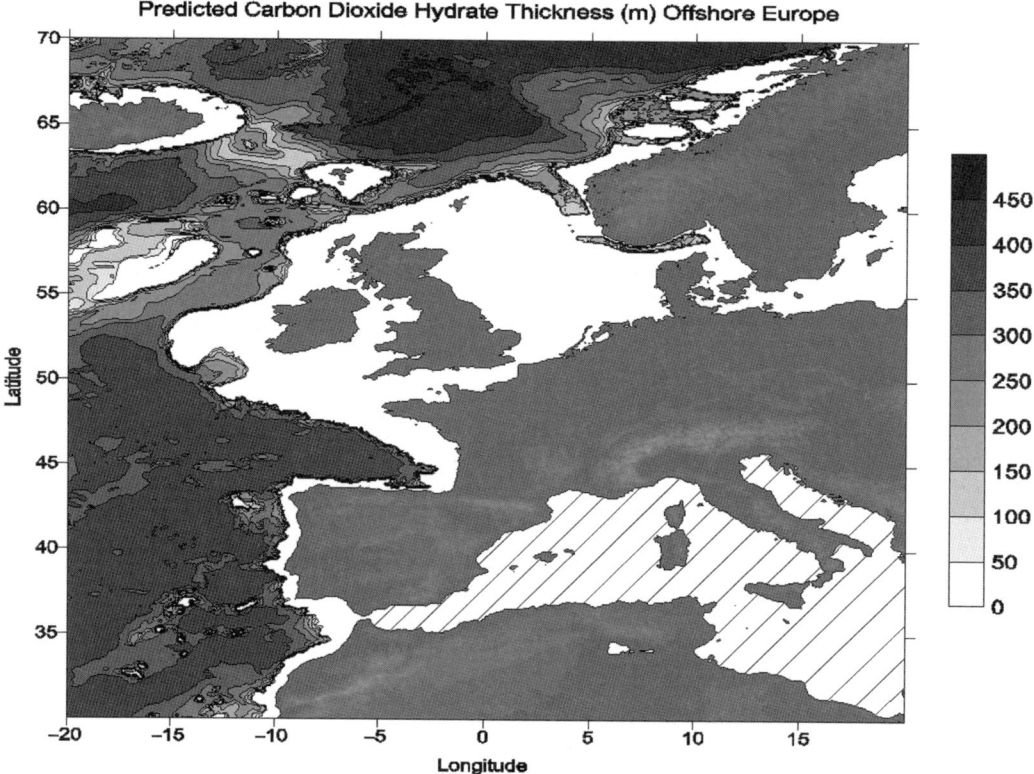

Fig. 3. Map of predicted thickness (m) of the CO_2 hydrate stability zone within seabed sediments for offshore Western Europe. Note that the Mediterranean Sea is not covered by this study.

CO$_2$ hydrate stability zone reaching a depth of up to about 450 m below the ocean floor. Given that the distribution of the hydrate stability zone shown in Figure 3 is largely controlled by the position of the continental slope, and given the scale of the map, the uncertainty over geothermal gradients mainly controls the thickness of the CO$_2$ hydrate zone rather than its spatial distribution. Nevertheless, the preliminary model does indicate that there is potential for the formation of a thick cap of CO$_2$ hydrate above a store of liquid CO$_2$.

The information in Figure 3 is also useful when considering the most appropriate locations for this type of storage methodology. For example, the relatively shallow seas around most of the UK preclude the formation of CO$_2$ hydrate in near-shore sediments (for all but the very NW of Scotland). As a consequence, this approach would necessitate considerable investment in pipelines to enable access to the deep cold sediments necessary for this type of storage. Conversely, Portugal, northern Spain, SW France and parts of Norway show greater potential where relatively deep waters lie close to shore. Although this does not identify whether suitable geological structures exist in these regions (or for that matter, whether large point sources of CO$_2$ exist close to these coastal regions), it can illustrate this approach to both industry and environmental policy makers this approach, and whether it is worthwhile considering it for inclusion in their portfolio of possible CO$_2$ management and mitigation strategies.

Hydrate within pore spaces – the small scale

Assuming porous sediments of suitable extent exist within the CO$_2$ hydrate stability zone, there is a need to know what impact CO$_2$ hydrate formation will have on the sediments and which sediments are most suitable. This might include whether hydrate will form in the centre of pores or on grain surfaces, whether it will cement grains together and make the sediment stronger, and/or whether precipitation will create an effective impermeable barrier to upward CO$_2$ migration. It will also be important to consider whether the origin of the CO$_2$ influences hydrate precipitation. For example, there may be differences in the nature of the hydrate when formed purely from dissolved CO$_2$ (i.e. in water-saturated rock *adjacent* to any CO$_2$ 'pool'), compared with that formed from within the CO$_2$-rich phase (i.e. within the CO$_2$ 'pool').

Laboratory-based studies can provide useful insights into processes controlling the above, but although there have been various studies on CO$_2$ hydrate, very few of them bring together the sediments, water of appropriate salinity and conditions appropriate to geological storage. As part of this study, CO$_2$ hydrate has been formed within synthetic sandy sediments under both seawater-saturated and seawater-poor conditions (Camps 2007). These reflect the conditions that may exist adjacent to, and within an underground store of CO$_2$. Damp sediments were used to represent CO$_2$-dominated conditions within the main storage region, it being assumed that most water would have been displaced from the pores except for a thin film on the grain surfaces. This provided an open pore network for CO$_2$ ingress. As a consequence of thin-water films and easy CO$_2$ ingress, hydrate formation was rapid and widespread throughout the sediment sample. The CO$_2$ hydrate replaced the water film completely and cemented the grains together (Fig. 4). In some pores larger hydrate crystals formed, reaching sizes of about 100 μm. Other parts of the pores remained open, and it is possible that CO$_2$ could still migrate through the sample, albeit in a restricted manner. The limiting factor for hydrate growth appears to have been the availability of water, which was all converted to hydrate (cf Gorman *et al.* 2002). An interesting consequence of this was halite precipitation (Camps 2007; Camps *et al.* 2009), which previous workers have predicted may form during hydrate growth in areas of restricted water availability (Harrison *et al.* 1995; Lorenz & Müller 2003).

Sediments saturated with water were used to represent conditions adjacent to the main storage area, it being assumed that CO$_2$ would either diffuse into the surrounding water, or CO$_2$-saturated porewater would migrate away from the CO$_2$–water interface. In the experiments only the upper part of the wet sediment was in contact with the CO$_2$. At the CO$_2$–water interface, precipitation of CO$_2$ hydrate was rapid, and some evidence of halite precipitation was again observed (Camps 2007). Further into the sediment, though still relatively close to the CO$_2$–water interface (in these experiments a few millimetres), hydrate filled all the intergranular pore space and cemented sediment grains together. The formation of this hydrate appears to have greatly reduced the transport of CO$_2$ into the rest of the sediment, leaving the majority of the remaining sediment uncemented. However, observations of the zone between the cemented and uncemented sediment (both visual and by SEM) suggest that hydrate tends to be restricted to the centres of pore spaces.

Cementation of sediment grains by CO$_2$ hydrate is advantageous for underground CO$_2$ storage because:

(1) It traps stored CO$_2$ as a solid phase.
(2) It makes the sediment more stable; this could be particularly important given that the

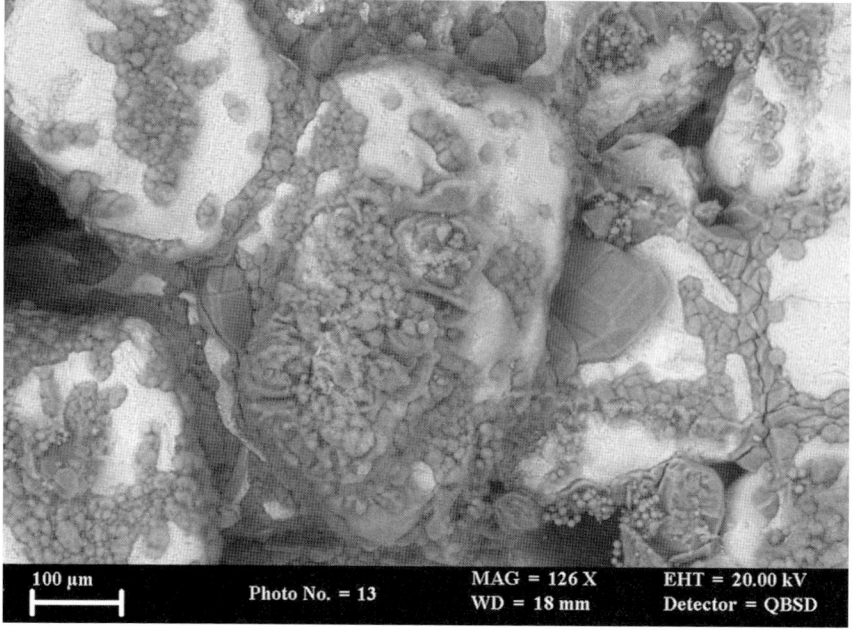

Fig. 4. Cryogenic SEM photomicrograph of sand grains (light grey) coated with a film of CO_2 hydrate (dark grey) that has been partly removed from the grains in places. Note larger crystals of CO_2 hydrate in the larger pore spaces – a good example is just to the right of the centre of the image. (Experiment using sand and synthetic seawater.)

sediments are expected to be relatively poorly consolidated.
(3) CO_2 migration through the sediments is reduced or, with sufficient hydrate formation, possibly even stopped.

The latter point is particularly important as it relates to the thickness of the hydrate 'cap' needed to contain a 'pool' of stored CO_2. In several of our simple laboratory experiments it was found that only a very thin (approximately 2 mm) hydrate layer was enough to restrict CO_2–water reaction and prevent further hydrate formation (at least over timescales of days to weeks). Similar observations have been made by other workers, who studied the release of liquid CO_2 into the deep ocean (e.g. Aya *et al.* 2000). More relevant however, are observations from complex laboratory experiments, which show that a relatively thin layer of rapidly formed CO_2 hydrate is capable of withstanding a significant differential pressure across it (Someya *et al.* 2006). Although more work is needed to ascertain the effects of other factors (such as hydrate strength, salinity, sediment mineralogy, etc.), the information currently available seems to suggest that a relatively thin 'cap' of CO_2 hydrate may be perfectly able to prevent a slightly buoyant 'pool' of stored liquid CO_2 from rising.

Other CO_2 trapping mechanisms operating near the hydrate stability zone

CO_2 hydrate will not be the only trapping mechanism for stored CO_2. In order to ascertain the overall potential for CO_2 storage it is important to consider the other mechanisms that will operate within, or close to, the CO_2 hydrate stability zone.

Density and viscosity

Most of the schemes currently being considered for underground CO_2 storage involve *in-situ* conditions above the critical point of CO_2 (i.e. >31.1 °C, >73.8 bar), where a supercritical phase is stable (Fig. 1). The density of this supercritical phase will vary with temperature and pressure, but may typically be of the order of 700 kg m^{-3}. As a consequence, a low permeability caprock (e.g. shale or evaporite) is needed to prevent the buoyant CO_2 rising towards the surface.

Within deep-water sediments or below permafrost regions, pressures may be equally as high as in a deep aquifer storage system, but temperatures may be much lower. Under these conditions the stable phase of CO_2 is likely to be a liquid, which is likely to have a higher density than that of supercritical CO_2 (e.g. Sasaki & Akibayashi 2000) (see Fig. 5). For example, at 10 MPa (100 bar) warm

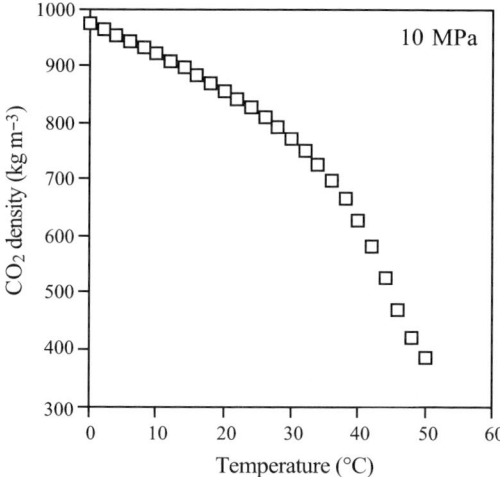

Fig. 5. Variation in CO_2 density over a range of temperatures, for an assumed hydrostatic head of 1 km (10 MPa) (prepared using a density model courtesy of Sintef). CO_2 density is about 20% greater at 10 °C compared with 30 °C.

CO_2 at 30 °C occupies more pore space than does cool CO_2 at 10 °C (in this case by about 20%). Thus, for similarly sized reservoirs, significantly greater quantities of CO_2 could be stored under cooler conditions.

There are other benefits from cool storage. As well as requiring less volume for storing the same weight of CO_2, the increased density would reduce buoyancy forces driving vertical migration. This could mean that a thinner caprock may be sufficient to contain the stored CO_2. Cooler temperatures would also increase the viscosity of CO_2. For example, at a pressure of 10 MPa (100 bar), data in Vesovic et al. (1990) indicates that CO_2 viscosity at 30 °C is approximately 70 μPa s^{-1}, but at 10 °C this increases to approximately 110 μPa s^{-1} (an increase of nearly 60%). As a consequence, vertical migration of cool CO_2 is likely to be slower than that of warm CO_2. Both increased density and increased viscosity are advantageous in reducing the potential for CO_2 to escape.

Solubility

Compared with many other gases, CO_2 is relatively soluble in water, and its dissolution into formation porewater will occur once it is injected underground. Indeed, previous studies have shown that a significant amount (>10%) of stored CO_2 can be trapped as a dissolved phase over intermediate timescales (e.g. Johnson et al. 2001, 2004; Wilson & Monea 2004). Once dissolved, the CO_2 will no longer be subject to the same buoyancy-driven upward migration as supercritical CO_2. Consequently, enhancing the amount of dissolved CO_2 will aid long-term storage.

The solubility of CO_2 increases with decreasing temperature up to the point where CO_2 hydrate is stable (Fig. 6). Therefore, porewaters adjacent to areas of CO_2 hydrate formation (i.e. just outside the hydrate stability zone) may also be able to store significant amounts of CO_2. For example, for seawater salinities and 10 MPa (100 bar) pressure, the solubility of CO_2 is approximately 25% greater at about 10 °C compared with that at 30 °C. In general terms, therefore, given that liquid CO_2 is stable at temperatures below that of supercritical CO_2, the storage of liquid CO_2 favours solubility trapping (and at sufficiently low temperatures, mineral trapping as a hydrate). Consequently, it appears to offer some benefits in terms of long-term containment.

CO_2 solubility is, however, also controlled by ionic strength (salinity), pressure and pH. It decreases with increasing ionic strength, increases with increasing pressure and increases with increasing pH. Overall CO_2 solubility is also controlled by the pH of the groundwater through linked equilibria such as:

$$CO_{2(aq)} + H_2O \Longleftrightarrow CO_{2(aq)} + H_2O$$
$$\Longleftrightarrow H_2CO_3^\circ \Longleftrightarrow HCO_3^- + H^+ \quad (2)$$

As a consequence, mineral assemblages that allow fluid–rock reactions to buffer pH at higher values

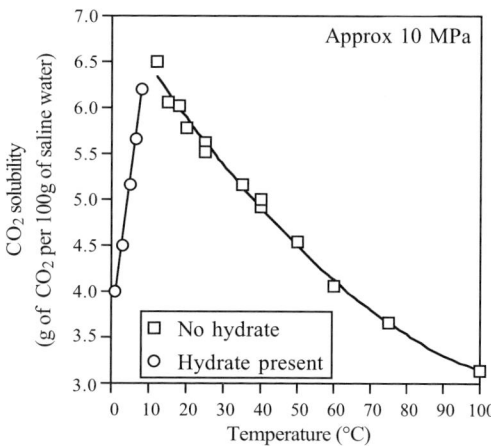

Fig. 6. Variation in CO_2 solubility over a range of temperatures, for an assumed hydrostatic head of 1 km (10 MPa) and for seawater-like salinities (prepared using data from Wiebe & Gaddy 1939, 1940; Wiebe 1941; Kuk & Montagna 1983; Enick & Klara 1990; King et al. 1992; Kojima et al. 2003).

will facilitate higher CO_2 solubility (Gunter et al. 1993; Rochelle et al. 2004). Assessment of the amount of CO_2 held in dissolved form therefore needs to be made on an individual site-by-site basis using appropriate in-situ temperature, pressure and fluid compositions.

There is one final advantage to dissolved CO_2. Formation water enriched in CO_2 is denser than its CO_2-free equivalent. It is possible, therefore (given a thick enough reservoir rock), that 'plumes' of CO_2-rich water may descend *slowly* from the CO_2–water interfaces of a storage scheme. This process facilitates further CO_2 dissolution through increased CO_2–water mixing. It also makes the trapped CO_2 descend further underground, as opposed to ascending as would occur if CO_2 were in its buoyant free-phase.

Mineral trapping

Although CO_2 hydrate is likely to form relatively rapidly at low temperatures, it is not the only solid phase that may form. Reactions of CO_2–water–rock may also produce a variety of secondary carbonate minerals that would enhance mineral trapping (e.g. Gunter et al. 1993, 1997; Bachu et al. 1994; Baker et al. 1995; Harrison et al. 1995; Rochelle et al. 2004). It is noted, however, that the rate and extent of such reactions is likely to be much slower under cool conditions than under warm storage conditions. Nonetheless, limited precipitation of carbonate minerals could still occur under conditions close to the hydrate stability zone, helping to trap CO_2. For example, Gunter et al. (1997) suggested that detrital Ca-rich feldspar might react to form calcite:

$$CaAl_2Si_2O_8 + CO_{2(aq)} + 2H_2O \Rightarrow$$
anorthite

$$CaCO_3 + Al_2Si_2O_5(OH)_4 \quad (3)$$
calcite   kaolinite

Similarly, Johnson et al. (2001, 2004) postulated that in saline solutions K-rich feldspar might react to form a different carbonate mineral, dawsonite:

$$KAlSi_3O_8 + Na^+ + CO_{2(aq)} + H_2O \Rightarrow$$
K-feldspar

$$NaAlCO_3(OH)_2 + 3SiO_2 + K^+ \quad (4)$$
dawsonite   quartz/chalcedony/cristobalite

Reactions such as these are likely to be relatively slow, however, and dependent upon the abundance and dissolution rates of the dissolving minerals. While they enhance the appeal of CO_2 sequestration through this approach, they are an added bonus, but may provide relatively little contribution in the short term compared with CO_2 hydrate formation.

Summary

The underground storage of CO_2 is increasingly seen as a possible method for reducing anthropogenic emissions of this greenhouse gas to the atmosphere without necessarily dramatically changing our energy-producing technologies. Most of the current approaches are aimed at storage within deep porous rocks below 800 m, where in-situ conditions of pressure and temperature are sufficient for injected CO_2 to exist as a buoyant *supercritical* phase. There is, however, an alternative approach to underground storage, using liquid CO_2 and associated CO_2 hydrate. This requires similarly high pressures, but would operate at lower temperatures, such as might be found beneath the floors of cold, deep oceans, or permafrost regions. Although this concept of 'cool storage' has received much less attention compared with that of 'warm storage', it does appear to offer certain advantages in terms of the mechanisms that may trap the stored CO_2:

(1) In terms of the free CO_2 phase, liquid CO_2 can have a significantly higher density compared with supercritical CO_2, so more of it can be stored in an equivalent volume of rock. Its lower buoyancy and higher viscosity would also help reduce the rate of vertical migration from the storage horizon.

(2) In terms of dissolved CO_2, its solubility increases significantly at lower temperatures (up to the point where CO_2 hydrate precipitates).

(3) In terms of storing CO_2 as solid phases, cool conditions would allow for the precipitation of CO_2 hydrate as well as carbonate minerals. CO_2 hydrate would be advantageous as it forms rapidly, only requiring the presence of water and CO_2. It could form via 2 routes; intentionally, as a primary storage mechanism, or unintentionally, as a secondary 'backup' storage mechanism (e.g. as a result of leakage of CO_2 from a deep, warm reservoir to shallower, cooler horizons). As a primary storage mechanism it may form an impermeable 'cap' above a 'pool' of liquid CO_2, enhancing the sealing properties of a natural caprock.

If CO_2 hydrate is to play a role in underground storage, we need to be able to predict where it will be stable on a regional scale. This has been done for offshore Western Europe and, although this does not identify local geological structures suitable for CO_2 storage, it does show that large regions have the potential for CO_2 hydrate formation in deep-water sediments. We also need to know about smaller-scale processes, such as the relationship between CO_2 hydrate and sediment grains at a pore scale, and how this influences the overall physical properties of the sediment. Results from laboratory experiments show that hydrate formation is rapid and that it can act as a cement to the sediment grains, although its morphology may differ if grown in water-saturated or CO_2-saturated conditions. Experimental results also indicate that even a relatively thin layer of hydrate can be effective at greatly retarding CO_2 migration rates. This suggests that even a relatively thin 'cap' of CO_2 hydrate may be able to prevent a slightly buoyant 'pool' of stored liquid CO_2 from rising.

In conclusion, there could be a role for CO_2 storage under cool conditions, and CO_2 hydrate could have an important part to play in this. However, in assessing the overall storage potential of an individual storage scheme, it will be important to consider all possible trapping mechanisms, and not just those involving CO_2 hydrate. To achieve this fully will require the close cooperation of those with in-depth knowledge of both hydrate phases and underground CO_2 storage. Much work remains to be undertaken to fully understand how CO_2 hydrate can best contribute to underground storage and the complex inter-relationship it may have with sediments. Sequestration of CO_2 as hydrate could provide another technique to add to the portfolio of strategies that could help reduce emissions of anthropogenic CO_2 to the atmosphere.

Ameena Camps acknowledges the Natural Environment Research Council for funding under grant NER/S/A/2003/11923. Erik Lindberg is thanked for the use of the Sintef CO_2 density model. Peter Miles and Mark Rodger are thanked for their useful comments that helped improve this paper. This paper is published with the permission of the Executive Director of the British Geological Survey (NERC).

References

ATKINS, P. W. 1982. *Physical Chemistry* (2nd edn). Oxford University Press, Oxford.

AUSTVIK, T. & LØKEN, K. P. 1992. Deposition of CO_2 on the seabed in the form of hydrates. *Energy Conversion and Management*, **33**, 659–666.

AYA, I., YAMANE, K. & KOJIMA, R. 2001. Simulation experiment of CO_2 storage at 3600 m deep ocean floor. *In*: WILLIAMS, D., DURIE, B., MCMULLAN, P., PAULSON, C. & SMITH, A. (eds) *Proceedings of the 5th International Conference on Greenhouse Gas Control Technologies GHGT-5*. CSIRO Publishing, Collingwood, 423–428.

BACHU, S., GUNTER, W. D. & PERKINS, E. H. 1994. Aquifer disposal of CO_2: Hydrodynamic and mineral trapping. *Energy Conversion and Management*, **35**, 269–279.

BAINES, S. J. & WORDEN, R. H. 2004a. *Geological Storage of Carbon Dioxide*. Geological Society, London, Special Publications, **233**.

BAINES, S. J. & WORDEN, R. H. 2004b. The long-term fate of CO_2 in the subsurface: Natural analogues for CO_2 storage. *In*: BAINES, S. J. & WORDEN, R. H. (eds) *Geological Storage of Carbon Dioxide*. Geological Society, London, Special Publications, **233**, 59–85.

BAKER, J. C., BAI, G. P., HAMILTON, P. J., GOLDING, S. D. & KEENE, J. B. 1995. Continental-scale magmatic carbon dioxide seepage recorded by dawsonite in the Bowen–Gunnedah–Sydney basin system, eastern Australia. *Journal of Sedimentary Research*, **A65**, 252–530.

BAKLID, A., KORBØ, L. R. & OWREN, G. 1996. Sleipner Vest CO_2 disposal, CO_2 injection into a shallow underground aquifer. *Society of Petroleum Engineers*, **36600**, 269–277.

BERGMAN, P. D., DRUMMOND, C. J., WINTER, E. M. & CHEN, Z.-Y. 1996. Disposal of power plant CO_2 in depleted oil and gas reservoirs in Texas. *Proceedings of the Third International Conference on Carbon Dioxide Removal*. Massachusetts Institute of Technology, Cambridge, MA, 9–11 September 1996.

BREWER, P. G., FRIEDERICH, G., PELTZER, E. T. & ORR, F. M., JR. 1999. Direct experiments on the ocean disposal of fossil fuel CO_2. *Science*, **284**, 943–945.

CAMPS, A. P. 2007. Hydrate formation in near surface ocean sediments. Unpublished PhD thesis, Leicester University.

CAMPS, A. P., LONG, D., ROCHELLE, C. A. & LOVELL, M. A. 2009. Mapping hydrate stability zones offshore Scotland. *In*: LONG, D., LOVELL, M. A., REES, J. G. & ROCHELLE, C. A. (eds) *Sediment-Hosted Gas Hydrates: New Insights on Natural and Synthetic Systems*. Geological Society, London, Special Publications, **319**, 81–91.

CZERNICHOWSKI-LAURIOL, I., SANJUAN, B., ROCHELLE, C., BATEMAN, K., PEARCE, J. & BLACKWELL, P. 1996a. Inorganic geochemistry. *In*: HOLLOWAY, S. (ed.) *The Underground Disposal of Carbon Dioxide*. Final Report of Joule II Project Number CT92-0031, Chapter 7.

CZERNICHOWSKI-LAURIOL, I., SANJUAN, B., ROCHELLE, C., BATEMAN, K., PEARCE, J. & BLACKWELL, P. 1996b. Analysis of the geochemical aspects of the underground disposal of CO_2. *In*: APPS, J. A. & TSANG, C.-F. (eds) *Deep Injection Disposal of Hazardous and Industrial Wastes, Scientific and Engineering Aspects*. Academic Press, London, 565–583.

ENICK, R. M. & KLARA, S. M. 1990. CO_2 solubility in water and brine under reservoir conditions. *Chemical Engineering Communications*, **90**, 23–33.

FREUND, P. & ORMEROD, W. 1996. Progress towards storage of CO_2. *Proceedings of the Third International Conference on Carbon Dioxide Removal*. Massachusetts Institute of Technology, Cambridge, MA, 9–11 September 1996.

GORMAN, A. R., HOLBROOK, W. S., HORNBACH, M. J., HACKWITH, K. L., LIZARRALDE, D. & PECHER, I.

2002. Migration of methane gas through the hydrate stability zone in a low-flux hydrate province. *Geology*, **30**(4), 327–330.

GUNTER, W. D., PERKINS, E. H. & MCCANN, T. J. 1993. Aquifer disposal of CO_2-rich gases: Reaction design for added capacity. *Energy Conversion Management*, **34**, 941–948.

GUNTER, W. D., WIWCHAR, B. & PERKINS, E. H. 1997. Aquifer disposal of CO_2-rich greenhouse gases: Extension of the time scale of experiment for CO_2-sequestering reactions by geochemical modelling. *Mineralogy and Petrology*, **59**, 121–140.

HARRISON, W. J., WENDLANDT, R. F. & SLOAN, E. D. 1995. Geochemical interactions resulting from carbon dioxide disposal on the seafloor. *Applied Geochemistry*, **10**, 461–475.

HAUGEN, H. A. & EIDE, L. I. 1996. CO_2 Capture and disposal: The realism of large scale scenarios. *Energy Conversion and Management*, **37**, 1061–1066.

HERZOG, H. J., ADAMS, E. E., AUERBACH, D. & CAULFIELD, J. 1996. Environmental impacts of ocean disposal of CO_2. *Energy Conversion and Management*, **37**, 999–1005.

HIRAI, S., OKAZAKI, K., TABE, Y. & HIJIKATA, Y. 1997. Numerical simulation for dissolution of liquid CO_2 droplets covered with clathrate film in intermediate depth of ocean. *Energy Conversion and Management*, **38**(suppl.), S313–S318.

HOLLOWAY, S. 1996a. An overview of the Joule II project 'The Underground Disposal of Carbon Dioxide'. *Energy Conversion and Management*, **37**, 1149–1154.

HOLLOWAY, S. (ed.). 1996b. *The Underground Disposal of Carbon Dioxide, Final Report of Joulle II Project Number CT92-0031*. British Geological Survey, London, 355pp. ISBN 0 85272 280 X.

HOLLOWAY, S. 1997. An overview of the underground disposal of carbon dioxide. *Energy Conversion and Management*, **38**(suppl.), S193–S198.

IEA GHG. 1998. Sleipner aquifer storage of CO_2. *Greenhouse Issues*, **34**, 4.

IEA GHG. 2000a. *Issues Underlying the Feasibility of Storing CO_2 as Hydrate Deposits*. IEA Greenhouse Gas R&D Programme Report **PH3/25**.

IEA GHG. 2000b. *Natural Gas and Methane Hydrates*. IEA Greenhouse Gas R&D Programme Report **PH3/27**.

IEA GHG. 2004. *Gas Hydrates for Deep Ocean Storage of CO_2*. IEA Greenhouse Gas R&D Programme Report **PH4/26**.

IMO. 1997. Convention on the prevention of marine pollution by dumping of wastes and other matter (London Convention 1972). *In*: *Compilation of the Full Texts of the London Convention 1972 and of the 1996 Protocol Thereto*. **LCo2/Circ.380**. International Maritime Organisation, London.

IOC, IHO & BODC. 2003. *Centenary Edition of the GEBCO Digital Atlas*, published on CD-ROM on behalf of the Intergovernmental Oceanographic Commission and the International Hydrographic Organization as part of the General Bathymetric Chart of the Oceans. British Oceanographic Data Centre, Liverpool.

IPCC. 1990. *Scientific Assessment of Climate Change*. Working Group 1 report of the Intergovernmental Panel on Climate Change. WNO, Geneva. World Wide Web Address: http://www.ipcc.ch/ipccreports/assessments-reports.htm.

IPCC. 2005. *Carbon Dioxide Capture and Storage*. METZ, B., DAVIDSON, O., DE CONINCK, H., LOOS, M. & MEYER, L. (eds) Working Group 3 report of the Intergovernmental Panel on Climate Change. Cambridge University Press, New York.

IPCC. 2007. *Fourth Assessment Report: Climate Change*. IPCC. World Wide Web Address: http://www.ipcc.ch/ipccreports/assessments-reports.htm.

JOHNSON, J. W., NITAO, J. J. & KNAUSS, K. G. 2004. Reactive transport modelling of CO_2 storage in saline aquifers to elucidate fundamental processes, trapping mechanisms and sequestration partitioning. *In*: BAINES, S. J. & WORDEN, R. H. (eds) *Geological Storage of Carbon Dioxide*. Geological Society, London, Special Publications, **233**, 87–106.

JOHNSON, J. W., NITAO, J. J., STEEFEL, C. I. & KNAUSS, K. G. 2001. Reactive transport modelling of geologic CO_2 sequestration in saline aquifers: The influence of intra-aquifer shales and the relative effectiveness of structural, solubility, and mineral trapping during prograde and retrograde sequestration. *Proceedings of the First National Conference on Carbon Sequestration*, Washington, DC, 14–17 May 2001.

JOIDES. 1992. JOIDES Pollution Prevention and Safety Panel, Ocean Drilling Program Guidelines for Pollution Prevention and Safety. *Joides Journal*, **18** (Special Issue 7).

KARL, T. R., KNIGHT, R. W. & BAKER, B. 2000. The record breaking global temperatures of 1997 and 1998: Evidence for an increase in the rate of global warming? *Geophysical Research Letters*, **27**, 719–722.

KING, M. B., MUBARAK, A., KIM, J. D. & BOTT, T. R. 1992. The mutual solubilities of water with supercritical and liquid carbon dioxide. *Journal of Supercritical Fluids*, **5**, 296–302.

KOIDE, H., TAKAHASHI, M. ET AL. 1997. Hydrate formation in sediments in the sub-seabed disposal of CO_2. *Energy*, **22**, 279–283.

KOJIMA, R., YAMANE, K. & AYA, I. 2003. Dual nature of CO_2 solubility in hydrate forming region. *In*: GALE, J. & KAYA, Y. (eds) *Greenhouse Gas Control Technologies*. Elsevier Science, Oxford.

KUK, M. S. & MONTAGNA, J. C. 1983. Solubility of oxygenated hydrocarbons in supercritical carbon dioxide. *In*: *Chemical Engineering at Supercritical Fluid Conditions*. Ann Arbor Science, Ann Arbor, 101–111.

LORENZ, S. & MÜLLER, W. 2003. Modelling of halite formation in natural gas storage aquifers. *Proceedings of the TOUGH Symposium 2003*, Lawrence Berkeley National Laboratory, Berkeley, CA, 12–14 May 2003.

MALIK, Q. M. & ISLAM, M. R. 2000. CO_2 injection in the Weyburn field of Canada: Optimization of enhanced oil recovery and greenhouse gas storage with horizontal wells. *Society of Petroleum Engineers*, **59327**.

MOBERG, R. 2001. The Weyburn CO_2 monitoring and storage project. *Greenhouse Issues*, **57**, 2–3.

NAKANO, S., YAMAMOTO, K. & OHGAKI, K. 1998. Natural gas exploitation by carbon dioxide from gas hydrate fields – high-pressure phase equilibrium for

an ethane hydrate system. *Proceedings of the Institution of Mechanical Engineers*, **212**(Part A), 159–163.
OSPAR. 1992. *Convention for the Protection of the Marine Environment of the North-East Atlantic*. OSPAR Commission, London.
PEARCE, J., CZERNICHOWSKI-LAURIOL, I. ET AL. 2004. A review of natural CO_2 accumulations in Europe as analogues for geological sequestration. *In*: BAINES, S. J. & WORDEN, R. H. (eds) *Geological Storage of Carbon Dioxide*. Geological Society, London, Special Publications, **233**, 29–41.
PEARCE, J., HOLLOWAY, S., WACKER, H., NELIS, M. K., ROCHELLE, C. A. & BATEMAN, K. 1996. Natural occurrences as analogues for the geological disposal of carbon dioxide. *Energy Conversion and Management*, **37**, 1123–1128.
PRUESS, K. 2003. Numerical simulation of leakage from a geologic disposal reservoir for CO_2, with transitions between super- and sub-critical conditions. *Proceedings of the TOUGH Symposium 2003*, Lawrence Berkeley National Laboratory, Berkeley, CA, 12–14 May 2003.
RCEP. 2000. *Energy – the Changing Climate*. Twenty-second report of the Royal Commission on Environmental Polution, Cm. **4749**.
READ, J. F., POLLARD, R. T & HIRST, C. 1991. *CTD Data from the North East Atlantic, April 1989, Collected on RSS Discovery Cruise 181*. Institute of Oceanographic Sciences, Deacon Laboratory, Report no. **285**.
RIESTENBERG, D., CHIU, E., GBORIGI, M., LIANG, L., WEST, O. R. & TSOURIS, C. 2004. Investigation of jet breakup and droplet size distribution of liquid CO_2 and water systems – implications for CO_2 hydrate formation for ocean carbon sequestration. *American Mineralogist*, **89**, 1240–1246.
ROCHELLE, C. & CAMPS, A. 2006. Underground storage of CO_2 as a liquid and solid hydrate. *Greenhouse Issues*, **82**, 8–9.
ROCHELLE, C. A., PEARCE, J. M. & HOLLOWAY, S. 1999. The underground sequestration of carbon dioxide: Containment by chemical reactions. *In*: *Chemical Containment of Waste in the Geosphere*. Geological Society, London, Special Publications, **157**, 117–129.
ROCHELLE, C. A., CZERNICHOWSKI-LAURIOL, I. & MILODOWSKI, A. E. 2004. The impact of chemical reactions on CO_2 storage in geological formations, a brief review. *In*: BAINES, S. J. & WORDEN, R. H. (eds) *Geological Storage of Carbon Dioxide*. Geological Society, London, Special Publications, **233**, 87–106.
SAKAI, H., GAMO, T. ET AL. 1990. Venting of carbon dioxide-rich fluid and hydrate formation in mid-Okinawa trough backarc basin. *Science*, **248**, 1093–1096.
SASAKI, K. & AKIBAYASHI, S. 2000. A calculation model for liquid CO_2 injection into shallow sub-seabed aquifer. *In*: HOLDER, G. D. & BISHNOI, P. R. (eds) *Gas Hydrates, Challenges for the Future*. Annals of the New York Academy of Sciences, **912**, 211–225.
SAUNDERS, P. M. & COOPER, S. 1987. *CTD Data on the Iberian Abyssal Plain*. Institute of Oceanographic Sciences, Deacon Laboratory, Report no. **247**.
SLOAN, E. D., JR. 1998. *Clathrate Hydrates of Natural Gases*. Marcel Dekker, New York.
SOMEYA, S., SAITO, K., NISHIO, M. & TSUTSUI, K. 2006. CO_2 sequestration under a sealed layer with clathrate hydrate in sediments. *Proceedings of Sediment-Hosted Gas Hydrates: New Insights on Natural and Synthetic Systems*, 25–26 January 2006, London. Geological Society, London, 34.
STUDLICK, J. R. J., SHEW, R. D., BASYE, G. L. & RAY, J. R. 1990. A giant carbon dioxide accumulation in the Norphlet Formation, Pisgah Anticline, Mississippi. *In*: BARWIS, J. H., MCPHERSON, J. G. & STUDLICK, J. R. J. (eds) *Sandstone Petroleum Reservoirs*. Springer, New York, 181–203.
VESOVIC, V., WAKEHAM, W. A., OLCHOWY, G. A., SENGERS, J. V., WATSON, J. T. R. & MILLAT, J. 1990. The transport properties of carbon dioxide. *Journal of Physics and Chemistry Refference Data*, **19**, 763–808.
WARZINSKI, R. P., LYNN, R. J. & HOLDER, G. D. 2000. The impact of CO_2 clathrate hydrate on deep ocean sequestration of CO_2 – experimental observations and modeling results. *In*: HOLDER, G. D. & BISHNOI, P. R. (eds) *Gas Hydrates, Challenges for the Future*. Annals of the New York Academy of Sciences, **912**, 226–234.
WIEBE, R. 1941. The binary system carbon dioxide-water under pressure. *Chemical Reviews*, **29**, 475–481.
WIEBE, R. & GADDY, V. L. 1939. The solubility in water of carbon dioxide at 50, 75 and 100°, at pressures to 700 atmospheres. *Journal of the American Chemical Society*, **61**, 315–318.
WIEBE, R. & GADDY, V. L. 1940. The solubility of carbon dioxide in water at various temperatures from 12 to 40° and at pressures to 500 atmospheres. Critical phenomena. *Journal of the American Chemical Society*, **62**, 815–817.
WILSON, M. & MONEA, M. (eds). 2004. IEA GHG Weyburn CO_2 Monitoring & Storage project Summary Report 2000–2004. *Proceedings of the 7th International Conference on Greenhouse Gas Control Technologies*, 5–9 September 2004, Vancouver. Petroleum Technology Research Centre, Regina, **3**.
WILSON, T. R. S. 1992. The deep ocean disposal of carbon dioxide. *Energy and Management*, **33**, 627–633.
ZHENG, L., WANG, S., LIAO, Y. & FENG, Z. 2001. CO_2 gas pools in Jiyang sag, China. *Applied Geochemistry*, **16**, 1033–1039.

Index

Note: Figures are shown in *italic* font, tables in **bold**.

AABW *see* Antarctic Bottom Water (AABW)
accretionary prism 103, 105
acoustic velocity 4
advanced piston corer (APC) 13
Advance Piston Corer Temperature (APCT) 13
AIW *see* Arctic Intermediate Water (AIW)
Alaskan permafrost wells 2
Alaska shelf seas 3
amplitude blanking 22
Angola *74*
anions 62–63
Antarctic Bottom Water (AABW) 83
Antarctic Peninsula 103
    BSR 4
Antarctic–Phoenix ridge 103
Anton Dorm seamount 82, *82*
AOI *see* areas of interest (AOI)
APC *see* advanced piston corer (APC)
APCT *see* Advance Piston Corer Temperature (APCT)
Archie's equations 4, 98, 123
Arctic coastlines 3
Arctic Intermediate Water (AIW) 82, *82*
areas of interest (AOI) 165
authigenic carbonates 65–66
    isotopes *66*
autonomous underwater vehicle (AUV) 33
    MC118 37
    MC798 *39–41*
AUV *see* autonomous underwater vehicle (AUV)

Barents Sea 131
base of the free gas reflector (BGR) 103
base of the hydrate stability zone (BHSZ) 31–32, 145
    seismic reflection imaging 33–34
bathymetry *see also* swath bathymetry
    British and Icelandic continental shelves *82*
    contoured *84*, *85*
    Faeroe–Shetland Channel *84*
    Northern Rockall Trough *85*
    Scotland *84*, *85*
Beam Line BW5 163, *163*
Bentheim sandstone 131
    MRI 133
BGR *see* base of the free gas reflector (BGR)
BHSZ *see* base of the hydrate stability zone (BHSZ)
bicarbonate 32
biogenic methanes 129
biostratigraphy
    MC118 *46*
    MC798 *46*
biosurfactants 32
Black Sea 23–24
Blake Ridge, South Carolina
    BSR **128**
    cell pressure **128**
    drilling campaigns 21
    gas evolution rate **128**
    hydrate zone **128**
    natural gas vents and seeps 23
    ODP Leg 164 21, 145
    seafloor 24
    sediments 124
    venting rate **128**
Blue Ridge 121
BODC *see* British Oceanographic Data Centre (BODC)
borehole
    data 98–99
    image logs 24, *99*
    laboratory medium modelling 93–102
    resistivity data 98–99
    seismic data 93–102
bottom simulating reflectors (BSR) 2, 12–13
    Antarctic Peninsula 4
    Blake Ridge **128**
    identification 31–32
    interpretation 21–22
    Namibe Basin 76–77
    phase reversal 103
    seismic line *74*
    site-survey seismic data 15
    South Shetland margin 103
    Walvis Ridge 73
Bragg slits 164, *164*
Bransfield Strait marginal basin 103
British continental shelves *82*
British Oceanographic Data Centre (BODC) 83
Brookhaven National Laboratory 124
BSR *see* bottom simulating reflectors (BSR)
bulk hydrate formation experimental setup *134*
Bullseye vent area 13
Bush Hill, Gulf of Mexico 4, 161–162
    crystallite size *167–168*
    fault zone natural gas vents and seeps 23–24

calcite-water carbon dioxide systems **140**
calcium-chloride ratios 63
Calyptogena bivalve shell 53, 55, 59, 65
Canada *see also* Mallik 2L-38 borehole, Canada
    shelf seas 3
    Weyburn 81–82
capillary inhibition 145–160
capillary pressure
    hydrate 3
    pore geometry *154*
    v. pore hydrate normalized volume fraction *156*
capillary-thermodynamic model 146
carbonates
    authigenic 65–66, *66*
    Derugin Basin 65
carbon dioxide
    calcite-water **140**
    Faeroe–Shetland Channel 90
    greenhouse gases offsetting storage program 103
    Rockall Trough 90
    sequestration 103

carbon dioxide hydrate 5–6, 131
   depth-temperature stability diagram *89*
   Faeroe–Shetland Channel 85, *86*
   methane hydrate conversion 131, *142*
   molecular dynamics simulation *137*
   predication program 87–88
   Rockall Trough 85, *86*
   Sloan's hydrate prediction model CSMHYD *88*
   underground carbon dioxide storage 171–184
Cascadia Basin 11
Cascadia Margin 4, 11, 145
   drilling transect 11–20
   IODP Expedition 311 11–20
Caspian Sea 3
cations 62–63
CDP *see* common depth point (CDP)
cell pressure **128**
cementation
   hydrate saturation 100
   models 122
Centigrade offset 1.1 degree 88–89
channel-overbank deposits 32
CHAOS *see* hydro-Carbon Hydrate Accumulations in Okhotsk Sea (CHAOS) International
Chevron Joint Industry Project 21
chirp sonar 33
   MC118 37
   MC798 *43*
chloride
   CHAOS venting site 67
   Hieroglyph venting site 68
   Hydrate Ridge, Oregon 66
   ion concentration **64**
   Kitami venting site 67, 68
   map *67*
   waters *62*
CIG *see* common image gather (CIG)
clathrate hydrates 131, 145
climate change 3
CMT *see* computed microtomography (CMT)
cold vent site U1328 17–18
common depth point (CDP) 108–109
   residual move-out 108–110
   residual NMO correction *109*
common image gather (CIG) 105, *106*
   picked horizons percentage differences *108*
   tomographic inversion *107*
computed microtomography (CMT) 129
   National Synchrotron Light Source at Brookhaven National Laboratory 129
conductivity-temperature-depth (CTD) 83
   Faeroe-Shetland Channel *84*
   Northern Rockall Trough *85*
Congo River 73
constant velocity field 106
contoured bathymetry
   Faeroe–Shetland Channel *84*
   Northern Rockall Trough *85*
contoured bottom water temperature *85*
Controlled Pore Glass (CPG) silica 148, *148*, 150
   dissociation loop PT data *149*
   methane hydrate growth *149*
   volume fraction pore hydrate v. capillary pressure *151, 152*

controlled source electromagnetic (CSEM) techniques 25
core samples analyses
   MC 118 41–42
   MC 798 41–42
Coulomb theory 24
CPG *see* Controlled Pore Glass (CPG) silica
cryogenic structures 54, *55*
crystallite size *see also* high-energy synchrotron radiation gas hydrate crystallite size
   Bush Hill region Gulf of Mexico *167–168*
   gas hydrate 161–162
   synthetic gas hydrate sample **166**
CSEM *see* controlled source electromagnetic (CSEM) techniques
CTD *see* conductivity-temperature-depth (CTD)

Davis Villinger Temperature Pressure (DVTP/P) tools 13
Debye–Scherrer cones *164*
Debye–Scherrer ring 163
deionized water (DI) 123, 126
DEM *see* differential effective medium (DEM)
density profiles *138*
depleted sediment methane hydrate formation-decomposition 121–130
   future work 129
   hydrate decomposition 124
   hydrate formation method 123–124
   materials and methods 123–124
   unit description 123
depth-temperature stability diagram *89*
Derugin Basin 52
   carbonates 65
DESY *see* Deutsche Elektronen Synchrotron (DESY)
Deutsche Elektronen Synchrotron (DESY) 163
DI *see* deionized water (DI)
differential effective medium (DEM)
   approach 93, 94, *95*
   cementation *99*
   load-bearing hydrate 96
   quartz-air mixture *97*
   theory 100
disseminated hydrate 3, 4
dissociation profile *127*
drilling strategy 13
DSDP drilling 122
   Namibe Basin 75–76
DVTP/P *see* Davis Villinger Temperature Pressure (DVTP/P) tools
dynamic mode 123

East North Atlantic Water (ENAW) 82, *82*, 83
echosounder record *53*
EDX *see* energy-dispersive X-ray microanalysis (EDX) detector
effective medium
   approach 98–99
   inversion 95–96
   models uncertainty 96
   theory 93, 94–95
electrical resistivity 4
electromagnetic methods 25–26
Elephant Island 105
ENAW *see* East North Atlantic Water (ENAW)

INDEX

energy-dispersive X-ray microanalysis (EDX) detector 162, 168
energy sources 1
environmental hazard 1
Expedition 311 12–13
    carbon gas profiles *14*
    gas hydrate concentration *15*
    gas hydrate examples *17*
    IW chlorinity profiles *14*
experimental gas 123
exploration methods 21–22
extended core barrel (XCB) systems 13

Faeroe–Shetland Channel 81, 82, *82*
    carbon dioxide hydrate 5, 85, *86*
    carbon dioxide storage potential 90
    compositional changes 89–90
    contoured bathymetry *84*
    CTD temperature profile *84*
    currents 89
    methane hydrate stability 84–85, *86*
    Mud diapirs 89
    natural methane hydrate evidence 89–90
    reflectors 89
Faeroe–Shetland Channel Bottom Water (FSCBW) 83
faults 23–24
    hydrate reservoirs morphology 24
Fermat's principle 105
FE-SEM *see* field-emission scanning electron microscope (FE-SEM)
field-emission scanning electron microscope (FE-SEM) 162
fine-grained sediments 32
FISH *see* flexible integrated study of hydrates (FISH)
flexible integrated study of hydrates (FISH) 121
    modifications 129
fluid compressibility 114, **115**
fluid/hydrate chamber *134*
fluid infiltration 66
fluid water (FW) 68
formal inverse approach 95
formation rate 42
fractures 23–24
    hydrate reservoirs morphology 24
FSCBW *see* Faeroe–Shetland Channel Bottom Water (FSCBW)
FW *see* fluid water (FW)

gas
    consumption as function of time *133*
    evolution rate **128**
    local production 22
    transport through faults and fractures 23–24
gas hydrate
    bearing sediment structures 53–55
    drilling transect across northern Cascadia 11–20
    Expedition 311 11–20, *17*
    forming fluids 51–72
    growth and dissociation 145–160
    host sediments 122
    images *166*
    induced structures 55
    marine exploration strategy 21–28
    narrow pore networks 145–160
    NE Sakhalin slope 51–72
    offshore deposits 21
    outcropping *33*
    reservoir accumulation mechanisms 22–24
    rock elastic properties 123
    Sea of Okhotsk 51–72
    venting structures 53
gas hydrate crystallite
    data processing 164–165
    experimental methods 162–164
    grain sizes 161–162
    high-energy synchrotron radiation 161–170
    results 165–168
gas hydrate stability zone (GHSZ) 21
    amplitude blanking 22
gas phase concentration *116*
    South Shetland margin 114
gas transport
    Green Canyon Block 185 24
    Hydrate Ridge, Oregon 23
GC *see* gravity-corer (GC)
GEBCO *see* Global Bathymetric Chart of the Oceans (GEBCO)
geochemical analysis 58–60
Geological Society Hydrocarbons Group 2006 meeting 2, 6
geothermal gradient 88–89
    histogram *88*
German–Russian KOMEX Project 52
GHSZ *see* gas hydrate stability zone (GHSZ)
Gibbs phase rule 131
Gibbs–Thomson equation 146, 151
Global Bathymetric Chart of the Oceans (GEBCO) 83
grain size distributions *167–168*
gravity-corer (GC) 53
Green Canyon Block 185 23–24
    gas transport 24
Green Canyon Block 204
    Gulf of Mexico 24
greenhouse gases
    emissions 81
    offsetting carbon dioxide storage program 103
Gulf of Mexico 2, 3, 21 *see also* Bush Hill, Gulf of Mexico; Mississippi Canyon area, Gulf of Mexico
    biogenic methanes 129
    gas hydrates 31–32
    Green Canyon Block 204 24
    hydrates 32–33
    hydrate stability zone 31
    natural gas vents and seeps 23–24
    seafloor mound *33*
    Z Zone 45
Gulf of Mexico Hydrates Research Consortium 46
    University of Mississippi 29

Hamburg Synchrotron Laboratory (HASYLAB) 163
HC *see* hydrocorer (HC)
Helgerud models 95
Hero fracture zone 103–105, *104*
Hieroglyph venting site
    chloride 68
    gas structure 53

Hieroglyph venting site (*Continued*)
  hydrated gases composition **65**
  structure *53*
high-energy synchrotron radiation gas hydrate crystallite size 161–170
  data processing 164–165
  experimental methods 162–164
  results 165–168
high-pressure set-up *147*
Hokkaido–Sakhalin dextral shear zone 52
HSZ *see* hydrate stability zone (HSZ)
hydrate(s) *see also* carbon dioxide hydrate; methane hydrates
  capillary pressure 3
  chamber *134*
  composition 65
  detailed velocity analysis quantifying 102–120
  disseminated 3, 4
  distribution 1–2
  domains 2–3
  dyke systems 24
  formation 3
  formation method 123–124
  fraction as function of time *135*
  global reserves 1
  Integrated Ocean Drilling Program Expedition 311 96
  kinetics and stability 131–144
  load-bearing 96
  marine gas 21–28
  massive 3
  morphology 3
  MRI *134*
  Namibe Basin 75–76
  natural 2
  natural gas 31–32, 145
  natural gas carbon dioxide 5
  nodular 3
  normalized induction times *45*
  reservoir detection 24–26
  sediment frame strength 93
  sediment grain interaction 4
  sediment-hosted gas 1–10
  sediment relationships research 6
  solid surfaces effects 131–144
  stability program construction 83–84
  structure marker 137
  thermodynamics 3
  vertical dykes 25
  zone **128**
hydrated gases composition **65**
  Chaos venting site **65**
  Hieroglyph venting site **65**
  Kitami venting site **65**
Hydrate Ridge, Oregon 3, 18
  chloride enrichments 66
  drilling campaigns 21
  gas transport 23
  hydrate saturation 22
  ODP Leg 204 4, 161–162
  seafloor 24
  sound propagation 25
  vent structure 25
hydrate saturation *94, 95*
  borehole resistivity data 98–99

Hydrate Ridge 22
  tomographic velocity model *96*
  velocity 96
hydrate stability zone (HSZ) 3, 145
  Scotland mapping 81–92
  seismic reflection imaging 33–34
hydrocarbon gas hydrates 31–32
  Mississippi Canyon area 29–50
hydrocarbon headspace gas measurements 15
hydro-Carbon Hydrate Accumulations in Okhotsk Sea (CHAOS) International 9
  chlorinity map *67*
  2003 Cruise *55*
  Research Project 53
  Russian–Japanese–Korean CHAOS Project 52
  side-scan sonar records *54*
  venting site **65**, 67
hydrocorer (HC) 53
hydrogen carbonate 32
hysteresis
  loops 149–151, 157
  origins 151–157
  phenomena 145–160
hysteresis phenomena 145–160

Icelandic continental shelf *82*
induction time 42
industrial-scale projects 5
infiltrating fluids 69–70
  isotopic model 68–69
infrared (IR) imaging 13
Integrated Ocean Drilling Program Expedition 311 11, 96
interstitial water (IW) 13
IR *see* infrared (IR) imaging
isobutane 15
isotopes
  authigenic carbonates *66*
  waters *63*, 63–64
IW *see* interstitial water (IW)

Japan
  CHAOS Project 52
  Nankai Trough 6, 21
Joint Industry Project (JIP) 21
  Atwater Valley site 22
Jolliet Reservoir 23
Juan de Fuca plate 11
Jurassic Louann Salt 31

Kitami venting site
  chloride 67, 68
  hydrated gases composition **65**
  structure *55*
KOMEX Project 52
Kyoto Protocol 81

laboratory data application 97–98
Labrador Sea Water (LSW) *82, 83*
Lennard–Jones model carbon dioxide model 139
lenticular-bedded structures 54, *55*
lithology
  MC118 43, *46*
  MC798 43, *46*
  Ocean Drilling Program Site 889 95

INDEX

lithostratigraphic units 13
load-bearing hydrates 96
location scan method 164
logging while drilling (LWD) 13, 16
low flux mechanisms 22–23
low resolution P wave tomographic study 95
LSW *see* Labrador Sea Water (LSW)
LWD *see* logging while drilling (LWD)

magnesium-chloride ratios 63
magnetic resonance imaging (MRI) 132
    Bentheim sandstone 133
    hydrate *134*
    methane hydrate *132*
Mallik 2L-38 borehole, Canada 98, 100
marine gas hydrate exploration strategy 21–28
marine sediments *see also* South Shetland margin case study regional
    free gas 102–120
    numerical modeling 122
marine seismic surveys 25
massive hydrates 3
material balance calculations 132
Mazurenko, Leonid vi
MC118 *see* Mississippi Canyon Area lease block 118 (MC118)
MC798 *see* Mississippi Canyon Area lease block 798 (MC798)
MDAC *see* methane-derived authigenic carbonate (MDAC)
mean crystallite size v. depth *169*
measurement while drilling (MWD) 13, 16
message passing interface (MPI) 139
methane, biogenic 129
methane-derived authigenic carbonate (MDAC) 32
methane hydrates *see also* depleted sediment methane hydrate formation-decomposition
    ambient temperature and pressure instability 122
    bearing sand samples 97
    conversion to carbon dioxide hydrate 131, *142*
    depth-temperature stability diagram *89*
    Faeroe–Shetland Channel 89–90
    growth and dissociation loop PT data *149*
    MRI 132, *132*
    natural 89–90
    natural systems 4–5
    Rockall Trough 89–90
    Sea of Okhotsk 52
    structure transition 122
    thermo-physical properties 122
methane hydrate stability
    Faeroe–Shetland Channel 84–85, *86*
    Rockall Trough 85, *86*
microbes 32
Minerals Management Service (MMS) 29
mini-basins 31
Mississippi Canyon area, Gulf of Mexico
    formation 31
    hydrocarbon gas hydrates 29–50
    quaternary geology 29–31
Mississippi Canyon Area lease block 118 (MC118) 29, *30*, 36, **36**, 36–38
    AUV 37
    biostratigraphy *46*
    chirp sonar profile 37
    core samples analyses 41–42
    lithology 43, *46*
    percentage sand *46*
    reflection polarity *38*
    seismo-acoustic profile *37*
    swath bathymetry 36, 47
Mississippi Canyon Area lease block 798 (MC798) 29, *30*, 38–41
    AUV *39–41*
    biostratigraphy *46*
    chirp sonar profile *43*
    contour map *42*
    core samples analyses 41–42
    lithology 43, *46*
    percentage sand *46*
    seismic reflection profiles 40–41
    seismo-acoustic profiles *43*
    swath bathymetry 38, *39–41*, 47
MMS *see* Minerals Management Service (MMS)
Molecular Dynamics simulations 138
molecular simulation snapshot *141*
Moving Area Detector Method 165, 168
MPI *see* message passing interface (MPI)
MRI *see* magnetic resonance imaging (MRI)
Mud diapirs 89
multibeam bathymetry map *12*
MWD *see* measurement while drilling (MWD)

NADW *see* North Atlantic Deep Water (NADW)
Namibe Basin 74
    BSR 76–77
    dip section *77*, *78*
    hydrate occurrences 73–80
    hydrates evidence 75–76
    ODP/DSDP drilling 75–76
    orthogonal features strike feature *77*
    regional setting 73–74
    seismic line *78*, *79*
    seismic surveys 76, 76–77
Namibian continental margin 3, 73–74
Nankai Trough, Japan 6, 21
narrow pore networks gas hydrate growth and dissociation 145–160
    confined geometries in phase behaviour 146–147
    experimental equipment and methods 147–148
    hysteresis origins 151–157
    pore blocking effects *155*, 155–157
    pore geometry 153–155
    results 148–151
    seaflow hydrate systems significance 157
National Synchrotron Light Source at Brookhaven National Laboratory 121
    computed microtomography technique 129
natural gas carbon dioxide hydrates 5
natural gas hydrates 31–32, 145
natural gas seeps 23–24
natural gas vents
    Black Sea 23–24
    Blake Ridge 23
    Bush Hill fault zone 23–24
    Gulf of Mexico 23–24
    Pacific Coast 23–24
natural hydrates 2

natural methane hydrates
  Faeroe–Shetland Channel 89–90
  Rockall Trough 89–90
natural systems 2–5
  characteristic hydrate domains 2–3
  methane hydrates stability 4–5
  natural hydrates setting 2
  sediment-hydrate interaction 3–4
NAW *see* North Atlantic Water (NAW)
NE Sakhalin slope, Sea of Okhotsk
  gas hydrate forming fluids 51–72
  water samples chemical and isotopic measurements **56–58, 59, 60, 61**
Niger Delta 73
NMR *see* nuclear magnetic resonance (NMR) cryoporometry
nodular hydrates 3
normalized induction times *45*
North Atlantic Deep Water (NADW) *82*, 83
North Atlantic Water (NAW) 82
Northern Rockall Trough
  contoured bathymetry *85*
  contoured bottom water temperature *85*
  CTD temperature profile *85*
North Sea 131 *see also* Sleipner gas field, North Sea
  Sleipner gas field 81–82, 131
Norwegian Sea Deep Water (NSDW) *82*, 83
Norwegian Sea Intermediate Water (NSIW) *82*, 83
Norwegian shelf 131
NSDW *see* Norwegian Sea Deep Water (NSDW)
NSIW *see* Norwegian Sea Intermediate Water (NSIW)
nuclear magnetic resonance (NMR) cryoporometry 148
numerical modelling 122

Ocean Drilling Program (ODP) 122
  Namibe Basin 75–76
Ocean Drilling Program (ODP) Leg 164 145
  Blake Ridge 21, 145
Ocean Drilling Program (ODP) Leg 204 164
  detector image *165*
  Hydrate Ridge 4, 161–162
Ocean Drilling Program (ODP) Site 889 95
ODP *see* Ocean Drilling Program (ODP)
offshore gas hydrate deposits 21
offshore Scotland mapping hydrate stability zones 81–92
Okhotsk Plate 52
Okhotsk Sea *see* hydro-Carbon Hydrate Accumulations in Okhotsk Sea (CHAOS) International
Okinawa Trough 81
OPR *see* out-of-plane reflections (OPR)
Oregon *see* Hydrate Ridge, Oregon
organic matter biogenic decay 2–3
Ormen Lange 131
out-of-plane reflections (OPR) 110
oxygen isotopes **64**

Pacific Coast natural gas vents and seeps 23–24
Parrish and Prausnitz model 87
PCS *see* pressure core sampler (PCS)
Peng–Robinson equation of state 42
permafrost 3
PFT *see* phase field theory (PFT) approach
phase-diagram and kinetic data 122

phase field theory (PFT) approach 132, 135–136
  results *139*
  simulations *140*
pipeline gas transportation 2
polyoxymethylene (POM) 133
POM *see* polyoxymethylene (POM)
pore blocking 157
pore geometry 153–155
  capillary pressure *154*
pore hydrate calculated volume fraction *150*
pore hydrate normalized volume fraction v. capillary pressure *156*
pore size/volume distributions (PSD) 156
pore-space gas-hydrates 123
porosity 95
potassium-chloride ratio 63
pressure
  carbon dioxide hydrate predication program 87–88
  methane hydrates 122
  v. time dissociation profile *127*
pressure core sampler (PCS) 13
pressure-temperature (PT) conditions 146
  primary growth and dissociation data *148*
pre-stack depth migration 106–108
Priest dataset 97
Priest resonant column measurement
  velocities 97, *98*
PSD *see* pore size/volume distributions (PSD)
PT *see* pressure-temperature (PT) conditions

quartz-air mixture 97
quaternary geology 29–31

RAB *see* resistivity at-the-bit (RAB) images
ray path analysis 25
real field datasets with known hydrate content 98–99
recovered core 13
reflection polarity *44*
  MC118 *38*
reflectors 89
reservoir accumulation mechanisms 22–24
residual salts 66–67
resistivity at-the-bit (RAB) images 13
resistivity-derived hydrate saturation velocities
  borehole image logs *99*
rock
  elastic properties 123
  physics approach 94
Rockall Trough 81–83, *82*
  carbon dioxide hydrate stability 85, *86*
  carbon dioxide storage potential 90
  contoured bathymetry *85*
  contoured bottom water temperature *85*
  CTD temperature profile *85*
  methane hydrate stability 85, *86*
  natural methane hydrate evidence 89–90
  surface waters 83
Rosemary Bank seamount 82, *82*
Russian–Japanese–Korean CHAOS Project 52

salt tectonics 31
SCA *see* self-consistent approximation (SCA) approach
scanning electron microscopy (SEM) 153

Scotland
  Bathymetric map *84, 85*
  mapping hydrate stability zones 81–92
  oceanographic setting 82–83
seabed
  geochemical sampling *79*
  moat *75*
  observatories 3
seafloor hydrate systems significance 157
seafloor mound
  Gulf of Mexico *33*
  mimic FISH unit *124*
Sea of Okhotsk 9 *see also* NE Sakhalin slope, Sea of Okhotsk
  geological setting 52–53
  methane hydrates 52
seawater (SW) 68
  sulfate 32
sediment
  Blake Ridge 124
  cool 5
  frame strength 93
  gas hydrate saturation 122
  grain interaction 4
  hosted gas hydrates 1–10, 5
  hosted hydrates 93–102
  hydrate interaction 3–4
  marine 122
  resistivity 100
  structure properties and permeability 122
  system complexity 122
  temperature v. run time trace *125–126*
  thermo-physical properties 122
sedimentation rates 47
seismic line
  BRS features *74*
  moat *76*
  Namibe Basin *78, 79*
seismic reflection
  BHSZ 33–34
  HSZ 33–34
  imaging 33–34
  MC798 40–41
  profiles 40–41
seismic surveys 25
  data for South Shetland margin 105
  marine 25
  Namibe Basin *76*, 76–77
seismic velocities 93
  anomalies 22
  South Shetland margin 105–108
seismo-acoustic profiles
  MC118 *37*
  MC798 *43*
self-consistent approximation (SCA) approach 93, 94, *95*
  cementation required 99
  load-bearing hydrate 96
  quartz-air mixture 97
SEM *see* scanning electron microscopy (SEM)
Shackleton fracture zone 103–106, *104*
  strike-slip fault 105
Shetland Shelf and Slope 84
Siberian permafrost wells 2
Siberia shelf seas 3

side-scan sonar records *54*
silica *see* Controlled Pore Glass (CPG) silica
site-survey seismic data 15
Site U1325 16
Site U1326 13–16
Site U1327 16
Site U1329 16–17
Sleipner gas field, North Sea 81–82, 131
Sloan's hydrate prediction model CSMHYD 86–87, *87*
  carbon dioxide hydrate pressure predictions *88*
  constants derived from 87
  graphical comparison *87*
smectite clay 32, 46
SMTZ *see* sulfate–methane transition zone (SMTZ)
Snell's law 106
sodium-chloride ratios 63
solid-state conversion 135
solid surfaces effects on hydrate kinetics and stability 131–144
  applied force fields 139–140
  experimental results 133–135
  interfacial system 140–142
  molecular dynamics simulation details 139–140
  practical implications 142–143
  techniques and approaches 132–133
  theoretical modeling 135–139
sound propagation 25
South Carolina *see* Blake Ridge, South Carolina
South Shetland margin
  BSR 103
  detailed velocity analysis 108–115
  gas phase concentration 114
  residual move-out of CDP 108–110
  residual move-out of CIG 110–113, *111, 112, 113*
  seismic data 105
  seismic velocity field 105–108
South Shetland margin case study regional 102–120
SSDR *see* surface-source-deep-receiver (SSDR) technique
static mode 124
Storegga slide 131
stratigraphic-type hydrate deposits 3
strike-slip fault 105
submarine slope failures 3
subsea biosphere 4
sulfate–methane transition zone (SMTZ) 32
surface energies **140**
surface-source-deep-receiver (SSDR) technique 33–36, *34*
surface waters 83
SW *see* seawater (SW)
swath bathymetry 33, *35*
  MC118 36, 47
  MC798 38, *39–41*, 47
S wave 95
synchrotron radiation 163–168, *164*, 168 *see also* high-energy synchrotron radiation gas hydrate crystallite size
synthetic gas hydrate sample 166–167
  crystal sizes **166**

temperature
  carbon dioxide hydrate predication program 87–88
  methane hydrates 122

temperature (*Continued*)
    Northern Rockall Trough *85*
    v. run time trace sediments *125–126*
    spike magnitude **128**
    time of appearance v. formation pressure **128**
three-phase Biot theory (TPB) 94
three-phase effective medium model (TPEM) 94
TOC *see* total organic carbon (TOC)
tomographic velocity model *96*
total organic carbon (TOC) 22
TPB *see* three-phase Biot theory (TPB)
TPEM *see* three-phase effective medium model (TPEM)
transect across margin 13

underground carbon dioxide storage 5
United States Department of Energy/Chevron Joint Industry Project 21
United States Department of Interior 29
University of Mississippi 29
Utsira formation 131

VAMP *see* velocity and amplitude anomalies (VAMP)
Vancouver Island
    formal inverse approach 95
    multibeam bathymetry map *12*
Vancouver Island margin 13
vein hydrate 3, 55
velocity
    acoustic 4
    constant field 106
    hydrate saturation 96
    methane hydrate-bearing sand samples 97
    Priest resonant column measurement *97, 98*
    resistivity-derived hydrate saturation *99*
    seismic 22, 93
    South Shetland margin 108–115
    South Shetland Margin case study 102–120
    tomographic model *96*

velocity and amplitude anomalies (VAMP) 25
venting
    evolution 24–26
    rate for Blake Ridge **128**
    structure for Hydrate Ridge 25
vertical seismic profiles (VSP) 13
    walkaway 25

Walvis Ridge *74*
    BRS 73
waters
    anions and cations correlation 62–63
    chloride contents 61–62, *62*
    chloride ion concentration **64**
    flow spatial patterns 67–68
    hydrate-forming **69**
    interstitial 13
    isotopes *63*, 63–64
    oxygen and hydrogen isotopic composition **64**
    samples chemical and isotopic measurements **56–58, 59, 61**
    surface 83
    Wyville–Thomson Ridge 90
weighted equation approach 94
West Africa 79
West Shetland Shelf and Slope 84
Weyburn, Canada 81–82
Wisconsin cycles 31
Wyville–Thomson Ridge 82, *82*, 83
    water overflow 90

XCB *see* extended core barrel (XCB) systems
X-rays 5

Young–Laplace equation 150

Z Zone, Gulf of Mexico 45